Solanum tuberosum – A Promising Crop for Starvation Problem

Edited by Mustafa Yildiz and Yasin Ozgen

Published in London, United Kingdom

IntechOpen

Supporting open minds since 2005

Solanum tuberosum – A Promising Crop for Starvation Problem
http://dx.doi.org/10.5772/intechopen.94735
Edited by Mustafa Yildiz and Yasin Ozgen

Contributors
María-Teresa Pino, Cristina Vergara, Belay Dereje, Nwankwo Stanley Chibuzo, Olusola Luke Oyesola, Olawole Obembe, Oluwadurotimi Aworunse, Onyemaechi Obiazikwor, Margaret Oniha, Olubunmi Atolagbe, Ayodele Sobowale, Jacob Popoola, Oluwakemi Bello, Peter VanderZaag, Tung Xuan Pham, Victoria Escobar Demonteverde, Cynthia Kiswa, Monica Parker, Shadrack Nyawade, Pieter Wauters, Alex Barekye, Rember Pinedo-Taco, Percy Rolando Egusquiza-Bayona, Dylan Anderson-Berens, Daniel P. Smith, Nathaniel T. Smith, Refik Bozbuga, Selman Uluisik, Mustafa Yildiz, Murat Aycan, Muhammet Cagri Oguz, Yasin Ozgen, Burak Onol, Elena Rakosy-Tican, Imola Molnar, Fulgence Waryoba, David Kinder, John Bamberg, Lisbeth Louderback, Bruce Pavlik, Alfonso Del Rio, Martin Raspor, Aleksandar Cingel, Sona S. Dev, Jini Joseph, Ligi Lambert D'Rosario, Tijjani Ahmadu, Khairulmazmi Ahmad, Adamu Abdullahi, Duraid K.A. AL-Taey, Rusul F. AL-Shmary, Prashant Kaushik, Navjot Singh Brar, Sat Pal Sharma

First published in London, United Kingdom, 2021 by IntechOpen
IntechOpen is the global imprint of INTECHOPEN LIMITED, registered in England and Wales, registration number: 11086078, 5 Princes Gate Court, London, SW7 2QJ, United Kingdom
Printed in Croatia

British Library Cataloguing-in-Publication Data
A catalogue record for this book is available from the British Library

Additional hard and PDF copies can be obtained from orders@intechopen.com

Solanum tuberosum – A Promising Crop for Starvation Problem
Edited by Mustafa Yildiz and Yasin Ozgen
p. cm.
Print ISBN 978-1-83969-166-9
Online ISBN 978-1-83969-167-6
eBook (PDF) ISBN 978-1-83969-168-3

For EU product safety concerns:
IN TECH d.o.o., Prolaz Marije Krucifikse Kozulić 3, 51000 Rijeka, Croatia, info@intechopen.com or visit our website at intechopen.com.

Meet the editors

Mustafa Yildiz obtained an MSc in Agricultural Sciences from Ankara University, Turkey, in 1996 with a thesis on the effect of cell structure on yield and sucrose concentration in sugar beet. In 1998, he studied plant biotechnology at Osaka Prefecture University, Japan, for five months. He received his Ph.D. from Ankara University in 2000 with a thesis on shoot regeneration and *Agrobacterium tumefaciens*-mediated gene transfer in flax. He is currently a professor in the Faculty of Agriculture, Department of Field Crops, Ankara University. Dr. Yildiz's research areas include plant tissue culture, plant biotechnology, molecular markers, *Agrobacterium tumefaciens*-mediated gene transfer, plant stress physiology, plant immune system, plant defense mechanism, and plant breeding. He has more than 170 scientific publications, two books, and nine book chapters to his credit. He was awarded First Place in the "International Sunflower Project Market" by the International Sunflower Association for his project titled "A New Environmental Friendly Production Method in Sunflower for High Seed and Crude Oil Yields.".

Yasin Ozgen obtained an MSc in 2015 with a thesis on the determination of chemical and morphological properties of different basil lines. He completed his Ph.D. in 2019 with a thesis on yield and agronomic characters of morphine and noscapine-type opium poppy hybrids. He is now working as a research assistant at the Faculty of Agriculture, Department of Field Crops, Ankara University, Turkey. Dr. Ozgen is currently researching industrial, medicinal, aromatic, and bulbous plants.

Contents

Preface **XIII**

Section 1
Potato Cultivation **1**

Chapter 1 **3**
Solanum tuberosum Yield for Selected Countries
by Fulgence Dominick Waryoba

Chapter 2 **19**
The Role of Crop Protection in Sustainable Potato (*Solanum tuberosum* L.) Production to Alleviate Global Starvation Problem: An Overview
by Tijjani Ahmadu, Adamu Abdullahi and Khairulmazmi Ahmad

Chapter 3 **53**
The Impact of Bio-Organic and N, P, K Fertilizers on the Growth and Yield of Potato
by Duraid K.A. AL-Taey and Rusul F. AL-Shmary

Chapter 4 **61**
Solanum tuberosum Cultivation Using Nitrogen Recovered from Local Wastewater
by Daniel P. Smith and Nathaniel T. Smith

Section 2
Origin and Wild Relatives **93**

Chapter 5 **95**
Solanum jamesii as a Food Crop: History and Current Status of a Unique Potato
by David Kinder, John Bamberg, Lisbeth Louderback, Bruce Pavlik and Alfonso Del Rio

Section 3
Nutrition for Increasing Population **105**

Chapter 6 **107**
Nutritional Composition and Biochemical Properties of *Solanum tuberosum*
by Belay Dereje and Nwankwo Chibuzo

Chapter 7 119
Red and Purple Flesh Potatoes a Healthy and Attractive Alternative
Associated with New Market Trends
by María-Teresa Pino and Cristina Vergara

Section 4
Potato Breeding 135

Chapter 8 137
Prospects for Genome Editing of Potato
by Sona S. Dev, Jini Joseph and Ligi Lambert D'Rosario

Chapter 9 157
Genetically Modified Potato for Pest Resistance: Thrift or Threat?
by Martin Raspor and Aleksandar Cingel

Chapter 10 195
Visiting Potato from a Breeding Perspective: Accomplishments
and Prospects
by Navjot Singh Brar, Sat Pal Sharma and Prashant Kaushik

Chapter 11 213
Impact and Management of Diseases of *Solanum tuberosum*
*by Olusola L. Oyesola, Oluwadurotimi S. Aworunse, Margaret I. Oniha,
Onyemaechi H. Obiazikwor, Oluwakemi Bello, Olubunmi M. Atolagbe,
Ayodele A. Sobowale, Jacob O. Popoola and Olawole O. Obembe*

Chapter 12 231
Molecular Host-Nematode Interactions and Tuber Development
by Refik Bozbuga and Selman Uluisik

Chapter 13 253
Gamma Radiation Effect on *Agrobacterium tumefaciens*-Mediated
Gene Transfer in Potato (*Solanum tuberosum* L.)
*by Murat Aycan, Muhammet Cagri Oguz, Yasin Ozgen, Burak Onol
and Mustafa Yildiz*

Chapter 14 265
Biotechnological Strategies for a Resilient Potato Crop
by Elena Rakosy-Tican and Imola Molnar

Section 5
Potato Seeds 303

Chapter 15 305
Use of Quality Potato Seeds in Family Farming Systems
in the Highlands Zones of Peru
*by Rember Pinedo-Taco, Percy Rolando Egusquiza-Bayona
and Dylan Anderson-Berens*

Chapter 16 321
Apical Rooted Cuttings Revolutionize Seed Potato Production by Smallholder Farmers in the Tropics
by Peter VanderZaag, Tung Xuan Pham, Victoria Escobar Demonteverde, Cynthia Kiswa, Monica Parker, Shadrack Nyawade, Pieter Wauters and Alex Barekye

Preface

It is estimated that the world population will increase to 11 billion and cultivated agricultural land per capita will decrease to 0.15 hectares in 2050. Every year parallel to the increasing population, agricultural fields are shrinking fast due to reasons such as misuse (settlement, road, factory, etc.), erosion, salinization, acidification, intensive agriculture, and overgrazing.

Every year, thousands of people in many parts of the world die due to malnutrition and hunger. For humankind to maintain its existence on earth, crop production needs to be increased. This can only be achieved by increasing the yield obtained per unit area. To achieve this, the genetic structure of plants should be improved and the agricultural techniques used in cultivation such as fertilization, irrigation, disease, and pest control should be applied. However, it has been observed that unconscious use of agricultural techniques adversely affects ecological balance in the long term.

Potato (*Solanum tuberosum* L.) is an annual plant in the Solanaceae family. It is reported that potato is the most consumed nutrition after cereals. In addition, it is the most produced plant in the world after maize, wheat, and rice. Among the reasons for such widespread consumption are its widespread production and consumption by almost all countries of the world due to it being cheap and easy to digest, as well as its high nutritional value and ability to grow in all kinds of climates allowing for obtaining more products per unit area. Considering that a significant number of people in the world are struggling with hunger, the importance of the potato plant is better understood. From this point of view, potato is one of the most important plants that can solve hunger and nutrition problems worldwide with their rich nutrient content. This book discusses these topics from all aspects. We hope this book will guide growers and researchers in solving problems in potato cultivation.

Dr. Mustafa Yildiz
Professor,
Faculty of Agriculture,
Department of Field Crops,
Ankara University,
Ankara, Turkey

Yasin Ozgen
Faculty of Agriculture,
Ankara University,
Turkey

Section 1

Potato Cultivation

Chapter 1

Solanum tuberosum Yield for Selected Countries

Fulgence Dominick Waryoba

Abstract

This chapter, aimed at analyzing potato yield among selected countries, has seven sections. The panel analysis of potato production and productivity has shown significant differences among countries. The main panel analysis of the random and fixed effect model indicates a negative influence of land size on yield and a positive influence on production. However, using multilevel mixed effect model, some country specific estimates deviate from main model results. In yield and output equations, the influence of land is positive for some countries and negative for others. Improvement of potato productivity is vital for hunger relief and starvation reduction. Even though, area specific analysis can bring in many determinants of potato production and productivity. A detailed analysis can give the right direction for policy makers in their effort to reduce hunger and starvation as well as improve the living standards of people.

Keywords: World potato share, regional potato production, potato yield, random effect model, fixed effect model, multilevel mixed effect model

1. Introduction

Spreading to about 160 countries of the world, *Solanum tuberosum* or Irish Potato or sometimes referred to as potato, is originated from the Andes of South America [1]. Potato is in the fourth order with respect to production and area harvested after maize, wheat and rice [2, 3]. Potato is one of the most world productive crop with high value as a balanced and nutritious food [2]. Being so important, potato is central to food security [4]. This implies that potato is an ideal crop for starvation related problems when weather is favorable. Smallholder farmers in developing countries like Kenya use almost 25 percent of their nearly 2.4 hectares farm area to grow potato for consumption and commercial purposes [5]. While fresh potato consumption has declined, it is observed that the consumption of processed products has continued to gain popularity [1].

This chapter analyzes potato productivity trend in some selected low and high income countries. In this analysis, the aim is to answer the following two questions. *Has there been an increase in potato production among selected countries? Is the increase in potato production a result of land expansion or yield improvement?* The question of increased potato yield is important because [6] increased population requires increased food production which is constrained by water and land availability. As a result, the increased potato production should be supported by increase in potato yield rather than land expansion. It is clear that with growing population and income, food crop yield must keep expanding to meet global food demand [6]. The rest of the study is organized as follows. Section two discusses about regional average potato production. Section three discusses about the contribution of potato in world food production. Section four discusses the productivity trend of potato. Section five concludes on the ability of the tuber crop in fighting against starvation.

2. World potato share of food crop

Potato belongs to the root and tuber crops group [7], such as sweet potato, yams, and cassava, which are among leading crops in the world [8]. The group of tuber crops as can be viewed in **Figure 1** comes fourth after cereals, sugar crops primary and sugar cane. However, as highlighted in section one, potato maintained a fourth position after maize, wheat and rice. This reality makes *Solanum tuberosum* (potato) the leading crop in the group of tuber crops because no other tuber crop has outweighed potato other than cereals. The average values provided in the figure, however, are computed from 1961 to 2029. Sugar cane is leading among crops standing on their own rather than in groups. Sugar consumption is very high from industrial consumption to domestic consumption. The crop can be eaten raw, processed to make juice and processed to make industrial and domestic sugar.

The demand for sugar is very high due to the fact that it is highly needed as an ingredient in other processed food. Most of the food people consume have sugar components [10], for instance cakes, bread, and other bites are all mixed with sugar. Juices from other fruits like orange, mango and others processed either domestically of at industrial levels are mixed up with sugar. There are a lot that can function with sugar, even fresh milk, ice cream, candy and chocolate that are favorites of children and people of all demographics must be mixed up with sugar contents. As a result, sugar cane production is highly favored for domestic consumption and industrial input for commercial purposes. It is found [11] for instance that US citizens' purchase is highly based on high processed foods with high contents of sugar. However, even with this high level of demand and high promotion of sugar cane production with large and new plantations being started, sugar cane cannot be promoted to fight against starvation. The crop cannot be consumed alone as food crop, but rather additional and cannot be taken in excess. As it is discussed in the book, "Sweeteners and Sugar alternatives in Food Technology", research is undertaken to find sugar alternatives for food in order to improve consumer health [12].

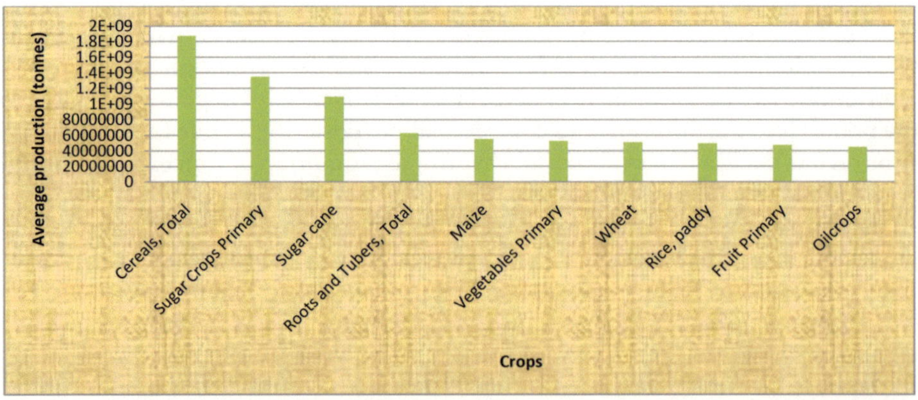

Figure 1.
World's most produced crops. Source: [9]. Note: The values used for the analysis are the average values of FAOSTAT computed from 1961 to 2029 [9].

Therefore, promotion of high yielding food crop is the most important decision among policy makers as it reduces hunger and fight against starvation.

3. Regional potato production

Potato production has spread worldwide. The regional production averages from 1961 to 2019 as shown from FAOSTAT in the table below, indicate a very high proportion of potato coming from Europe. The average is very high for the period mentioned showing that Europe is the leading region compared to other regions. For the period under consideration, Asian region comes second followed by the Americas, that is, North and South America combined with 11 percent. Africa which is one of the least developed regions comes with only 4 percent of the world potato share. The shares are approximated, for instance in the upper right chart of panel (b), Oceania region has a 0.5 percent but due to round off, the figure turned into 0 percent. The region produces on average, about 1736491.5 tones of potato for the period from 1994 to 2019. The periods have been randomly selected to check on consistence of potato production dominance.

In panel (b), by changing the period of analysis, the world average share of potato production by region also changes. Using the full period data to get the average favors Europe compared to other regions. The European region is over-weighed by the Asian region in the period starting from 1994 to 2019. A simulation analysis on the impact of climate change on potato [13], indicate a significant reduction of potato production in Eastern Europe and Northern America. Even though, African region remained in the same position with a very low share of production. The change can be attributed to the improvement in agricultural production technology in Asia. The average production in Asia increased but declined in Europe. For instance, the average production in Europe was 166767397.02 tons in the 1961–2019 period and was 125436231.88 tons in the 1994–2019 periods. While the average production in Asian increased from 87285656.59 to 140597815, respectively. This clearly indicates an improvement in average potato production in the Asian region compared to the average potato production in the European region. Nevertheless, the increased average potato production in Asia is likely to result from both land expansion and productivity improvement. Accordingly [14], China is the world leading potato producer due to land expansion and increase in potato yield (**Figure 2**).

The average share of potato production in panel (b) indicates the Asian dominance. This dominance comes from a great improvement because as it is shown in panel (d), the periods before 1994 a large proportion of world potato came from Europe. About 72 percent of world potato came from Europe and Asia, through that period, produced only 16 percent of world potato. It is clear evidence that Asia worked very hard to reach to the later period's level of potato production. All the periods later after the 1990s, Asian region has dominated the rest of regions in potato production. Even taking the 2000s' average production in panel (c), we still witness potato production dominance by the Asian region compared to other regions. Asia's awake in potato production is very important in increasing the world food security. With improved potato processing technology, potato is becoming a promising food crop suitable for fighting against starvation. Nevertheless, the crop can also be a driver for industrial development particularly in developing countries.

Figure 3 provides a list of countries leading in potato production using the average values computed from 1961 to 2019. The list includes even the Soviet Union which collapsed [15] in 1991. That is why Russian Federation has a shorter span of time from 1991 to 2019.

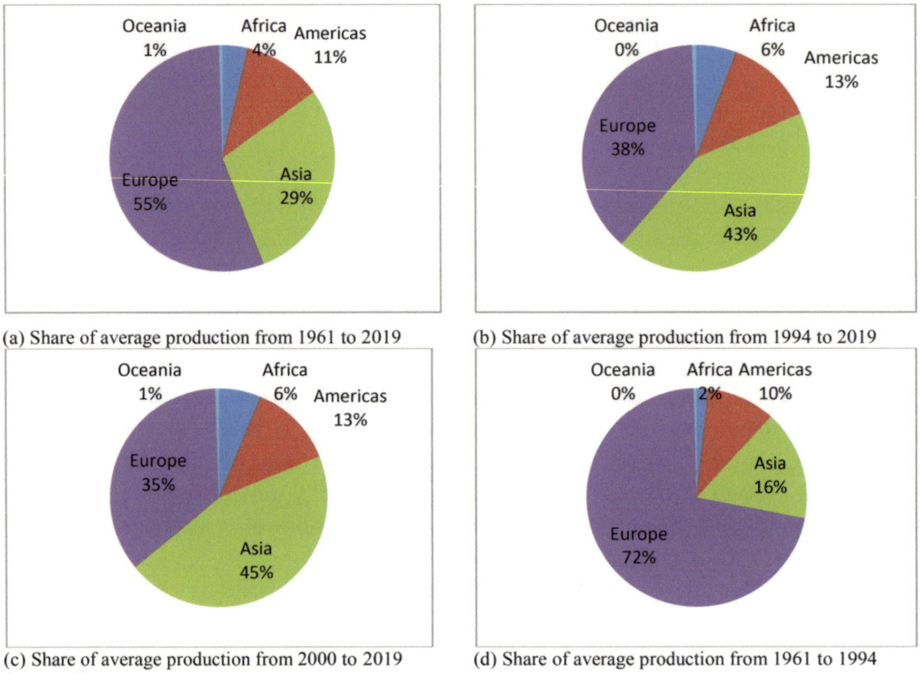

(a) Share of average production from 1961 to 2019

(b) Share of average production from 1994 to 2019

(c) Share of average production from 2000 to 2019

(d) Share of average production from 1961 to 1994

Source: (9)

Figure 2.
Region production shares of potato. Note: Due to approximations, Oceania seems to have zero share of potato in the world. However, Oceania has in every panel some quantities of potato produced.

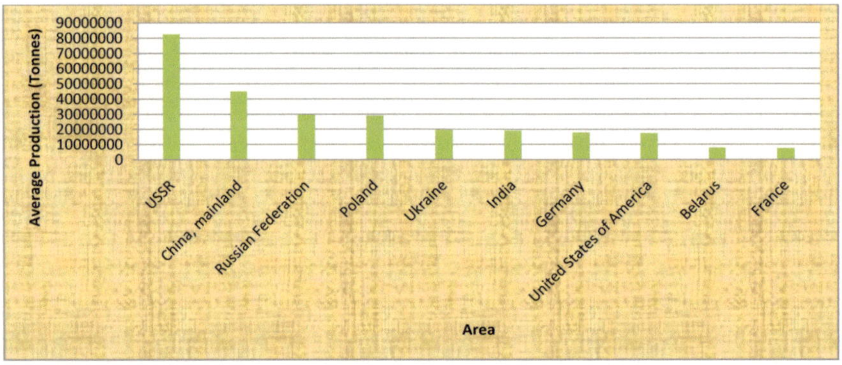

Figure 3.
Top 10 in potato production. Source: [9].

The shorter time period of Russian Federation is likely to be the reason for lower average potato production compared to other countries included in the analysis. The Soviet Union is leading in the average potato production since 1961 to 2019 even though it collapsed in 1991 because the union involved many countries. It is a group of countries rather than an individual country. For individual countries, China mainland is leading in average potato production from 1961 to 2019. Even in the trend analysis, China has a very high level of potato production which is again trending upward at a very high speed. The slope of the production trend is roughly steeper than those of other countries. As a result, even if China's potato productivity is almost half the potato productivity of the USA, total potato production is far larger than that of the USA. This is highly influenced by the increased land under potato production, but also potato productivity [14] in China. The increased land devoted for potato production in China is due to increased processing of coarse starch which is the most important component of potato processing industry in China. But also other processing industries such as crisps, and French friesare expanding [14].

4. Potato harvestedarea for some selected countries

As shown in the previous section, Africa's potato share has remained stagnant and lagging behind other regions except Oceania. The region is large in geographical size, but with a lower average production compared to other small sized regions like Europe. This is due to the fact that most agriculture practices in Africa is under subsistence farming where farmers grow crops in small plots with poor farming implements. In Kenya for instance [5] farmers cultivate on a land of less than 2.4 hectares with diseases constraint. In this part, six African countries that are, Nigeria, Rwanda, Senegal, Uganda, Tanzania, and Zambia have been selected to represent the region. For Europe, Sweden and Romania have been used as representatives. China, Korea, Japan and Philippines have been used to represent the region of Asia. Nevertheless, Peru, Uruguay, Canada and the United States have been selected to stand for the Americas region. The Oceania has been represented by Australia. As the list shows, the representation is not even but only to provide some light on the production status of the region. The countries as we all know have different characteristics to become a regional representative. But, for analytical purposes, the sample is still worth of knowledge generation. A clear specific country trend analysis can be provided on request. But due to space limitation, the overlay graphical analysis is used to highlight important results.

The graphical analysis of area harvested, quantity produced and the land productivity is provided in this section. It is important to have production area expansion especially in low income countries due to low agricultural mechanization. The increased technology leads to increased production without necessarily expanding the area under potato cultivation. From appendices, the trend shows that on average almost every country increased the area under potato production although with some variations of up and downs. For instance, in the late 1990s, Nigeria expanded potato production area which had previously been almost constant from 1961. An expansion from less than 50,000 hectares to about 200,000 hectares is significant in increasing production quantities. The potato production area trends in Rwanda, Tanzania, and Zambia are shown with hump shaped structure reaching peaks in late 2000s and dropping in around 2010s. The likely explanation of this drop is poor weather condition because the area provided here is that which has been harvested with potato. So, it is likely that the cultivated area did not shrink but due to unfavorable weather conditions, the harvested area declined.

Potato production in Europe has also been declining in terms of the world average potato share. For the selected countries, however, Israel is also termed as a European country in the analysis here. Israil cannot be accepted in Europe due to its geographic position. But for analytical purpose, it still works better to place Israel in Europe. An important point to stress here is that we have the ability to talk about Israel potato production regardless of the region we place the country. The analysis just takes few of the selected countries to show how area harvested has been trending throughout the analysis period. The selection is not based on any scientific reasoning but rather a random selection made discretionarily.

The analysis shows a quit unique trending for each of the countries involved. However, unlike other European countries which have shown some hump shaped curves, Sweden has reduced the land size for potato production by time. Even

Romania has also decreased the land size for potato production. Although, it is the area harvested with potato, the reduction of land size can be an appropriate explanation for harvested area shrinking. The reason comes from the fact that Europe has a high agricultural technology which improves productivity thereby reducing the area to be cultivated with crops. The reduction in area under potato harvest explains why world potato share of Europe has been declining by time.

The harvested area for Canada has been low but then increased shortly just to return to the lower levels making a hump shaped curve. Peru has shown increasing interests after dropping down in the 1990s then rising throughout. The United States of America and Uruguay have almost the same potato harvest pattern of fluctuations but continually dropping down. Even though, as it can be seen, the two have a significant difference in terms of absolute potato harvest area. The Uruguay's harvested area is almost ten times less than the one of the USA. So, even if the harvested areas of both countries have been declining in the same trending pattern, the USA produces large quantities of potato compared to Uruguay and the rest of included countries in the regional analysis.

After we have traced the trend of potato harvested areas for Americas, Africa, and Europe, it is also important we slightly analyze the case for some Asian countries. In this case, China, Japan, Korea, and Philippines have been used to provide a glimpse of the Asian region. However, as noted before, they are not used as reflecting what is happening in Asia but rather their analysis presents a representative of the region.

In the harvested area, China has been expanding the size annually, the pattern which is also experienced with Philippines. Even though, China's area is far larger than Philippines' harvested area. The rest of the included countries, that is, Japan and Republic of Korea have continually experienced potato harvested area shrinking. The increased technology in the countries allow for the reduction of planted area with either increased or constant potato output. In the next section, the trend

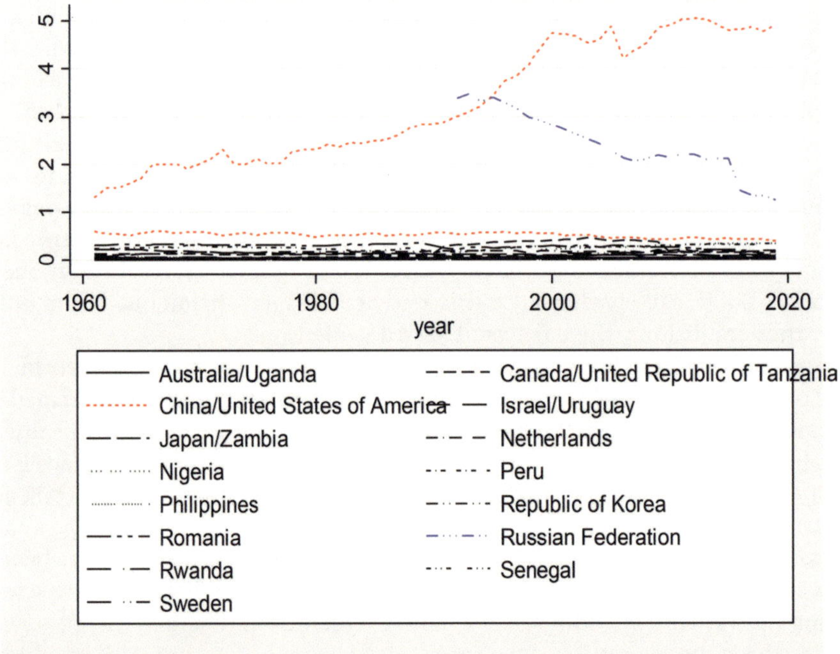

Figure 4.
Area harvested with potato. Source: [9].

analysis of potato land productivity is provided. This is the focus of the chapter which however becomes more informative after the harvested area trend analysis. Increased potato productivity is at the center of starvation and poverty reduction.

From **Figure 4** above, the red colors are for China and the USA with larger sizes being recorded for China. The upper red line is for China while the lower red line is for the USA showing a very huge difference between two countries. After Soviet Union collapsed, Russian Federation started operating independently with potato allocated land of the size of China. But the size of potato cultivated land in Russia has been continually declining, whilst that of China's has been continually increasing. However, in terms of land productivity, China is far less than the USA as the analysis in the following section shows.

5. Potato productivity for selected countries

Before we consider productivity, it is important to have a total production analysis. As it is shown in **Figure 5** below, total production trend almost resembles area under potato production in **Figure 4** above. The red lines again stand for China and USA, and the blue line is for Russia. Just like for land under potato production, total potato production when Russian Federation started operating independently was the same as that of China. However, the production for Russia has been downward trending following potato production land shrinking. Potato production for the USA has been lower than the production for China. Apart from China, Russia and the USA, potato production trends for the remaining countries lie below 20 million tones.

But, contrary to total production analysis, potato yield trend analysis shows a different picture. The productivity analysis shows very different outcomes whereby countries with increasing harvested areas have experienced lower productivity than countries with declining harvested area. In other words, countries which have

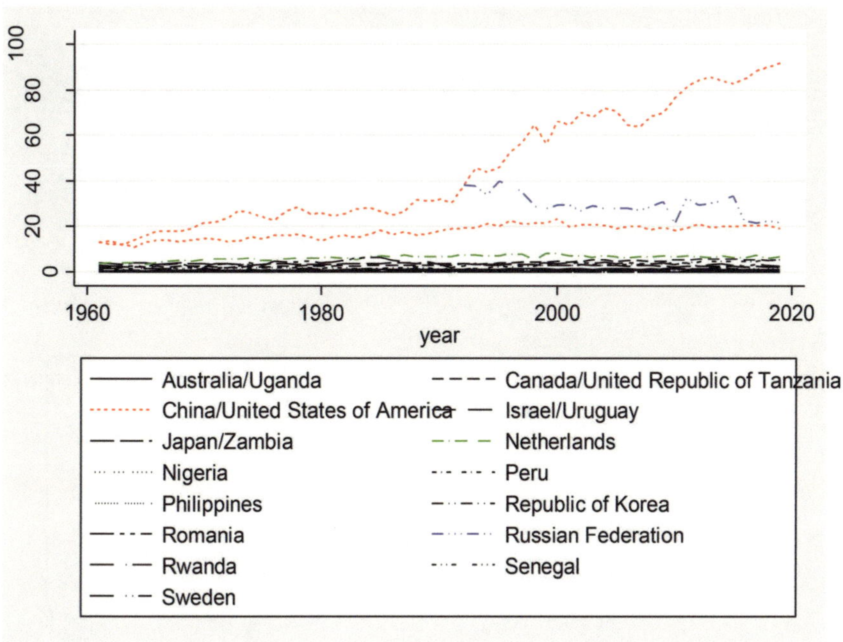

Figure 5.
Potato production among selected countries. Source: [9].

experienced increasing productivity, have discretionarily reduced the size of their potato farming land. For instance, the red lines in **Figure 6** again stand for China and the USA. But this time, it is the lower line which represents China and the upper line standing for the USA. As it can be seen, from almost 2012 or later, the USA is leading in terms of potato productivity. For low income countries like Uganda which is represented by the blue line, productivity is very low. Uganda and Australia which are represented by the blue lines have a very big difference with Australia's potato productivity being far larger than that of Uganda.

The country specific trend analysis in appendix shows that the USA and Uruguay have been experiencing a growth in potato productivity which led into a declining potato harvested area. However, the productivity level of the USA is far larger than the productivity of Uruguay. Japan with a declining harvested area has very high productivity levels compared to China for instance which has experienced increasing harvested area. Productivity in Japan has gone up to higher levels than 30 tons per hectare while that of China has not managed to reach even 20 tons per hectare. As shown in **Figure 3**, China is leading in tuber crop production. Nevertheless, in appendix B, China has the highest level of potato production far larger than the rest of the countries involved in the analysis. It is therefore, clear that the increased production level has been influenced by land expansion. An improvement in China's productivity level to that of the USA for instance without shrinking the potato cultivable land will improve food security to a large extent. As a result, increasing productivity is necessary for development as the remaining land can be utilized for other development purposes. The increased potato and other crop yield is important [6] to spare land for nature conservation.

The next section provides panel regression analysis on the relationship between output and harvested area, as well as yield and harvested area. The importance of this analysis is its ability to tell on the statistical significance of the relationship between output or yield and harvested area. Nevertheless, the panel estimation is

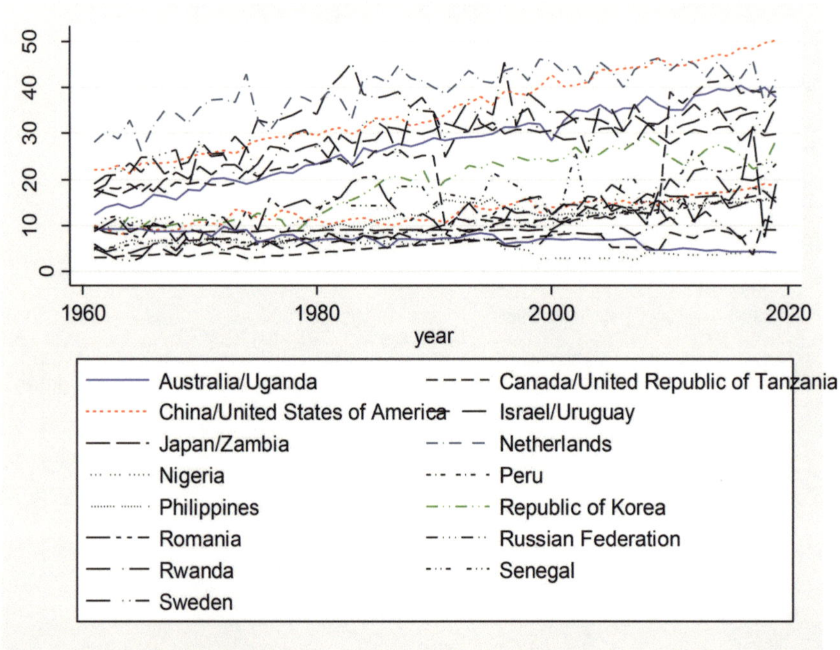

Figure 6.
Potato yield for selected countries. Source: [9].

followed by the post estimation test for random effect. This helps to identify the appropriate model for the coefficient estimation in the study.

6. The effect of land on potato production

6.1 Post-estimation test

After running the random effect model, the test for a zero variance hypothesis is imperative. This is because the random effect assumes a zero correlation between the error term and the independent variable. The zero correlation hypotheses are tested using Breusch and Pagan lagrangian multiplier test for random effects. From the tests, we reject the null hypothesis or zero variance because the test is statistically significant at all levels of significance. Nevertheless, the null hypothesis of zero correlation between the independent variable and the error term is rejected for both output and yield equation. The procedure involves testing whetherthe following equations holds or not.

$$Yield[Country, t] = Xb + u[Country] + e[Country, t] \quad (1)$$

$$Output[Country, t] = Xb + u[Country] + e[Country, t] \quad (2)$$

Where, X is the natural logarithm of land which is the only independent variable used in the analysis, b is the land elasticity of yield in the first equation or output in the second equation. The variance of the model due to country specific differences is represented by u[Country], and the variance of the model due to country differences and time differences is represented by e [Country, t].

In the random effect model, the assumption is that the equation does not hold because there is no influence of country specific differences. The crucial difference [16] between random effect and fixed effect models is whether the unobservable individual elements are correlated with the regressors in the model. However, the results show clearly that country specific differences influence the variation of the dependent variable since $Var(u) \neq 0$ for all the equations estimated, that is yield equation as well as output equation.

6.2 Estimation results

Estimates of both random effects and fixed effects models are provided in **Table 1** below. The difference in the magnitude of their coefficients, however, is not significant. Potato yield is negatively affected by land size because land is in the denominator of the ratio. This implies that any increase in the denominator affects the yield ratio negatively. The results contend with the yield trend analysis where countries with high yield growth rates have also reduced potato harvested area significantly. However, land expansion is better for higher levels of potato output.

Something worth noting in the table above is the similarity of standard errors for each model. That is, random effect model estimation provides similar standard errors for both equations and the same applies for fixed effect model. Nevertheless, even the intercept coefficient is similar for that matter. Similarities are also noted in the variance parameters provided in the lower part of **Table 1**. This similarity tells us on the similarities of the equations due to the fact that the yield equation has both output and land as it is for the output equation.

	Random Effect Model		Fixed Effect Model	
Variable	**Output**	**Yield**	**Output**	**Yield**
Land	.922*** (.014)	−.078*** (.014)	.916*** (.015)	−.084*** (.015)
Constant	3.50*** (.209)	3.50*** (.209)	3.56*** (.156)	3.56*** (.156)
sigma_u	.628	.628	.678	.678
sigma_e	.328	.328	.328	.328
rho	.786	.786	.810	.810

Source: [9].
*Note: *** implies statistically significant at 1 percent levels of significance and standard errors are in parentheses.*

Table 1.
Random effect and fixed effect estimates.

Since we have involved many countries of different characteristics in terms of mechanization, it is important to have estimates for each of the countries in the sample size. Nevertheless, it is true that these countries differ in terms of potato production level, productivity and the size of land apportioned to potato production also differ from country to country. The multilevel mixed effect model is applied to get country specific effect. The approach uses overall data to get estimates for a specific country.

So, even if a country or unit of analysis has lower number of observations, no estimation problem will result due to insufficient number of observation. For instance, for the sample used in the analysis, Russian Federation has 28 observations which is less than the minimum requirement of 30 observations. But, due to the application of multilevel mixed effect model, estimates for Russia have no observational problem just like the rest of the countries which have each 59 observations. The results show differences in intensity of influence of the land size on potato output or yield level.

The country specific estimates show differences in the influence of land on either yield or output. Some countries show results contrary to the main model estimation results in **Table 1**. For instance, yield is negatively affected by land size in the main model estimation. But, country specific estimations show some positive effects of land on potato yield. Potato yields for countries like China, Israel, Netherlands, Peru, Philippines, Senegal, Tanzania, and Zambia, are positively influenced by land size. For these countries, an increase in land size helps to increase potato yield. It can be interpreted that for these countries, although there has been expansion of land under potato production, production of potato has increased at a higher rate than the land expansion rate. Therefore, the pulling down effect of land expansion has been outweighed by the pushing up effect of production increase. It can therefore, be argued that potato yield has been growing at a higher rate than the land expansion rate. An investigation in Yunnan province [17], suggests the use of mixed cropping for developing countries where farming is dominated by small-scale farming. Their findings revealed an increased crop yield between 33.2 percent and 84.7 percent for the same season due to mixed cropping.

For the rest of the countries, potato yield is negatively affected by land size whereby increased potato yield has led to a reduction in the size of land under potato production. It is best to interpret the results that way rather than saying that in order to increase potato yield the area under potato production must be reduced. It is of course the increased potato productivity that influences a particular country to reduce the size of land under potato production. The findings contend with [18] where increased food crop yield corresponds to reduced food crop grown area.

Alternatively, these countries have reached the saturation levels of potato productivity that increases in land size increases potato production but at a rate lower than expansion rate. But, as we have seen in the trend analysis, most of these countries have reached to a point where no improvement in productivity or productivity has been fluctuating around a constant acting as if stationary without increasing or decreasing trend. But, some of the countries have shown increasing yield trend while declining trend of land under potato production.

On the output side, some countries have their potato production being negatively affected by land size. From **Table 2**, it is clear that potato productions in the USA and Australia are negatively influenced by increases in land under potato cultivation. The implication here is that, even with a shrinking cultivated land size, total production in these two countries has increased. As a result, their potato productivity is very high as comparedto other countries. High agricultural technology in these countries, has led into high improvement in potato production that reduction in the area under cultivation does not reduce potato production. When all countries reach to this level of potato productivity, hunger and starvation will remain history in the world.

	Yield		Output	
Country	**Coefficient**	**Constant**	**Coefficient**	**Constant**
Australia	-1.46^{***} (.333)	18.7^{***} (3.50)	$-.463^{***}$ (.333)	18.7^{***} (3.50)
Canada	$-.609^{***}$ (.070)	10.3^{***} (.848)	$.391^{***}$ (.070)	10.3^{***} (.848)
China	$.484^{***}$ (.035)	-4.69^{***} (.515)	1.48^{***} (.035)	-4.69^{***} (.515)
Israel	$.122^{**}$ (.048)	2.36^{***} (.433)	1.12^{***} (.048)	2.36^{***} (.433)
Japan	$-.543^{***}$ (.042)	9.64^{***} (.486)	$.457^{***}$ (.042)	9.64^{***} (.486)
Netherlands	$.818^{***}$ (.155)	-6.12^{***} (1.85)	1.82^{***} (.155)	-6.12^{***} (1.85)
Nigeria	$-.211^{***}$ (.010)	3.85^{***} (.106)	$.789^{***}$ (.011)	3.85^{***} (.106)
Peru	$.694^{***}$ (.201)	-6.42^{***} (2.50)	1.69^{***} (.201)	-6.42^{***} (2.50)
Philippines	$.801^{***}$ (.045)	-4.45^{***} (.384)	1.80^{***} (.045)	-4.45^{***} (.384)
Republic of Korea	$-.996^{***}$ (.073)	13.2^{***} (.759)	$.004^{**}$ (.073)	13.2^{***} (.759)
Romania	$-.488^{***}$ (.184)	8.65^{***} (2.31)	$.512^{***}$ (.184)	8.65^{***} (2.31)
Russian Federation	$-.515^{***}$ (.066)	10.1^{***} (.964)	$.485^{***}$ (.066)	10.1^{***} (.964)
Rwanda	$.284^{***}$ (.038)	-1.10^{***} (.412)	1.28^{***} (.038)	-1.10^{***} (.412)
Senegal	$.252^{***}$ (.087)	$.975^{***}$ (.568)	1.25^{***} (.087)	$.975^{***}$ (.568)
Sweden	$-.421^{***}$ (.049)	7.78^{***} (.518)	$.579^{***}$ (.049)	7.78^{***} (.518)
Uganda	$-.176^{***}$ (.049)	3.75^{***} (.511)	$.824^{***}$ (.049)	3.75^{***} (.511)
United Republic of Tanzania	$.248^{***}$ (.031)	-1.02^{***} (.332)	1.25^{***} (.031)	-1.02^{***} (.332)
United States of America	-1.63^{***} (.191)	24.9^{***} (2.51)	$-.628^{***}$ (.191)	24.9^{***} (2.51)
Uruguay	$-.830^{***}$ (.046)	10.1^{***} (.436)	$.170^{***}$ (.046)	10.1^{***} (.436)
Zambia	$.240^{***}$ (.048)	$.781^{***}$ (.314)	1.24^{***} (.048)	$.781^{**}$ (.314)

Source: [9].
Note: ***, **, and * are levels of significance at 1, 5, and 10 percent respectively. Standard errors are given in parentheses.

Table 2.
Multilevel mixed effects model estimates.

7. Conclusion

The chapter aimed at answering two questions that is whether there has been an increase in potato production among countries involved in the analysis, and whether this increase is due to productivity or land expansion. From the findings, it is clear that, on average, potato production has been increasing from 1961 to 2019. However, potato production increment for some countries has been due to productivity increase, while for others, production increase has been a result of land expansion. For some countries like Tanzania, however, potato yield has been positively influenced by potato cultivable land expansion. For almost all countries except Australia and the USA, potato production has been positively influenced by potato cultivable land expansion. As it has been emphasized [19], potato production increase should come from yield increase rather than land expansion.

The crop is highly produced in developed countries compared to developing countries. Potato productivity for most developed countries has been increasing compared to potato productivity among developing countries. Higher per hectare productivity in developed countries is a result of agricultural mechanization. The application of appropriate farming technologies, which are more advanced, influences more output coming from one hectare. Low potato productivity in developing countries can be explained by low farming technology which forces countries to expand farming land in order to increase production. Given the availability of improved technology and farming techniques in developed countries, developing countries can adopt the technology to increase their own potato productivity. With high technology adoption among low income countries, both potato production and yield are expected to increase in the near future at a higher rate thereby fighting against hunger and starvation.

Potato is an ideal food crop to fight against hunger and starvation especially in low income countries. However, in order for potato to help reducing starvation, countries particularly low income countries must invest in advanced farming technologies to increase potato productivity. Higher potato productivity will avail food at a lower cost, increase employment from industrial processing and therefore improve the living standard of people. Nevertheless, for developed countries with high productivity, potato production land reduction decisions should be revised. This means that since productivity is increasing in developed countries, more land should be available for potato production to ensure food security even in countries with low productivity. Countries with low productivity can access food from high income countries at an affordable price. So, reduction of land under potato production does not match with food security improvement strategy. Even though, if all countries reach to a level of technology where land size reduction leads to increased production and higher potato productivity, then it will be optimal to reduce potato cultivable land for other uses. At this stage, starvation will be something of the past.

This chapter has analyzed potato production and productivity in relation to potato harvested area which is the approximation of potato grown area. There are more factors which influence potato production such as labor, machineries, irrigation, fertilizer application, and spacing which are not included in the current chapter. Future, studies should incorporate these factors to make a detailed analysis. Nonetheless, focusing the analysis on a small area is more appealing as it can include social and economic characteristics of the farm manager which are more important in influencing potato production. From these specific analyses, policy recommendations can be more useful for farmers which are mainly smallholders in developing countries to improve food security by applying more advanced farming technology.

Author details

Fulgence Dominick Waryoba
Economics Department, St. Augustine University of Tanzania, Mwanza, Tanzania

*Address all correspondence to: fuldominick@yahoo.com

References

[1] Camire ME, Kubow S, Donnelly DJ. Potatoes and Human Health. Crit Rev Food Sci Nutr. 2009 Dec;49(10):823-40.

[2] Spooner DM. Solanum tuberosum (Potatoes). In: Brenner's Encyclopedia of Genetics. Elsevier; 2013. p. 481-3.

[3] Razzaghi F, Zhou Z, Andersen MN, Plauborg F. Simulation of potato yield in temperate condition by the AquaCrop model. Agric Water Manag. 2017 Sep;191:113-23.

[4] Xu X, Pan S, Cheng S, Zhang B, Mu D, Ni P, et al. Genome sequence and analysis of the tuber crop potato. Nature. 2011 Jul;475(7355):189-95.

[5] Jane Muthoni J, Shimelis H, Melis R. Potato Production in Kenya: Farming Systems and Production Constraints. J Agric Sci. 2013;

[6] van Ittersum MK, Cassman KG, Grassini P, Wolf J, Tittonell P, Hochman Z. Yield gap analysis with local to global relevance—A review. F Crop Res. 2013 Mar;143:4-17.

[7] Sanginga N. Root and Tuber Crops (Cassava, Yam, Potato and Sweet Potato) Roots. An Action Plan African Agric Transform. 2015;

[8] Liu Q, Liu J, Zhang P, He S. Root and Tuber Crops. In: Encyclopedia of Agriculture and Food Systems. Elsevier; 2014. p. 46-61.

[9] Food and Agricultural Organization of the United Nations. FAOSTAT Crops [Internet]. Crops. 2020. p. last access: May,2020. Available from: http://ec.europa.eu/eurostat/statistics-explained/index.php?title=Agricultural_production_-_crops%0Ahttp://ec.europa.eu/eurostat/statistics-explained/index.php/Agricultural_production_-_crops

[10] Baker P, Friel S. Processed foods and the nutrition transition: evidence from Asia. Obes Rev. 2014 Jul;15(7):564-77.

[11] Poti JM, Mendez MA, Ng SW, Popkin BM. Is the degree of food processing and convenience linked with the nutritional quality of foods purchased by US households? Am J Clin Nutr. 2015 Jun;101(6):1251-62.

[12] Mitchell H. Sweeteners and Sugar Alternatives in Food Technology. Mitchell H, editor. Sweeteners and Sugar Alternatives in Food Technology. Oxford, UK: Blackwell Publishing Ltd; 2006.

[13] Raymundo R, Asseng S, Robertson R, Petsakos A, Hoogenboom G, Quiroz R, et al. Climate change impact on global potato production. Eur J Agron. 2018 Oct;100:87-98.

[14] Jansky SH, Jin LP, Xie KY, Xie CH, Spooner DM. Potato Production and Breeding in China. Potato Res. 2009 Feb;52(1):57-65.

[15] Borjas GJ, Doran KB. The Collapse of the Soviet Union and the Productivity of American Mathematicians*. Q J Econ. 2012 Aug;127(3):1143-203.

[16] Torres-Reyna O. Panel Data Analysis Fixed & Random Effects. Princet Univ. 2014;

[17] Li C, He X, Zhu S, Zhou H, Wang Y, Li Y, et al. Crop Diversity for Yield Increase. Fuller DQ, editor. PLoS One. 2009 Nov;4(11):e8049.

[18] EWERS RM, SCHARLEMANN JPW, BALMFORD A, GREEN RE. Do increases in agricultural yield spare land for nature? Glob Chang Biol. 2009 Jul;15(7):1716-26.

[19] West PC, Gibbs HK, Monfreda C, Wagner J, Barford CC, Carpenter SR, et al. Trading carbon for food: Global comparison of carbon stocks vs. crop yields on agricultural land. Proc Natl Acad Sci. 2010 Nov;107(46):19645-8.

Chapter 2

The Role of Crop Protection in Sustainable Potato (*Solanum tuberosum* L.) Production to Alleviate Global Starvation Problem: An Overview

Tijjani Ahmadu, Adamu Abdullahi and Khairulmazmi Ahmad

Abstract

Among food crops in terms of consumption, potato ranks fourth, most important and valuable crop worldwide in terms of production and area harvested after maize, wheat and rice. In the coming years, potato production must keep pace with global population expansion nutritiously and sustainably which can partially be achieved by reducing the yield losses caused by the destructive pest and disease activities to the crop. The challenge of 70–80% total microbial crop yield loss posed by pathogens must be addressed for sustainable potato production in order to properly alleviate the global starvation problem. Potato as a food security crop can help to achieve the four food security requirements: food availability, quality, accessibility and stability. Health benefits of potato have shown the presence of phytochemicals as well as resistant starch which serve as anticancer and antidiabetic. The role of potato in the global food security should not be over emphasized, hence in this chapter we want to give an overview on the global hunger and food security at present, and the role played by potato as a food security crop. In addition, potato yield losses caused by pests and diseases especially phytopathogens, their etiology and the role of crop protection in sustainable potato production to alleviate global starvation problem will be discussed.

Keywords: Crop protection, food security, potato, starvation, yield losses

1. Introduction: global hunger and agricultural growth at present

Mankind cannot survive without food which is one of the three basic necessities of life. Potato (*Solanum tuberosum*) ranks fourth most important food crop in the world after maize, wheat and rice in terms of human consumption (**Figure 1**) [1]. Potato as a food crop can help to achieve the four food security requirements: food availability, quality, accessibility and stability. At present, agri-food systems do not sufficiently provide nutritious food in a sustainable and eco-friendly way to the growing global population [2, 3]. Potatoes continue playing a very important role in

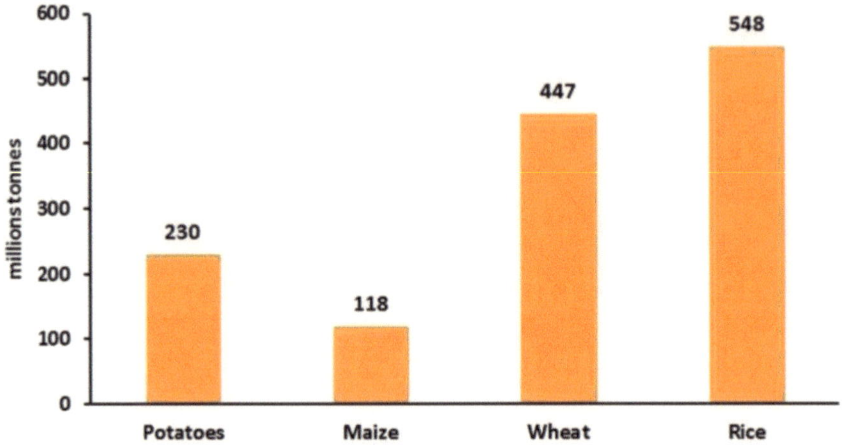

Figure 1.
Global consumption of maize, wheat, rice and potato. Source: FAO [1].

feeding the human population. According to [4], by 2050 an estimated global population of 9.7 billion people will demand 70% more food than is consumed today and feeding this expanded population both sustainably and nutritiously will require substantial improvements to the global food system. The food system should be one that provides livelihoods for farmers as well as nutritious products to consumers while conserving the environment and passing it *in situ* to the coming generations [5]. The Global Hunger Index (GHI) as reported by [6], showed substantial progress in terms of hunger reduction for the developing world and the GHI ranks countries on a 0–100-point scale with 0 being the best score (no hunger) and 100 being the worst. Whereas the 2000 GHI score for the developing world was 29.9, the 2017 GHI score is 21.8, with 27% reduction. Although, there are pronounced disparities in hunger at all levels and progress has been uneven. Poverty is the clearest manifestation of societal inequality and this supported the GHI report of 2017 that emphasizes the fact that inequality and hunger are extremely linked and both are rooted in uneven power relations that often are exacerbated and perpetuated by laws, policies, attitudes, and practices. GHI in 2013 showed that fifty-six countries are at alarming levels of hunger as published by the International Food Policy Research Institute [7]. The GHI aggregates three equally weighted indicators: (i) Prevalence of underweight in children (ii) Proportion of undernourished (iii) Mortality rate of children under five. Today more than 850 million people are suffering from hunger in addition to the several hundred million children categorized as malnourished children or as "hidden hungry".

Growth in agricultural sector can particularly be effective in reducing hunger effect, starvation and malnutrition problem because most of the extremely peasant poor farmers depend solemnly on agriculture and other related activities for their livelihoods. At the 2014 World Economic Forum (WEF), Shenggen Fan, Director of the IFPRI, advanced that tackling hunger and malnutrition is not only a moral issue but also one that makes economic sense as mentioned in a debate on "Rethinking Global Food Security". The world loses Gross Domestic Product (GDP) of 2–3% per year because of hunger, while investing US$1 in tackling hunger that yields a return of US$30. Additionally, it was mentioned in a debate on Rethinking Global Food Security by Ajay Vir Jakhar, Farmers' Forum Chairman (Bharat Krishak Samaj) in India that farmers think on food security at their household level, but not global level. Globally, if small-scale farmers were supported, they could become self-sufficient and also food insecurity problem would be solved by 40 to 60%. Therefore policies should be geared towards localized solutions, worldwide issues and solutions, and motivation from the private and public groups required to help the huge number of peasant farmers who are cultivating small areas of land and which have an important role to play in the chain of food production and social development. The International Year of Family Farming declared in 2014 supported the acknowledgement made by the United Nations on the importance of family farming in improving worldwide food security and poverty reduction. Hence localized, technical, and commercial solutions with the support of both public and private sectors are needed in combination with global food security policies. An important way forward to design research in agriculture is to understand where hunger and poverty are converged. Potato is produced in poor areas globally including China and the Andes of South America; hence, innovations particularly on potato science can be a very important tool for targeting the poor and hungry as part of a broader set of research and development activities.

2. Potato as a food security crop

According to [8, 9] "Food security exists when all people, at all times, have physical, social and economic access to sufficient, safe and nutritious food to meet their dietary needs and food preferences for an active and healthy life." Food security has four major key dimensions: (a) food availability (b) quality and use (c) stability and (d) accessibility to food.

- Availability of food implies that supply of food at both levels (regional and national) that determines the ultimate price of food should be improved in order to reduce food insecurity and hunger.
- Accessibility to food implies the ability for one to buy or produces his food, which has to do with having the purchasing power to do so.
- Quality of food and use implies the level of nutrition obtained from food intake (consumption) at a nutritional, sanitary, sensory, and sociocultural point of view.
- Stability of food implies the idea of having food accessibility at all times thus incorporating issues such as price stability and securing incomes for affected populations [9].

With this breakdown, efforts in research may likely assist food security in the categories below:

- Access: Encouraging farm production competition, farmers' income improvement and other agri-food systems.
- Availability: Rising agricultural production via good cropping systems, integrated pest control for loss minimization and genetic improvement.
- Food use and quality: Quality food safety as well as food quantity by value addition to traditional local products.
- Stability: Improving agriculture and food production through sustainable management of such natural resources such as soil, water, and biodiversity.

Some peculiar features of potato crop such as adaptation range coupled with high nutritional value and production ease has aroused the interest of many people to embark on its cultivation which has led to the steady increases in potato production and consumption in many developed and developing countries. In the last few years, there is an increase in production of potato and its demand in Asia, Africa, and Latin America from less than 30 million tons to more than165 million tons. Today, the biggest potato producer is China followed by India. According to FAO, potato yields more food per unit of cropland in less time than any other major crop [10]. Millions of farmers depend on potatoes for food as well as cash income. Potato is a highly reliable food security crop that can help ease future turmoil in world food supply and demand [10]. Potato cropping systems help improve resilience especially among smallholder farmers by providing direct access to nutritious food, increasing household incomes, and reducing their easiness to food price volatility (**Figure 2**).

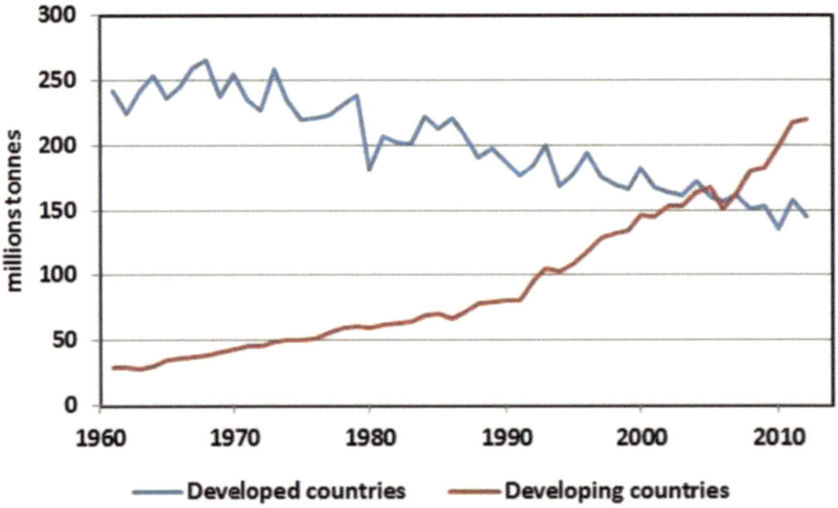

Figure 2.
Shift in potato production. Source: FAO [1].

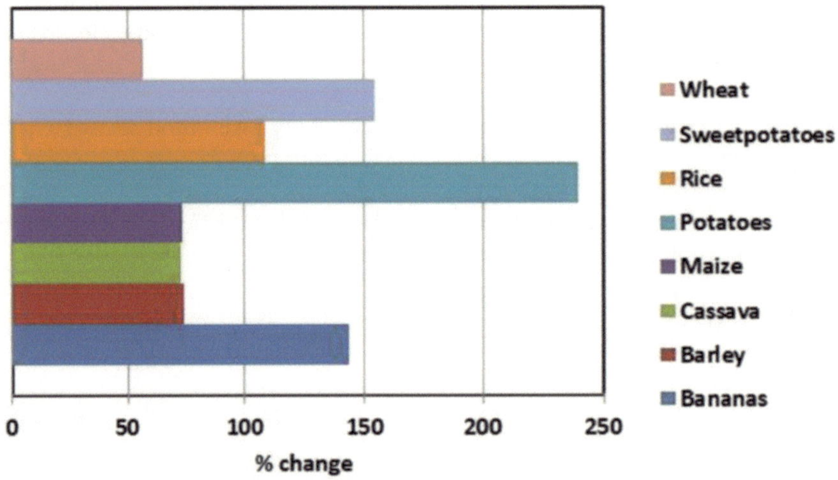

Figure 3.
Percent change in crop production of staple food crops in sub-Saharan Africa, 1994–2011. Source: FAO [1].

Farmers in Africa have responded to increased demand for food by increasing the production area for numerous crops that include banana, potato, sweet potato, and rice. This production increase was as a result of an increase in the production area of potato, which has doubled from 1994 to 2011 and now exceeds that of the Caribbean and Latin America (**Figure 3**) [1].

3. Otato in the global food system: production and demand trends by region

At present the global stands for potato production is 378 million tons on an estimated 19 million hectares of farmland worldwide (**Table 1**). Temperate area of northern hemisphere is where potato is mostly produced during the summer period (frost-free period). In these regions, potato is cultivated mainly as a cash crop and an important income source. Potato is significant in the Rift valley of the tropical regions of the African highlands, the highlands of the Andes, and the volcanic mountains of West Africa and Southeast Asia, where production is both for cash and food [12]. The crop is cultivated at the heat-free period as a winter crop in the subtropical regions such as in the southern China, Mediterranean region and North India. The crop is not considered as a staple crop in the lowlands of the tropics due to the high temperatures in the areas that do not favor potato growth and development [13]. **Figure 4** illustrates the recent pattern of the potato distribution worldwide [14, 15].

In order to boost the impact on the lives of the peasant farmers on investment in potato-related research and innovation, it is important to identify who are the peasant farmers, where they are dwelling, and that potato crop is crucial in the local food chain. In areas where production of the crop is in existence with poor income, there is a good chance to use potato as a tool to slashed poverty. In order to get optimal potential effect on living welfare of peasant farmers, International Potato Center (CIP) has prioritized its programs based on a pro-poor research-for-development paradigm where scientific research adhere to specific needs to address the peasant poor, other than a science driven paradigm which generates outputs of research, that may or may not adhere to real demands, and hands them to the partners.

Across landscapes of the globe, the adaptability of the potato combined with increases in its cultivation in different countries in the pass decades is dissimilar, even though this rise has been generated primarily by land expansion followed by improvements in yields. Statistics of the global potato production indicated a shift

Continent (Region)	**2014–2016**		
	Production %	**Yield %**	**Area %**
Europe	119,551	21.6	5547
Latin American and Caribbean	18,334	17.9	1023
North American	24,430	32.0	763
Asia	190,617	19.1	9975
Africa	25,270	14.4	1756
World	378,202	19.8	19,063

Source: [11].

Table 1.
Indicators of global potato production.

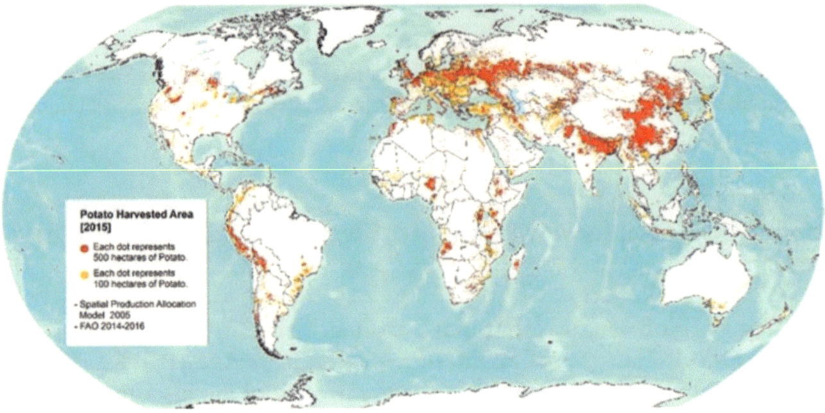

Figure 4.
Potato global distribution and harvested area. Source: You et al. and FAOSTAT [11, 14].

towards developing countries virtually with strong rise in production in Africa, more especially in East Africa. More so, the developed world's potato production is below that of developing world for the first time in 2005 [1] and this confirms the rising value of potatoes as a source of food, income and employment in Africa, Asia and Latin America. As world population levels are predicted to show the greatest rise in Africa in the coming decades, increased contribution of potato to local food systems in this region is of considerable importance [16]. Taking into consideration, production of potato in Latin American and Caribbean (LAC) over some years (at least 60 years), the average annual potato domestic production has risen from 7.2 million tons in the 1961–1963 periods up to 19.6 million tons in 2011–2013, which represents an average annual growth rate of 2%. By way of comparison, growth rates for potato production in Asia and Africa averaged over 4% for a similar period, i.e. more than double those of LAC [17]. Most of the production is oriented towards human consumption (74%, maintaining this trend throughout the period) and it highlights a relatively low processing level of 1% [11].

The fifth largest potato producer in the world is the United States in the global potato statistics with more than 420,000 ha harvested in 2013 and a total output of nearly 20 million tons [18]. Although in the United States potato is no longer the traditional staple of the past, it is nevertheless gaining increased appreciation by nutritionists because of its nutrient density and its contribution to a more balanced diet [19]. Potato yields in the United States have more than doubled over the last 50 years, rising from 22 tons ha^{-1} in 1961 to 49 tons ha^{-1} in 2016 as a result of improvements in the management practices [20]. This therefore will meet the demand of the agro- processing industry to produce chips and frozen French fries for consumers in the markets. In Asia, China became the world's largest potato producer in 1993 and currently accounts for almost one quarter of global potato production and about 28% of total cultivated areas [18] and used is mainly for food, both in processed forms and as a vegetable [21]. In India potato is cultivated mainly in the plain called Indo-Gangetic plain, intercropped in rotation with maize, rice and/or wheat or as singled crop (monoculture); and it is considered as important cash and a staple crop. Rising in potato production volumes, its yields have significantly increased in India at an average of 2% per annum, as a result of successful quality seed systems, breeding programs, and storage infrastructure that have decrease post-harvest losses [21].

Regarding Europe; France, Germany, Netherlands, Belgium and the United Kingdom are together the largest producers of potato in the European Union (EU), as potato yields more than 40 tons ha^{-1} in this area of north Western Europe and to the strong links of production with the dynamic European potato processing industry. Potato is also versatile in Eastern European countries, particularly in Ukraine, Russia, and Poland where per capita consumption has traditionally exceeded 100 kg annually. Future prospect and trends by region indicate a major production increase in Asia and Africa as compared to other regions. Taking into consideration some assumptions like increase in population, economic growth pathways, and climate change, a decline in population in China and growth of per capita (GDP) was projected by the UN which will subsequently influence their diet composition at long run. Hence, the supply of potato in China in future will not continue to grow faster as it was in the past. According to [22], it is in India where potato supply will almost triple because of the very high population growth, especially under certain socioeconomic scenarios.

4. Benefits of potato in the diet and health

4.1 Benefits in the diet

To a larger or lesser extent, the contributions of potato in the diet and health of human being for thousands of years should not be overemphasized. Proteins, fibers, carbohydrates, vitamins, lipids and minerals are present as food for human diet. Other benefits include its contribution as antioxidants, anticancer, anti-inflammatory, hypocholesterolemic, anti-obesity, and antidiabetic etc. Like other plant foods, the nutritional contribution and composition of potatoes is affected by many factors including bioavailability, bio-accessibility and cooking as well. For example cooking, the most important nutrient compounds found in potato that includes minerals, dietary fibers, and proteins are well retained after cooking as well as anthocyanins and carotenoids [23]. Vitamins C and B_6 are significantly reduced after cooking from the food matrix. Vitamin E is also contained in potato tubers at moderate amount [24]. Potatoes are mainly eaten as boiled and provide between 28 and 38% of the recommended total energy requirements for women [25]. The energy provided by 100 g of boiled tubers of potatoes varies from 96.33 to 123.17 kcal [25], which is similar to the energy provided by 100 g of cooked rice (130 kcal) but lower than the energy provided by 100 g of wheat (361 kcal), 100 g of cooked cassava (160 kcal) and soybeans (173 kcal) [26]. Total lipids in potatoes are found in low quantities ranging from 0.1 to 0.5 g–100 g^{-1} FW, consisting mainly of galactolipids and glycol (22%) and phospholipids (47%) that are structurally elements of biological membranes as well as neutral lipids (21%) like free fatty acids and acylglycerols [27]. More than 94% of the tuber lipids contain esterified fatty acids. According to [28], the protein content of potatoes generally ranges from 1 to 1.5 g/100 g^{-1} FW depending on the potato cultivar. Also [25] reported higher levels of protein in cooked tubers of Peruvian floury landraces (1.76–2.95 g-100 g^{-1} FW). Potassium is the most abundant mineral in potato with concentrations varying from 150 to 1386 mg-100 g^{-1} FW [29]. Potassium functions as an important electrolyte in the nervous system. High intake levels of potassium can help control high blood pressure and may decrease the risk of stroke [30]. Adequate Intake (AI) of boiled potatoes (100 grams) can contribute potassium recommended for adults (4700 mg per day). Phosphorus, iron, magnesium and zinc are also present in potato in small quantities ranging from 42 to 120 mg-100 g^{-1} FW for phosphorus and from 16 to 40 mg-100 g^{-1} FW for magnessium, respectively [31]. Again, potatoes are excellent source of diet in the form of iron especially in the highlands of Andean due to their high consumption and where accessibility to meat is little and levels of anemia and malnutrition are high. Typical example is that of Huancavelica, in the highlands of Peruvia, where women and children on average consume 200 to 840 g of potato per day [25].

4.2 Benefits in the health

Regarding health benefits, health-promoting effects attributed by potatoes were observed in human cell culture, human clinical studies, and experimental animals, including anti-inflammatory, anti-cancer, hypocholesterolemic, anti-diabetic and anti-obesity features. Compounds such as phenolics compounds, fiber, anthocyanins, and starch as well as compounds regarded as anti-nutritional compounds like lectins, glycoalkaloids, and proteinase inhibitors are considered to be attributed to the health benefits of potatoes. Many compounds found in potato are good in health promotion although some could be beneficial or detrimental to human depending on specific circumstances. Studies geared to investigate the association between potato consumption and diabetes, cardiovascular disease, obesity, and cancer while controlling for fat intake are needful [32]. As a key dietary source of potassium, vitamin C, and dietary fiber, potatoes contribute significantly to nutrients with defined roles in promoting cardiovascular health [33]. It was mentioned earlier that potato contains high amount of potassium and intake of potassium-rich foods has been shown to protect people against risk of stroke [34]. It was also reported that gelatinized potato starch containing a high level of phosphate reduced concentrations of serum-free fatty acids and triglycerides and liver triglycerides [35]. Potato consumption has often been associated in cohort studies with elevated risk of type 2 diabetes [36] and obesity [37], which has been attributed to a relatively high glycemic index in some potato varieties and processed potato products containing added saturated and trans fats. A major confounding factor in such studies is typical Western dietary patterns associated with increased disease risk typically include potato consumption along with high intake of red and processed meat, refined grains, high-fat dairy products, fried foods and sugar [38]. Reddivari et al. [39] showed that α-chaconine exhibited potent anti- proliferative properties and increased cyclin-dependent kinase inhibitor p27 levels in two prostate cancer cell lines, LNCaP and PC3. More recently, it has been reported that α-solanine, has a positive effect on the inhibition of pancreatic cancer cell growth *in vitro* and *in vivo*. Sun et al. [40] demonstrated that α-solanine inhibited cancer cell growth through caspase 3-dependent mitochondrial apoptosis and that the expression of tumor metastasis-related proteins, MMP-2 and MMP-9, was also decreased in the cells treated with α-solanine.

5. Pests and diseases of potato

5.1 Pests of potato

In addition to the present global climate change that used to worsen the situation, the potato's vulnerability to numerous pests such as tuber moth of potato (*Phthorimaea operculella*) [41], the leaf miner fly of potato (*Liriomyza huidobrensis*) [42], Guatemalan potato tuber moth (*Tecia solanivora*) [43]; the White flies (*Bemisia tabaci*), Andean potato tuber moth (*Symmetrischema tangolias*) [44] and *Trialeurodes vaporariorum* [45]. Pests, especially insects, are the major living factors affecting potato yield and tuber quality. Globally, losses are estimated on average at 16% [46]. It has been estimated that about 30–70% loss in tuber yield and quality can occurred for various pests, if pest infestation not routinely controlled [47, 48].

Common name	Scientific name	Order	Family	Origin	Distribution	Reference
Potato Tuber Moth	*Phthorimaea operculella*	Lepidoptera	Gelechiidae	South America	North, Central, and South America, Africa, Asia, Australia, and Europe North America, New Zealand. Australia, and Indonesia	[49]
	Symmetrischema tangolia					[48]
Andean potato tuber moth	*Tecia solanivora*			South America (Peru and Bolivia)	Colombia, Ecuador and Contrai America,	[50]
Guatemalan potato tuber moth				Guatemala		
Potato Leaf miner	*Liriomyza huidobrensis*	Diptera	Agromyzidae	South America	Different countries around the world	[42, 51]
Andean Potato Weevils	*Premnotrypes suturicallus* *P. vorax* *P. latithorax*	Coleoptera	Curculionidae	Andean region	Venezuela and Argentina	[52, 53]
Potato Psyllid	*Bactericera cockerelli*	Hemiptera	Triozidae	North America	United States, Nevada Colorado, Arizona, New Mexico, Mexico, El Salvador Honduras, and Nicaragua	[54–56]
Bud Midge	*Prodiplosis longlifila*	Diptera	Cecidomyiida	Americas	North America (Florida and Virginia), South America (Ecuador, Peru and Colombia)	[57, 58]
White Grubs	*Phyllophaga spp. Melolontha melolontha Anomala spp. Popillia japonica Holotrichia javana Holotrichia oblit Anomala oriontalis*	Coleoptera	Scarabaeidae		worldwide	[59–61]
Flea Beetles	*Epitrix tuberis* *E. papa* *E. cucumeris* *E. yanazara*	Coleoptera	Chrysomelidae	North America Peru	Spain and Portugal	[52, 62, 63]
Colorado Potato Beetle	*Leptinotarsa decemlineata*	Coleoptera	Chrysomelidae	Mexico	Central America, Canada, Europe, and parts of Asia	[64, 65]

Common name	Scientific name	Order	Family	Origin	Distribution	Reference
European Corn Borer	*Ostrinia nubilalis*	Lepidopetera	Crambidae	Europe	United States, Canada, China, India, Iran, Syria, Indonesia, North Africa (Algeria, Egypt, Libya, Morocco and Tunisia)	[66, 67]
Aphids	*Myzus persicae*	Hemiptera	Aphididae	China	North America, Europe and Asia	[68, 69]
White files	*Bemisia tabaci*	Homoptera	Aleyrodidae	Brazil or Mexico	Many counties in the world	[70]
Potato ladybirds beetles	*Henosepilachna vigintioctomaculata H. vigintioctopunctata*	Coleoptera	Chrysomelidae		Japan, Korea, China, Russia, Pakistan, Pacific islands, New Zealand and Autralia	[71–73]
Armyworms	*Spodoptera frugiperda* *S. eridania* *S. ornithogalli*	Lepidoptera	Noctuidae		North, Central, and South America, Africa, India and Europe	[74, 75]
Thrips	*Frankliniella occidentalis Thrips tabaci*	Thysanoptera	Thripidae	Western North America	Tropical, Subtropical, and Temperate Regions	[76, 77]

Table 2.
General description of major and minor Pest of pot.

Disease	Incitant	Symptoms	Reference
Fungal diseases			
Late blight	*Phytophthora ifestans*	• water-soaked light to dark brown spots on leaves • brown spots on stems • (slightly depressed areas with reddish-brown color on tubers)	[2, 80–82]
Early blight (EB)	*Alternaria solani* *A. alternate*	• dark brown to black necrosis on the lowest oldest leaves • a series of dark concentric rings are visible within the ring • The symptoms of EB on tubers are dark, slightly sunken lesions • It is not possible to distinguish between the different *Alternaria spp*	[83, 84]
Black scurf	*Rhizoctonia solan*	• It affects roots, stolen, stems and tubers • formation of sclerotia on the surface of the tubers • girdling on the stem with brown color • upward rolling of the leaves	[85]
Wart	*Synchytrium endobioticum*	• small greenish warts on the top of plants: stem, foliage and in extremely conditions on inflorescences • the typical symptoms of the disease on tubers are the proliferating warts which may vary markedly in form but are primarily spherical to irregular • the color of the vary with the variet	[86, 87]
Powdery Scab	*Spongospora subterranean*	• it infect all underground organs of porato (i.e. stolons, tubers, and roots) • purplish brown lesions are observed as initial symptoms on tubers • infection can be susceptible to root or stolon gall production	[88]
Bacterial diseases			
Bacterial wilt	*Ralstonia solanacearum*	• Wilting is the common symptoms	[89, 90]
Bacterial Blackleg and Tuber Soft Rot	*Pectobacterium aroidearum P. atrosepticum* *P. betavasculorum* *P. brasiliense* *P. cacticida* *P. carotovorum* *P. odoriferum* *P. parmentieri* *P. peruviense* *P. polaris* *P. punjabense* *P. wasabiae*	• stem necrosis • pith of the stem is often decayed • Infected plants produce few or no tubers • Plant leaves may turn bright yellow and the plant will eventually and die	[91–93]

Disease	Incitant	Symptoms	Reference
Potato Ring Rot	*Clavibacter michiganensis*	• Young infected leaves expand more slowly in the infected zones and become distorted • Leaves affected by xylem blockages further down the stem often develop chlorotic, yellow to orange, interveinal areas- • Leavers and tubers may simply be reduced in size and occasionally whole plants can be stunted	[94]
Common Scab	*Streptomyces spp.*	• necrosis on all underground parts of a potato • pitted scab, erumpent scab, and mild netted scab on the tuber	[95, 96]
Zebra chips	*Liberubacter spp.*	• severe Zebra chips on both the foliage and the tubers • Tuber development slows or ceases in symptomatic plants, resulting in yield losses • Infected tubers either do not sprout or have only hair sprouts	[97]

Table 3.
Summary of fungal and bacterial diseases of potato.

In this book chapter we provided some major and minor insect pests present in tropical, subtropical and temperate regions of the world (**Table 2**). Many pests have their evolution in the potato centre of origin, and farmers in the Andean region are confronted by numerous insect pests than those in Asia or Africa. Some species like the leaf miner fly (*Liriomyza huidobrensis*) and potato tuber moth (*Phthorimaea operculella*) has become highly invasive pests in many tropical and subtropical regions. In contrast, the strong adaptation of Andean potato weevils (*Premnotrypes spp.)* to the climate of the Andean region and its monophagous feeding habitat on potato and its wild relatives has restricted its distribution. Similarly, bud midge (*Prodiplosis longifilia*) presently with a distribution restricted in Florida, Virginia and South America (Peru, Colombia, and Ecuador) could be an invasive pests adapted by its polyphagous feeding habit. The Colorado potato beetle (*Leptinotarsa decemlineata*) native to Mexico, has spread across most of the United States, and was introduced into France in the 1920s from where it spread further reaching also parts of China [64]. Farmers in tropical and subtropical countries must contend with a higher number of pest species, and with some exceptions, a minimum of 2–4 pests often reach pest status requiring the application of control methods [52].

5.2 Diseases of potato

A disease is series of harmful physiological processes caused by continuous irritation of the host by a primary agent called a pathogen and exhibited as morbiphic cellular activity known as symptoms [78]. Potato diseases can be caused by fungi, bacteria, and viruses. Globally, the major potato diseases are late blight caused by *Phytophthora infestans*, early blight caused by *Alternaria solani*/, *A. alternata*, Fusarium dry rot caused by *Fusarium spp.*, Potato common scab caused by pathogenic *Streptomyces spp.* Black leg of potato caused by *Erwinia spp.* and bacterial wilt

Distribution	Genus/family	Virus	Transmission
Southern Andean region	Tepovirus. Betaflexiviridae	Potato virus T (PVT)	Contact, seed
Andean region, Brazil	Comovirus. Secoviridae	Andean potato mottle virus (APMoV)	Beetles
Peru	Nepovirus. Secoviridae	Potato black ringspot virus (PBRSV = TRSV-Ca)	true seed, nematodes
Peru		Potato virus U (PVU)	Nematodes
Peru		Potato virus B (PVB)	Nematodes
Europe		Cherry leaf roll virus (CLRV)	Nematodes, TPS, pollen?
Australia, Europe and New Zealand		Tomato black ring virus (TBRV)	Nematodes
Europe, North and South America		Lucerne Australian latent virus (LALV)	Unknown
Worldwide	Polerovirus	Potato leaf roll virus (PLRV)	Aphids
Worldwide	Potexvirus, Alphaflexiviridae	Potato virus X (PVX)	Contact
		Potato aucuba mosaic virus (PAMV)	
Worldwide	Cartavirus, Betaflexiviridae	Potato virus S (PVS)	Contact, Aphids
China		Potato virus H (PVH)	Unknown
Argentina and Brazil		Potato virus P	
Worldwide		Potato virus M (PVM)	Aphids
North America		Potato latent virus (PotLV)	
Europe. South America	Potyvirus, Potyviridae	Potato virus V (PW)	Aphids
Worldwide		Potato virus A (PVA)	
		Potato virus Y (PVY)	
Andes, only reported in wild potatoes		Wild potato mosaic virus (WPMV)	
Andean region	Tymovirus, Tymoviridae	Andean potato latent virus (APLV)	Beetles
		Andean potato mild mottle virus (APMMV)	
Carribean	Begomovirus. Geminiviridae	Potato yellow mosaic virus (PYMV)	Whiteflies
North America	Nucleorhabdovirus, Rhabdoviridae	Potato yellow dwarf virus (PYDV)	Leafhoppers
Worldwide	Pospiviroid, Pospiviroidao	Potato spindle tuber viroid (PSTVd)	Aphids, contact
Americas, Europe, Asia in cool and humid environments	Pomovirus, Virgaviridae	Potato mop-top virus (PMTV)	Spongospora
Colombia		Colombian potato soil-borne virus (CPSbV)	

Table 4.
Summary for viral diseases of potato.

caused by *Ralstonia solanacearum* [79] etc. The summary of the diseases caused by both fungal and bacterial pathogens is presented in **Table 3** and viral diseases in **Table 4**. Annual losses have been estimated for late blight (*Phytophthora infestans*) alone to be about €6.1 billion with resulting effects on food security, especially in developing countries. Disease symptoms can be noticed in the leaves as spots with light to dark brown water-soaked appearance. In the stems spots are usually brown in color and tubers appeared with slightly depressed areas with reddish-brown color. Mild temperatures and high humidity are requisite for disease development and, under optimal conditions; the disease can destroy a field in a few days.

6. Crop protection and sustainable potato production

According to [98], sustainability is the development that meets the needs of the present generation without compromising the ability of future generations to meet their own needs. The growing global population needs to be satisfied with food availability and accessibility through an intensive agricultural production system which signifies the need for various green revolutions. At present, our practices that involve indiscriminate use of synthetic chemicals, chemical fertilizers, and high utilization of non-renewable energy source have led to a large threat to environmental sustainability. For example use of agrochemicals to increase crop production is one way of adding unwanted substances to the environment, which eventually contributes to the emission of greenhouse gases and subsequent environmental alterations. These harmful practices can be reduced if appropriate crop protection measures are used stewardly in agricultural activities for vital approaches of improving potato crop production. Although continuous increase in the world population at an alarming rate requires more food for nutritional security [99], but the world is now facing a great challenge to adopt sustainable measures, green technologies, sustainable science, and cleaner production such that the generations to come may be able to benefit from the earth's ecology at its conserved form [100]. Conservation of the planet becomes necessary as "We don't have a Plan B, since there is no Planet B" [101]. All key processes in the biosphere and related human activities are quite interdependent, interconnected, and hence should be steered through a mutual systems approach [102]. Food security is one of the three most pressing super challenges of the twenty-first century, after climate change and overdependence on petroleum importation, and microbes are good enough in meeting out these challenges [103]. The aims or goals of sustainable production or development are People, Planet, Profit (Prosperity), Peace, and Partnerships (**Figure 5**) [105] and if good crop protection measures such as used of microbes are utilized judiciously they can make a significant contribution in the achievement of these goals [100, 106]. Microorganisms as part good crop protection measures are much of our past and our future, pivotal agents of ecosystem and planet's functioning hence are key parts of the stewards committee of planetary health and sustainability.

6.1 Crop protection and pests Management in Sustainable Potato Production

6.1.1 Chemical control of potato pests

At present, the most important challenge facing professionals in agriculture worldwide is ensuring sustainability in potato production [107]. Insect pests are major biotic constraints affecting potato tuber quality and yield. Global losses are estimated on average at 16% [46]. Locally, if not routinely controlled, reductions

Figure 5.
The five goals of sustainable production and development. Source: Tijjani and Khairulmazmi [104].

in tuber yield and quality can be between 30 and 70% for various pests [47, 48]. Likewise in other cultivated plants, control of insect pests in potato is achieved predominantly via application of pesticides. By some estimates, potatoes are the most chemically dependent crop in the world [107]. Even though insecticides have been largely successful in keeping potato production successfully going, there are well-known and serious concerns about long-term sustainability of this approach. Chemical control of pests involves the use of synthetic chemicals which have a long-standing reputation in agriculture and ensures produce protection. They produce instant effects on the pests because they are fast-acting biocides, resulting in the arrest of pest infestations [108]. Negative effects of insecticides on numerous organisms, including health risks to farmers and beneficial insects, gained considerable notoriety since 1960s [109]. Development of resistance and environmental concerns are the major reasons that lead to phasing out of many insecticides. Numerous cases are recorded for Potato pest species that are most prone to evolving resistance to a wide variety of chemicals. For example, the (2018) Arthropod Pesticide Resistance Database lists 300 cases of Colorado potato beetle (*Leptinotarsa decemlineata*) resistance to a total of 56 active ingredients; 469 cases of green peach aphid (*Myzus persicae*) resistance to a total of 80 active ingredients; 501 cases of two-spotted spider mite (*Tetranychus urticae*) resistance to 95 active ingredients; 111 cases of greenhouse whitefly (*Trialeurodes vaporariorum*) resistance to rather impressive 27 active ingredients [2].

It is a difficult and expensive task to develop replacement insecticides, and it is highly questionable that a plethora of new active ingredients will regularly appear on the market in perpetuity [110]. Therefore, good stewardship of existing chemicals is imperative and, whenever possible, their replacements with nonchemical control alternatives become an increasingly important business strategy for the pesticide industry and potato farmers. According to [110] development of resistance by insects could be managed by preventing the situation when only highly resistant homozygotes survive in a population and this can be achieved by doing the followings:

- Avoiding applications of the same or related products repeatedly throughout a growing season.

- Monitoring insecticide efficacy.

- The use of insecticide applications should not be the first option, but when the ultimate control option in an IPM approach after all other management options could not prevent to keep a specific pest population under the economic threshold.

- Whenever possible, leaving parts of the field untreated to allow susceptible pests to survive and interbreed with resistant pests.

- Applying insecticides at rates that are not lower than a recommended minimum. Otherwise, heterozygotes will survive and breed with each other.

- Applying insecticides only when pest populations are sufficiently high to cause economically important damage.

6.1.2 Integrated Pest management (IPM)

IPM is defined as an "ecosystem approach for crop production and protection that combines different management strategies and practices to grow healthy crops and minimize the use of pesticides. Kogan [111] also defined IPM as "a decision support system for the selection and use of pest control tactics, singly or harmoniously coordinated into a management strategy, based on cost/benefit analyses that consider the interests of and impacts on society, producers and the environment". It means "a careful consideration of all available pest control techniques and subsequent integration of appropriate measures that discourage the development of pest populations and keep pesticides and other interventions to levels that are economically justified and reduce or minimize risks to human health and the environment. IPM emphasizes the growth of a healthy crop with the least possible disruption to agro-ecosystems and encourages natural pest control mechanisms" [112].

The integration of control methods will ensure quality and safety and also provide retailers with desired extended shelf life. Other benefits of IPM include dramatic slowdown of evolution of pesticide resistance. Again, simultaneous adaptations to diverse and unrelated management techniques will require statistically unlikely genetic changes in pest populations [110]. Moreover, integrating various measures to control pests, it may possibly reduce our over-dependence on the "pesticide treadmill" of constantly replacing longstanding chemicals. Additionally, with all IPM advantages, it should be made close and available to farmers across all potato-growing regions of the world. For example in the management of leafminer fly, an IPM measure based on the use of seed treatment, action threshold, trapping devices, and steward application of insecticides showed a higher efficiency in the control of potato pests including *Liriomyza huidobrensis* rather than the conventional application of insecticides by farmers in the Canete valley of Peru. IPM decreased the total amount of pesticides used per season by 56% compared to the conventional management, representing a decrease of 69.2% in the environmental effect. Furthermore, IPM achieved 35% of higher marketable potato yield rather than conventional management [113].

6.1.3 Cultural practices

A number of cultural practices commonly used for the management of potato tuber pest are many and should not be overemphasized. Such practices including deep planting, the use of pest-free seed tubers, regular irrigation to avoid soil cracking, timely harvest, high hilling to protect tubers, clearing tubers after harvest exposed in the field for a long time (especially throughout the night), i.e., immediate harvest and storage; and removal of leftover tubers to reduce the overwintering field population are all common practices. Also, early maturing varieties can contribute for reduced risk of infestation. For example, weeding and removal of alternative and overwintering hosts such as wild mustards (*Brassica spp.*), use of wheat straw or white plastic as mulch, and intercropping with onion, garlic or coriander (*Coriandrum sativum* L.) have shown to reduce aphid populations.

6.1.4 Biopesticides/biological control

It can be an effective strategy in all those regions in which the pests have been unintentionally introduced and where natural enemies of the pests are absent to keep the pest population below economic threshold. The endoparasitoids *Halticoptera arduine* Walker (Pteromalidae), *Phaedrotoma scabriventris* (Nixon) (Braconidae) and *Chrysocharis flacilla* Walker (Eulophidae) were successfully introduced and established in three agro-ecological regions (low, middle, and high altitude) in Kenya [42, 114]. Also in the Andes, predators like carabids are widespread and affect the weevil population. Most common species are *Blennnidus sp.*, *Notiobia schnusei* (Van Endem) and *Harpalus turmalinus* Er. Additionally, fungi like *Beauveria bassiana* (entomopathogenic) and nematodes (*Heterorhabditis sp.*, and *Steinernema sp*.) have been identified and used to develop biocontrol strategies [52, 53, 115]. Biopesticides such as spinosins and abamectins generally provide excellent control of Colorado potato beetle (*Leptinotarsa decemlineata*) pest (but see cautionary note on insecticide resistance below). Bacterial insecticides based on delta endotoxin of bacterium *Bacillus thuringiensis* subsp. *tenebrionis* are also effective, but they must be applied against the first two instars. Plant extracts from such plants including leaves of *Nerium oleander* L., Melia azedarach L. fruits, neem leaves and seeds, *Bassia muricate* (L.) Asch., *Parthenium sp.*, *Lantana sp.*, *Hyptis sp.*, Tephrosia nubica (Boiss.) Baker, *Ipomoea carnea* Jacq., *Bidens pilosa* L. and *Rumex nepalensis* Spreng, roots have been shown to demonstrate an excellent level of toxicity to the larvae of *Agrotis ipslon* [74].

6.1.5 Physical control

It is essentially a good method used to control pests in the field or at storage. For example yellow attracts most insects; therefore, yellow sticky traps can effectively reduce the leafminer fly adult population. In the Cañete valley of Peru, a cumulative capture of up to seven million adults ha^{-1} by using fixed and mobile yellow sticky traps which resulted in a reduction of the control costs by 55.5% compared with chemical control, and an average use of six adulticide applications per season [116].

6.1.6 Resistance plant varieties

The most valuable and effective strategies to manage some pests like zebra chip is to discourage vector feeding by using plants that are resistant to psyllid feeding or less preferred by the psyllid.

6.2 Crop protection and diseases management in sustainable potato production

Control of disease in plants is defined as keeping disease severity below the level at which it may become economically significant [78]. In a bid to control these fungal pathogens causing losses to valuable crops at present, chemical control have been identified as the most common, popular and most effective strategy for managing plant diseases but public opinion demands a reduction in the use of chemical [117–119]. In addition to chemical control, there are a number of strategies including physical, biological, cultural, use of resistant varieties and in recent time plant-based pesticides that are enabling and instrumental to manage potato diseases and extend their shelf life without pollution to the environment and risk to the public health. This part of the chapter highlights the different techniques that are used to manage myco-induced potato diseases and other perishable produce viz.

6.2.1 Biological control

Biological control is the inhibition of infection, growth, survival and activity of one pathogen (organism) via the use of another organism with the result that there is a reduction in the evidence of the disease caused by the pathogen [118, 120, 121]. Biocontrol strategy can be a matter of harnessing any form of biological agent that exists in the environment or introduction of exotic species. The most important microoganisms causing serious losses annually in agriculture are the fungal plant diseases [122] but some of the fungal diseases including postharvest diseases of fruits and vegetables caused by fungal pathogens such as *P. infestans* and other disease causing organisms have been successfully controlled via the use of biocontrol agents [121–123]. The first experiment in biological control with antagonists was conducted by GB Sandford in Canada [2]. The mechanism of activity of these biocontrol agents may be by antibiosis (secretion of antibiotics as a result of an interaction with microorganisms, which at low concentration poisons or kills other microorganisms); by competition for space and nutrients; by metabolite production (production of cell wall lytic enzymes that can breakdown polymeric compounds, including chitin, cellulose, DNA, hemicellulose and proteins; by parasitism in form of hyperparasitism (in which the pathogen is directly attacked by a specific biological control agent that kills it or its propagules or mycoparasitism, that is microbial predation that results in the reduction of the pathogenicity of the pathogen [117, 121, 123, 124]. Combining biocontrol agent or antagonist with other postharvest treatments could increase the efficacy of the biocontrol agents [125, 126] Vesicular-arbuscular mycorrhizae (VAM) and Plant Growth Promoting Rhizobacteria (PGPR) are well known to reduce plant diseases and increase crop yield. Biocontrol applications on potato plants require a better understanding of the symbiotic fungal partners. Numerous bio-agents in the phyllosphere are antagonistic to *P. infestans*, which included the yeasts *Acetobacter spp.*, *Sporobolomyces spp.*, isolates of *Bacillus spp.* and *Pseudomonas spp.* [127, 128]. Various naturally occurring microorganisms, that is, *Trichoderma viride*, *P. aurantiogriseum*, and *Penicillium viridicatum*, *Chaetomium brasiliense* [129], *Acremonium strictum* [130], *Myrothecium verrucaria* and *P. aurantiogriseum* [131], showed antagonistic effect against *P. infestans*. Application of *P. fluorescens* at 0.5% was found effective against early blight disease of potato for decreasing the intensity of the disease under field conditions [132]. The biological control agents *T. harzianum* and *P. fluorescens* (seed treatment + foliar spray) were effective in decreasing the intensity of early blight disease of potato and also increase tuber yield [85].

6.2.2 Use of resistant varieties

Identification of new resistance sources and functional resistance or susceptibility genes has been recently greatly accelerated by modern techniques, such as effectoromics and resistance gene enrichment sequencing technologies. After the discovery of the Mexican wild species *Solanum demissum* as an excellent source of resistance, eleven major genes were introduced in cultivated tetraploid potato breeding lines [133, 134]. Although some of these genes can be considered defeated, others, for example R8, are still effective against current pathogen populations [135]. Over 50 R genes have been identified from wild *Solanum* species as detailed by [136], and the research field remains active with a growing list of genes available for potato breeding programs [135, 137–139]. However, due to crossing barriers and linkage drag, there are only few successful cases where R genes have been introduced into improved tetraploid breeding lines by classical breeding [140]. Introduction of a single R gene from wild germplasm is a lengthy procedure as demonstrated by the examples of commercial varieties Bionica and Toluca that contain Rpi-blb2 originating from *Solanum bulbocastanum*, and were released almost 50 years after the first crosses were made [141]. However, recently it was shown that R genes can also have quantitative effects. The potato cultivar Sarpo Mira contains at least four R genes that confer complete resistance against incompatible isolates and a quantitative R gene, Rpi-Smira1 that confers broad-spectrum field resistance [142]. Durability of quantitative resistance will, however, continue to depend on the size of the cultivation area of a variety as well as the dynamics of the pathogen population.

6.2.3 Chemical control

The application of chemical fungicides continues to be the most common strategy for the control of most disease causing phytopathogens, for example making late blight one of the top drivers for pesticide use in the world. The demand for weekly applications generates a billion-dollar business globally every year [143]. Chemical control involves the use of synthetic chemicals to control the pathogens which have a long-standing reputation in agriculture and ensures produce protection. They produce instant effects on the pathogens because they are fast-acting biocides, resulting in the arrest of disease epidemics [118, 144]. Various synthetic fungicides that have a broad spectrum of application in the field, transit, markets or storage houses have been used for controlling postharvest fruit rot diseases of tomato caused by many fungi [118, 145]. For example, [146] reported that the use of low-weight chemical compounds of sulfur dioxide (SO_2), ozone, and acetic acid as fumigants used for postharvest protection of produce especially fruits have proved to be effective in eradicating most of the rot-inducing pathogens. To optimize the use of fungicides, it is important to know the efficacy and type of activity of the active ingredients. The frequency and timing of fungicide applications may depend on the foliar resistance of the cultivar, fungicide characteristics, rate of growth of new foliage, weather conditions, irrigation, and incidence of blight in the region [147]. The most common chemicals used include diphenyl, dichloran, sodium-*o*-phenyl phenate, 2-amino-butane, benomyl, thiabendazole, imazalil, thiophanate-methyl, triforine, iprodione, captan, vinclozolin, borax and soda ash [148]. They have fungicidal properties and are used as wash treatments and are highly effective when used "hot" at temperatures in the range from 28 to 50°C depending on the crop susceptibility to the hot injury [148]. Fungicides like biphenyl, acetaldehyde vapors dichloran, and some ammonia-emitting or nitrogen trichloride-forming chemicals are used as supplementary volatile in pakage of fungistats impregnated in paper sheets during storage and transit. Some strains fungi are resistant to one or more of the synthetic fungicides therefore broad spectrum fungicides should be used in their control [148].

6.2.4 *Cultural control*

Cultural control includes all the measures undertaken as agronomic activities to change the microclimate, condition of the host and also the behavior of the pathogen in order to interfere with the activity of the pathogen i.e. reproduction, dispersal and survival [149]. Include in these cultural practices are the use clean certified seed, use of adequate inter and intra-row spacing, hilling, crop rotation, destruction of plant debris, harvesting at hot conditions and when tubers are matured [82, 149]. Crop rotation is one cultural practice that influences the occurrence of many pathogens. For example *A. solani* that causes early blight disease, the fungus persists as spores or mycelium in plant debris or soil in the field from one potato-growing season to the other or next. Therefore, the practice crop rotation, including the control of host plants such as weeds (black shadow) in the nonhost crops, reduces the initial soil born inoculum. A short crop rotation with host crops (tomato, potato) results in an earlier and more severe early blight epidemic [150]. In addition, the removal or burning of infected plant debris reduces the inoculum level. The fungus does not directly infect intact periderm, and so allowing tubers to fully mature before harvest reduced the risk of tuber infection. Additionally, wounding prevention at harvest and providing good storage conditions to promote wound healing can also reduce tuber infection [151]. The use of disease- and virus-free seed potatoes is the basis for an economical potato production. Virus-infected potato plants are more susceptible to most pathoges than healthy plants. Another important thing is the legislation related primarily to prohibit the importation of infected potatoes from one country to another. Disease avoidance using uncontaminated seed in uninfested soil represents the best method of disease prevention. The relative importance of soil inoculum level in causing disease on tubers was conclusively demonstrated by [152] who showed that when arbitrary soil inoculum threshold values of 0, <10 and > 10 sporosori-g^{-1} soil were set, it was observed that the number of crops developing powdery scab increased with the level of inoculum quantified in the field soil preplanting. In field trials carried out to investigate the link between the amount of inoculum added to the soil and disease development, disease incidence and severity on progeny tubers was found to be significantly ($P < 0.01$) greater in plots with increasing levels of inoculum.

6.2.5 *Integrated disease management*

Integrated disease management implies the integration of two or more control methods to benefit from their additive or synergistic effects and improve the efficacy of each method in order to tailor a complete disease management [153]. The combination of various methods may provide a more durable, sustainable and practical solution to the producers who utilize the available methods to eliminate the menace of pathogens [117]. The integration of control methods will ensure quality and safety and also provide retailers with desired extended shelf life. Amalgamation of compatible and complementary approaches will lead to efficient disease control. The combination could be bio control agent with physical treatment, bio-control agents (BCAs) with chemical at low doses, plant product with soil amendment and BCAs, BCAs with another BCAs, and fungicide with natural waxes, etc. Many researchers have reported the synergistic effect of combining different control methods together for the control of postharvest decay of potato tubers. For example integrated disease management to control early blight requires the implementation of several approaches. The disease is primarily controlled by the use of cultural practices (to reduce the soil born inoculum), less susceptible cultivars and the use of pesticides. *Trichoderma spp.* are beneficial fungi in the rhizosphere of plants in which some species are reported to act as BCAs either by directly antagonizing other pathogens or indirectly by inducing ISR [154]. When applied in alteration with a fungicide, the latter does not have impact the growth of the BCAs, and performance in disease control is enhanced. Also when *B. cereus* is applied as seed treatment, it induces systemic resistance that could reduce the number of sprays of another non-systemic fungicide like chlorothalonil, to manage early blight caused by *Alternaria solani* in potato and tomato [155]. The number of fungicide sprays therefore could be scaled down from 10 to 20 applications while the yield was unaffected over a 90-day field study, confirmingthelong-lasting effect of inducers of resistance on plant defense mechanisms.

7. Concluding remarks

The role of crop protection in environmental management and sustainable potato production should not be overemphasized as it offers countless benefits. The food security challenge is to produce just as much, but waste less through better pre and post-harvest management. Pre-harvest and post-harvest management in potato, including pests and diseases management; storage, processing and value chain efficiency, is a much larger problem than cereals and deserves special attention. Reduction of food losses appears as a key opportunity. The main causes of losses are poor crop and harvest management, infested tubers by pest and diseases, high percentage of small tubers and weather conditions: frost and heavy rains etc. Since potato is a major crop for humankind, it has a global distribution and it is attacked by pests which can substantially reduce its productivity and its quality. The increasing awareness about the nutritional, agronomic, and cash creating advantages potato provides is likely to further increase its status as a global crop, particularly in developing subtropical and tropical countries. The development, adaptation and use of integrated pest management will be an important area of future research crucial for a sustainable and more resilient and economic profitable potato production in all potato growing regions worldwide. Emphasis should be given to develop and use biological approaches in pest management. This will reduce the dependence on insecticides as well as will reduce the risk that insect populations develop resistance against insecticides. Diseases of potato have remained an economically significant disease worldwide. Farmers lose millions of dollars annually due to activities of diseases. However, considering the perspective of climate change, effective utilization of crop protection measures can provide better chance of their vast application in environmental as well as agricultural sustainability.

Acknowledgements

The authors are grateful and sincerely appreciate the valuable contributions of Dr. Khairulmazmi and Malam Adamu Abdullahi. Their inputs, opinions, and suggestions have greatly contributed to improve this chapter.

Conflict of interest

The authors declare no conflict of interest.

Author details

Tijjani Ahmadu[1,2*], Adamu Abdullahi[1] and Khairulmazmi Ahmad[1,3]

1 Department of Plant Protection, Faculty of Agriculture, Universiti Putra Malaysia, Selangor Darul Ehsan, Malaysia

2 Department of Crop Production, Faculty of Agriculture and Agricultural Technology, Abubakar Tafawa Balewa University, Bauchi, Bauchi State, Nigeria

3 Institute of Plantation Studies, Universiti Putra Malaysia, Selangor Darul Ehsan, Malaysia

*Address all correspondence to: tijjaniahmadu72@gmail.com

References

[1] FAO (2019) FAO statistical databases FAOSTAT. http://faostat3.fao.org.

[2] Folberth C, Khabarov N, Balkovic J, Skalský R, Visconti P, Ciais P, Janssens IA, Peñuelas J, Obersteiner M. The global cropland-sparing potential of high-yield farming. Nat. Sustain. 2020;3:281-289.

[3] Wu W, Yu Q, You L, Chen K, Tang H, Liu J. Global cropping intensity gaps: increasing food production without cropland expansion. Land Use Policy. 2018; 76:515-525.

[4] FAO, IFAD, UNICEF, WFP and WHO. The state of food security and nutrition in the world 2018. Building climate resilience for food security and nutrition. FAO, Rome. 2018.

[5] Foley JA. Solution for a cultivated planet. Nature. 2011;478:337-342.

[6] Von Grebmer K. Global hunger index: getting to zero hunger. Welthungerhilfe/ International Food Policy Research Institute/Concern Worldwide, Bonn/Washington, DC/ Dublin. https://doi.org/10.2499/9780896292710. 2017.

[7] FPRI. 2013. Global food security index, the challenge of hunger: building resilience to achieve food and nutritionsecurity. The International Food Policy Research Institute (IFPRI), Concern Worldwide, and Welthungerhilfe. 66 pp. http://www.ifpri.org/sites/default/files/publications/ghi13.pdf. Accessed: 2014-Aug-12.

[8] FAO. The state of food insecurity in the world 2001, Rome: 2002.p.4-7.

[9] FAO. Food security. Policy brief. FAO, Rome. 2006a.

[10] FAO. International year of the potato 2008: new light on a hidden treasure. http://www.fao. org/potato-2008/en/events/book.html. 2009.

[11] FAOSTAT. Food and agriculture data. http://www.fao.org/faostat/en/#data/QCinfo.2017.

[12] Muthoni J, Mbiyu MW, Nyamongo DO. A review of potato seed systems and germplasm conservation in Kenya. J Agric Food Inform. 2010;11: 157-167.

[13] Haverkort A, de Ruijter FJ, van Evert FK, Conijn JG, Rutgers B. Worldwide sustainability hotspots in potato cultivation. 1. Identification and mapping. Potato Res. 2014;56:343-353

[14] You L, Wood-Sichra U, Fritz S, Guo Z, See L, Koo J. Spatial Production Allocation Model (SPAM) 2005 v3.2. http://mapspam.info. 2014.

[15] FAO. Seed security assessment: a practitioner's guide, Rome: 2016.

[16] Birch PRJ, Bryan GJ, Fenton B, Gilroy EM, Hein I, Jones JT, Prashar A, Taylor MA, Torrance L, Toth IK. Crops that feed the world 8: potato: are the trends of increased global production sustainable? Food Secur.2012;4:477-508. https://doi.org/10.1007/s12571-012-0220-1

[17] Scott G. Growth rates for potatoes in Latin America in comparative perspective: 1961-07. Am J Potato Res. 2011;88:143-152.

[18] FAO. Household seed security concepts and indicators. Discussion paper. 2015a;10 pp.

[19] Bohl WH, Johnson SB. Commercial potato production in North America. The Potato Association of America Handbook. http://potatoassociation.org/wp-content/ uploads/2014/04/A_

ProductionHandbook_Final_000.pdf. 2010.

[20] Douches DS, Maas D, Jastrzebski K, Chase RW. Assessment of potato breeding progress in the USA over the last century. Crop Sci. 1996;36(6): 1544-1552.

[21] Scott GJ, Suarez V. Growth rates for potato in India and their implications for industry. Potato J. 2011;38:100-112.

[22] Rosegrant MW, Sulser TB, Mason-D'Croz D, Cenacchi N, Nin-Pratt A, Dunston S, Zhu T, Ringler C, Wiebe K, Robinson S, Willenbockel D, Xie H, Kwon H-Y, Thomas TS, Wimmer F, Schaldach R, Nelson GC, Willaarts B. Quantitative foresight modeling to inform the CGIAR research portfolio. Project report for USAID. International Food Policy Research Institute (IFPRI), Washington, DC. 2017.

[23] Tian J, Chen C, Ye X, Chen S. Health benefits of the potato affected by domestic cooking: a review. Food Chem. 2016;202:165-175.

[24] Chitchumroonchokchai C, Diretto G, Parisi B, Giuliano G, Failla ML. Potential of golden potatoes to improve vitamin A and vitamin E status in developing countries. PLoS One. 2017;12(11):e0187102.

[25] De Haan S, Burgos G, Liria R, Rodriguez F, Creed-Kanashiro H, Bonierbale M. The Nutritional Contribution of Potato Varietal Diversity in Andean Food Systems: a Case Study. Am J Potato Res. 2019;96:151. https://doi.org/10.1007/s12230-018-09707-2.

[26] King J, Slavin J. White potatoes, human health, and dietary guidance. Adv Nutr. 2013;4:393S–401S. https://doi.org/10.3945/an.112.003525.

[27] Ramadan M, Oraby H. Fatty acids and bioactive lipids of potato cultivars: an overview. J Oleo Sci. 2016;65(6): 459-470. https://doi.org/10.5650/jos.ess16015.

[28] Camire ME, Kubow S, Donnelly DJ. Potatoes and human health. Crit Rev Food Sci Nutr. 2009;49:823-840

[29] Nassar AM, Sabally K, Kubow S, Leclerc YN, Donnelly DJ. Some Canadian-grown potato cultivars contribute to a substantial content of essential dietary minerals. J Agric Food Chem. 2012;60:4688-4696. https://doi.org/10.1021/jf204940t.

[30] Bethke P, Jansky S. The Effects of Boiling and Leaching on the Content of Potassium and Other Minerals in Potatoes. J Food Sci. 2018;73(5): H80–H85.

[31] Bonierbale M, Burgos G, zum Felde T, Sosa P. Composition nutritionnelle des pommes de terre. Cahiers de nutrition et dietétique. 2010;45:S28–S36.

[32] Visvanathan R, Jayathilake C, Chaminda Jayawardana B, Liyanage R. Health-beneficial properties of potato and compounds of interest. J Sci Food Agric. 2016; 96:4850-4860. https:// doi.org/10.1002/jsfa.7848. Epub 2016 Jul 7. Review

[33] McGill C, Kurilich A, Davignon J (2013) The role of potatoes and potato components in cardiometabolic health: a review. J Ann Med 45(7):467-473.

[34] Adebamowo SN, Spiegelman D, Willett WC. Association between intakes of magnesium, potassium, and calcium and risk of stroke: 2 cohorts of US women and updated metaanalyses. Am J Clin Nutr. 2015;101:1269-1277.

[35] Kanazawa T, Atsumi M, Mineo H, Fukushima M, Nishimura N, Noda T. Ingestion of gelatinized potato starch containing a high level of phosphorus

decreases serum and liver lipids in rats. J Oleo Sci. 2008;57:335-343.

[36] Muraki I, Rimm EB, Willett WC, Manson JE, Hu FB, Sun Q. Potato consumption and risk of type 2 diabetes: results from three prospective cohort studies. Diabetes Care. 2016;39(3):376-384. https://doi.org/10.2337/dc15-0547.

[37] Borch D, Juul-Hindsgaul N, Veller M, Astrup A, Jaskolowski J, Raben A Potatoes and risk of obesity, type 2 diabetes, and cardiovascular disease in apparently healthy adults: a systematic review of clinical intervention and observational studies. Am J Clin Nutr. 2016;104:489-498.

[38] Pastorino S, Richards M, Pierce M, Ambrosini GL. A high-fat, high-glycaemic index, low- fibre dietary pattern is prospectively associated with type 2 diabetes in a British birth cohort. Br J Nutr. 2016;115:1632-1642.

[39] Reddivari L, Vanamala J, Safe SH, Miller JC. The bioactive compounds α-chaconine and gallic acid in potato extracts decrease survival and induce apoptosis in LNCaP and PC3 prostate cancer cells. Nutr Cancer. 2010;62:601-610.

[40] Sun H, Lv C, Yang L, Wang Y, Zhang Q, Yu S, Kong H, Wang M, Xie J, Zhang C, Zhou M. Solanine induces mitochondria-mediated apoptosis in human pancreatic cancer cells. BioMed Res Int. 2014;805926.

[41] Kroschel J, Sporleder M, Carhuapoma P. Potato tuber moth, *Phthorimaea operculella* (Zeller). In: Kroschel J, Mujica N, Carhuapoma P, Sporleder M, editors. Pest distribution and risk atlas for Africa—potential global and regional distribution and abundance of agricultural and horticultural pests and associated biocontrol agents under current and future climates. International Potato Center, Lima, Peru: 2016b.p.7-23.https://doi.org/10.4160/9789290604761. (online publication: http://cipotato.org/riskatlasforafrica/).

[42] Mujica N, Carhuapoma P, Kroschel J. Serpentine leafminer fly, *Liriomyza huidobrensis* (Blanchard 1926). In: Kroschel J, Mujica N, Carhuapoma P, Sporleder M, editors. Pest distribution and risk atlas for Africa—potential global and regional distribution and abundance of agricultural and horticultural pests and associated biocontrol agents under current and future climates. International Potato Center (CIP), Lima, Peru:2016.p.114-125. https://doi.org/10.4160/9789290604761 (online publication: http://cipotato.org/riskatlasforafrica/).

[43] Schaub B, Carhuapoma P, Kroschel J. Guatemalan potato tuber moth, *Tecia solanivora* (Povolny 1973). In: Kroschel J, Mujica N, Carhuapoma P, Sporleder M, editors. Pest distribution and risk atlas for Africa—potential global and regional distribution and abundance of agricultural and horticultural pests and associated biocontrol agents under current and future climates. International Potato Center (CIP), Lima, Peru: 2016.p.24-38. https://doi.org/10.4160/9789290604761. (online publication: http://cipotato.org/riskatlasforafrica/).

[44] Sporleder M, Carhuapoma P, Kroschel J. Andean potato tuber moth, *Symmetrischema tangolias* (Gyen 1913). In: Kroschel J, Mujica N, Carhuapoma P, Sporleder M, editors. Pest distribution and risk atlas for Africa—potential global and regional distribution and abundance of agricultural and horticultural pests and associated biocontrol agents under current and future climates. International Potato Center (CIP), Lima, Peru: 2016.p.39-53. https://doi.org/10.4160/9789290604761. (online publication: http://cipotato.org/riskatlasforafrica/).

[45] Gamarra H, Mujica N, Carhuapoma P, Kreuze J, Kroschel J. Sweetpotato white fly, *Bemisia tabaci* (Gennadius 1989) (Biotype B). In: Kroschel J, Mujica N, Carhuapoma P, Sporleder M, editors. Pest distribution and risk atlas for Africa—potential global and regional distribution and abundance of agricultural and horticultural pests and associated biocontrol agents under current and future climates. International Potato Center (CIP), Lima, Peru:2016a.p.85-99. https://doi.org/10.4160/9789290604761 (online publication: http://cipotato.org/riskatlasforafrica/).

[46] Oerke EC, Dehne HW, Schönbeck F, Weber A. Crop production and crop protection. Estimated losses in major food and cash crops. Amsterdam: Elsevier Science;1994.p.808.

[47] Mujica N, Kroschel J. Pest intensity-crop loss relationships for the leafminer fly *Liriomyza huidobrensis* (Blanchard) in different potato (*Solanum tuberosum* L.) varieties. Crop Prot. 2013; 47:6-16.

[48] Kroschel J, Schaub B. Biology and ecology of potato tuber moths as major pests of potato. In: Giordanengo P, Vincent C, Alyokhin A. editors. Insect pests of potato: global perspectives on biology and management. Oxford: Academic Press; 2013.p. 165-192.

[49] Jeger M, Bragard C, Caffier D, Candresse T, Chatzivassiliou E, Dehnen-Schmutz K, Gilioli G, Gregoire J-C, Jaques Miret JA, Navajas Navarro M, Niere B, Parnell S, Potting R, Rafoss T, Rossi V, Urek G, Van Bruggen A, Van der Werf W, West J, Winter S, Gardi C, Bergeretti F, MacLeod A. Scientific opinion on the pest categorisation of *Tecia solanivora*. EFSA J. 2018;16:5102.

[50] Kroschel J, Sporleder M, Carhuapoma P. Potato tuber moth, *Phthorimaea operculella* (Zeller). In: Kroschel J, Mujica N, Carhuapoma P, Sporleder M, editors. Pest distribution and risk atlas for Africa—potential global and regional distribution and abundance of agricultural and horticultural pests and associated biocontrol agents under current and future climates. International Potato Center, Lima, Peru: 2016b.p.7-23. https://doi.org/10.4160/9789290604761. (online publication: http://cipotato.org/riskatlasforafrica/).

[51] CABI. 2012. *Liriomyza huidobrensis* (Blanchard). Invasive species compendium: datasheets, maps, images, abstracts and full text on invasive species of the world.http://www.cabi.org/isc/ datasheet/30956. Accessed: 2018-Feb-2.

[52] Kroschel J, Mujica N, Alcazar J, Canedo V, Zegarra O. Developing integrated pest management for potato: experiences and lessons from two distinct potato production systems of Peru. In: He Z, Larkin RP, Honeycutt CW, editors. Sustainable potato production: global case studies. UK: Springer; 2012.p.419-450.

[53] Alcázar J, Kroschel J. Andean potato weevil. In: Wale SJ, Platt HW, Cattlin ND, editors. Diseases, pests and disorders of potatoes: a color handbook. Boston: Academic Press;2008.p.129-131.

[54] CABI. 2016.*Bactericera cockerelli* (tomato/potato psyllid). http://www.cabi.org/isc/datasheet/45643. Accessed: 2018-Feb-20.

[55] OIRSA. El psílido de la papa y tomate *Bactericera* (*Paratrioza*) *cockerelli* (Sulc) (*Hemiptera: Triozidae):* ciclo biológico; la relación con las enfermedades de las plantas y la estrategia del manejo integrado de plagas en la región del OIRSA. Editorial Tauro, Mexico: 2016.p. 48.

[56] Rehman M, Melgar J, Rivera C, Urbina N, Idris AM, Brown JK. First report of *Candidatus Liberibacter psyllaurous* or Ca. *Liberibacter*

solanacearum associated with severe foliar chlorosis, curling, and necrosis and tuber discoloration of potato plants in Honduras. Plant Dis. 2010;94:376.

[57] Hernandez LM, Guzman YC, Martínez-Arias A, Manzano MR, Selvaraj JJ.The bud midge *Prodiplosis longifila*: damage characteristics, potential distribution and presence on a new crop host in Colombia. Springer Plus. 2015;4:205. https://doi.org/10.1186/s40064-015-0987-6

[58] EPPO. *Prodiplosis longifila (Diptera: Cecidomyiidae)*. In: EPPO technical document No. 1068, EPPO study on pest risks associated with the import of tomato fruit. Paris: EPPO;2015.p.133-134.

[59] Pathania M, Chandel RS. Life history strategy and behaviour of white grub, *Brahmina coriacea* (Hope) (Coleoptera: Scarabaeidae: Melolonthinae) an invasive pest of potato and apple agro-ecosystem in northwestern India. Oriental Insects. 2016;51:46-69.

[60] Visser D, Stals R. *Temnorhynchus coronatus* (Fabricius) (Scarabaeidae: Dynastinae), potentially a pest of sweet potato in South Africa. Afr Entomol.2012; 20:402-407.

[61] Visser D. Beetles and their grubs that damage potato tubers. Technical news. ARC-Roodeplaat. 2012.

[62] Malumphy C, Everatt M, Eyre D, Giltrap N. Potato flea beetle, *Epitrix* sp. Plant pest factsheet. Department for Environment Food & Rural Affairs, United Kingdom: 2016.p.1-6.

[63] Furth DG, Savini V, Chaboo CS. Beetles (Coleoptera) of Peru: a survey of the families. Chrysomelidae: Alticinae (flea beetles). J Kans Entomol Soc. 2015; 88:368-374.

[64] CABI. 2017a. *Leptinotarsa decemlineata* (Colorado potato beetle) data sheet. In: Invasive species compendium. CAB International, Wallingford, UK. http://www.cabi.org/isc/datasheet/30380. Accessed: 2018- Feb-25.

[65] Jacques RL, Fasulo TR. Colorado potato beetle *Leptinotarsa decemlineata*. DPI Entomology Circular 271. Florida Department of Agriculture and Consumer Services, Division of Plant Industry and University of Florida. http://entnemdept.ufl.edu/creatures/veg/leaf/potato_beetles.htm. 2015.

[66] EPPO (2014) PQR database. European and Mediterranean Plant Protection Organization, Paris, France. http://www.eppo.int/DATABASES/pqr/pqr.htm

[67] CABI. 2017b. *Ostrinia nubilalis* (European maize borer). In: Invasive species compendium. CAB International, Wallingford, UK.http://eol.org/pages/464566/details. Accessed: 2018-Feb-4.

[68] CABI. 2017c. *Myzus persicae* (green peach aphid). In: Invasive species compendium. CAB International, Wallingford, UK. http://www.cabi.org/isc/datasheet/35642. Accessed: 2018-Feb-20.

[69] Powell W, Pell JK. Biological control. In: Van Emden HF, Harrington R, editors. Aphids as crop pests. 2nd ed. Wallingford: CABI Publishing; 2017.p.469-514.

[70] CABI. 2017d. *Macrosiphum euphorbiae* (potato aphid). In: Invasive species compendium. CAB International, Wallingford, UK. http://www.cabi.org/isc/datasheet/32154. Accessed: 2018-Feb-5.

[71] Katakura H. Classification and evolution of the phytophagous ladybirds belonging to *Henosepilachna vigintioctomaculata* complex (Coleoptera: Coccinellidae) [Thesis]. Hokkaido University, Japan. 1980.

[72] Naz F, Inayatullah M, Rafi MA, Ashfaque M, Ali A. *Henosepilachna vigintioctopunctata* (Fab.) (Epilachninae; Coccinellidae); its taxonomy, distribution and host plants in Pakistan. Sarhad J Agric. 2012;28:421-427.

[73] Xu R, Zhou J, Chen Z, Jia Y, Xu K. Phototactic Behavior of *Henosepilachna vigintioctomaculata* Motschulsky (Coleoptera: Coccinellidae). The Coleopterists Bulletin. 2013;69:806-812.

[74] CABI. 2018a. *Agrotis ipsilon* (black cutworm) datasheet. In: Invasive species compendium. Wallingford, UK: CAB International. https://www.cabi.org/isc/datasheet/3801. Accessed: 2018-Dec-10.

[75] CABI. 2018d. *Mamestra configurata* (bertha armyworm). In: Invasive species compendium. CAB International, Wallingford, UK. https://www.cabi.org/isc/datasheet/8492. Accessed: 2018-Dec-10.

[76] Learmonth S. Thrips: potato pest in Indonesia and Western Australia. Department of Primary Industries and Regional Development. https://www.agric.wa.gov.au/potatoes/ thrips-potato-pest-indonesia-and-western-australia. 2017.

[77] CABI. 2017i. *Frankliniella occidentalis* (western flower thrips). In: Invasive species compendium. CAB International, Wallingford, UK. https://www.cabi.org/ISC/datasheet/24426. Accessed: 2018-Feb-5.

[78] Roberts, D. A. and Boothroyd, C. W. Fundamentals of plant pathology. 2nd ed. New York: W.H. Freeman and Company; 1972.p.432.

[79] Chaube HS, Pundhir VS. Crop diseases and their management. New Delhi: Prentice Hall of India Private Limited; 2005.P.703.

[80] Li Y, Shen H, Zhou Q, Qian K, van der Lee T, Huang S. Changing ploidy as a strategy: The Irish Potato Famine pathogen shifts ploidy in relation to its sexuality. Mol Plant-Microbe Interact. 2017;30:45-52.

[81] Domazakis E, Lin X, Aguilera-Galvez C, Wouters D, Bijsterbosch G, Wolters PJ, Vleeshouwers VG. Effectoromics-based identification of cell surface receptors in potato. Methods Mol Biol. 2017:337-353

[82] Perez W, Forbes G. Potato late blight: technical manual. International Potato Center (CIP), Lima. http://www.cipotato.org/publications/pdf/005446.pdf. 2010.

[83] Fairchild KL, Miles TD, Wharton PS. Assessing fungicide resistance in populations of *Alternaria* in Idaho potato fields. Crop Prot. 2013;49:31-39.

[84] Ayad D, Leclerc S, Hamon B, Kedad A, Bouznad Z, Simoneau P. First report of early blight caused by *Alternaria protenta* on potato in Algeria. Plant Dis. 2017 101:836.

[85] Mane MM, Lal AA, Zghair NQ, Sobita S. Efficacy of certain bioagents and fungicides against early blight of potato (*Solanum tuberosum* L.). Int J Plant Protect. 2014;7:433436.

[86] Obidiegwu JE, Flath K, Gebhardt C. Managing potato wart: a review of present research status and future perspective. Theor Appl Genet. 2014;127:763-780.

[87] Przetakiewicz J. Manual of security sensitive microbes and toxins: *Synchytrium endobioticum*. Liu D, ed. Boca Raton: CRC Press; 2014a.p.823-829.

[88] Falloon R. Control of powdery scab of potato: towards integrated disease management. Am J Potato Res. 2008;85:253-260.

[89] Prior P, Ailloud F, Dalsing BL, Remenant B, Sanchez B, Allen C (2016) Genomic and proteomic evidence supporting the division of the plant pathogen Ralstonia solanacearum into three species. BMC Genomics 17(1):90

[90] Salgon S, Jourda C, Sauvage C, Daunay MC, Reynaud B, Wicker E, Dintinger J. Eggplant resistance to the *Ralstonia solanacearum* species complex involves both broad-spectrum and strain-specific quantitative trait loci. Front Plant Sci. 2017;8:828

[91] Dees MW, Lebecka R, Perminow JIS, Czajkowski R, Motyka A, Zoledowska S, Śliwka J, Lojkowska E, Brurberg MB. Characterization of Dickeya and Pectobacterium strains obtained from diseased potato plants in different climatic conditions of Norway and Poland. Eur J Plant Pathol. 2017a;148(4):839-851.

[92] Sarfraz S, Riaz K, Oulghazi S, Cigna J, Sahi ST, Khan SH, Faure D. *Pectobacterium punjabense* sp. nov., isolated from blackleg symptoms of potato plants in Pakistan. Int J Syst Evol Microbiol. 2018;68:3551-3556

[93] Waleron M, Misztak A, Waleron M, Franczuk M, Wielgomas B, Waleron K Transfer of *Pectobacterium carotovorum* subsp. carotovorum strains isolated from potatoes grown at high altitudes to *Pectobacterium peruviense* sp. nov. Syst Appl Microbiol. 2018;41:85-93.

[94] Van Vaerenbergh J, De Paepe B, Hoedekie A, Van Malderghem C, Zaluga J, De Vos P, Maes M. Natural infection of *Clavibacter michiganensis* subsp. sepedonicus in tomato (*Solanum Lycopersicum*). New Dis Rep. 2016;33:7– 7.

[95] Han L, Dutilleul P, Prasher SO, Beaulieu C, Smith DL. Assessment of common scab- inducing pathogen effects on potato underground organs via computed tomography scanning. Phytopathology. 2008;98:1118-1125.

[96] Hiltunen LH, Kelloniemi J, Valkonen JPT. Repeated applications of a nonpathogenic Streptomyces strain enhance development of suppressiveness to potato common scab. Plant Dis. 2017;101:224-232.

[97] Rashed A, Workneh F, Paetzold L, Rush CM. Emergence of '*Candidatus Liberibacter solanacearum*'-infected seed potato in relation to the time of infection. Plant Dis.2015;99:274-280.

[98] Brundtland GH. What is sustainable development. Our common future. 1987.p. 8-9. http://www. un-documents. net/wced-ocf.htm

[99] Kumar A, Bahadur I, Maurya BR, Raghuwanshi R, Meena VS, Singh DK, Dixit J Does a plant growth-promoting rhizobacteria enhance agricultural sustainability? J Pure Appl Microbiol. 2015a;9:715-724

[100] Akinsemolu AA. The role of microorganisms in achieving the sustainable development goals. J Clean Prod. 2018;182:139-155.

[101] Ki-moon B. 2016. N Secretary-general's Remarks at COP22 Press Conference [WWW Document]. United Nations Secr. Statements. https://www.un.org/sg/en/content/sg/press-encounter/ 2016-11-15/ un-secretary-generals-remarkscop22-press-conference. Accessed: 2017-July-20.

[102] United Nations. United Nations Summit on Sustainable Development 2015. Informal Summary. 70th Session of the General Assembly. 25-27 September 2015 United Nations Headquarters, New York. 2015.

[103] Mishra J, Prakash J, Arora NK Role of beneficial soil microbesin sustainable agricultureand environmental

management. Clim Chang Environ Sustain. 2016; 4:137-149.

[104] Tijjani A, Khairulmazmi A. Global Food Demand and the Roles of Microbial Communities in Sustainable Crop Protection and Food Security: An Overview. In: Gamini S, Junaida SJ, Role of Microbial Communities for Sustainability. Singapore: Springer Nature Ltd; 2020.p.81-108.ISBN 978-981-15-9911-8. https://doi.org/10.1007/978-981-15-9912-5

[105] Costanza R, Fioramonti L, Kubiszewski I. The UN sustainable development goals and the dynamics of well-being. Front Ecol Environ. 2016;14: 59. https://doi.org/10.1002/fee.1231.

[106] Kuhad R. Microbes and their role in sustainable development. Indian J Microbiol. 2012;52:309-313.

[107] Vincent C, Alyokhin A, Giordanengo P. Potatoes and their pests—setting the stage. In: Giordanengo P, Vincent C, Alyokhin A, editors. Insect pests of potato: global perspectives on biology and management. Oxford: Academic Press; 2013.p.3-8.

[108] Oerke, E.C. Crop losses to pests. J. Agric. Sci. 2019;44:31-43.

[109] Orozco FA, Cole DC, Forbes G, Kroschel J, Wanigaratne S, Arica D. Monitoring adherence to the International Code of Conduct—highly hazardous pesticides in central Andean agriculture and farmers' rights to health. Int J Occup Environ Health. 2009;15:255-268.

[110] Alyokhin A, Mota-Sanchez D, Baker M, Snyder WE, Menasha S, Whalon M, Dively G, Moarsi WF. Red Queen on a potato field: IPM vs. chemical dependency in Colorado potato beetle control. Pest Manage Sci. 2015;71:343-356.

[111] Kogan M. Integrated pest management: historical perspectives and contemporary developments. Annu Rev Entomol. 1998;43:243-270.

[112] FAO. 2018. http://www.fao.org/agriculture/crops/core-themes/theme/pests/ipm

[113] Kaya H, Alcazar J, Parsa S, Kroschel J (2009) Microbial control of the Andean potato weevil complex. Fruit Veg Cereal Sci Biotechnol 3(Special Issue 1):39-45

[114] Muchemi SK, Ngichop FC, Fiaboe KKM. Report of the establishmnet of the released hymenoptera *Phaedrotoma scabriventris* Nixon (Braconidae), *Halticpotera arduine* (Walker) (Pteromalidae) and *Chrysocharis flacilla* (Walker) (Eulophidae) in vegetables production systems of Kenya. Submitted to Kenyan Plant Health Inspectorate Service (KEPHIS). International Centre of Insect Physiology and Ecology (icipe): 2014.13.p.

[115] Kaya H, Alcazar J, Parsa S, Kroschel J. Microbial control of the Andean potato weevil complex. Fruit Veg Cereal Sci Biotechnol. 2009;3:39-45.

[116] Mujica N, Fonseca C, Suarez F, Fabian F, Marchena M, Cisneros F. Reducción del uso de insecticidas en el control de la mosca minadora, *Liriomyza huidobrensis* Blanchard, por medio del uso de técnicas etológicas. In: Arning G, Lizárraga A, editors. Control Etológico. Uso de feromonas, trampas de colores y luz para el control de plagas en la agricultura sostenible. Lima, Peru: 2000.

[117] Abhishek, T., Sharma, N., Sharma, V and Afroz, A. A review on conventional and non-conventional methods to manage postharvest diseases of perishables. Researcher. 2013;5:6-19.

[118] Enyiukwu, D. N., Awurum, A. N. and Nwaneri, J. A. Efficacy of

plant-derived pesticides in the control of myco-induced postharvest rots of tubers and agricultural products : A review, Net J Agr Sc. 2014;2:30-46.

[119] Tijjani, A., Adebitan, S. A., Gurama, A. U., Haruna, S. G. and Safiya, T. Effect of some selected plant extracts on *Aspergillus flavus*, a causal agent of fruit rot disease of tomato (*Solanum lycopersicum*) in Bauchi State. Int J Biosc. 2014;4: 244-252.

[120] Cook KA, Ratcliffe ST, Gray ME, Steffey KL. Potato leafhopper (*Empoasca fabae* (Harris)). Insect Fact Sheet, University of Illinois, Department of Crop Sciences. https:// ipm.illinois.edu/ fieldcrops/insects/ potato_leafhopper.pdf.2004.

[121] Heydari A, Pessaraki MA. A review on biological control of fungal plant pathogens using microbial antagonists. J Biol Sc. 2010;10(4): 237-290.

[122] Agrios GN. Plant Pathology. 3rd ed. New york: Academic Press Inc;1998.

[123] Bhattacharjee R, Dey U. Biofertilizer, a way towards organic agriculture: A review. Afr J Microbiol Res. 2014; 8:2332-2343.

[124] Obidiegwu JE, Flath K, Gebhardt C. Managing potato wart: a review of present research status and future perspective. Theor Appl Genet.2014;127:763-780.

[125] Nihorimbere V, Ongena M, Smargiassi M, Thonart P. Beneficial effect of the rhizosphere microbial community for plant growth and health. Biotechnol Agron Soc Environ. 2011; 2:327-337.

[126] Mouria B, Ouazzani-touhami A, Mouria A, Benkirane R, Douira A. Effect of compost and antagonistic fungi on suppression of Tomato Grey Mold. Int Quat J Biol Life Sc. 2015;3: 378-390.

[127] Ramos L, Ciampi L, Gonzales S. Biological control of *Phytophthora infestans* in potato plants. Simiente. 1993;63:53-54.

[128] Sanchez V, Bustamante E, Shattock R. Selection of antagonists for biological control of *Phytophthora infestans* in tomato. Manejo Integrado de Plagas. 1998;48:25-34.

[129] Gupta H, Singh BP, Mohan J. Biocontrol of late blight of potato. Potato J. 2004;31:39-42.

[130] CPRI. 1999. Annual Progress Report. Central Potato Research Institute, Shimla, India.

[131] Roy S, Singh BP, Bhattacharyya SK. Biocontrol of late blight of potato. *Phytophthora* Newslett. 1991;17:18.

[132] Yadav R, Pathak SP. Management of early blight of potato through fungicides and botanicals and bioagents. Plant. 2011;11:1143-1145.

[133] Black W, Mastenbroek C, Mills WR, Peterson LC. A proposal for an international nomenclature of races of *Phytophthora infestans* and of genes controlling immunity in *Solanum demissum* derivatives. Euphytica.1953;2:173-179.

[134] Malcolmson JF, Black W (1966) New R genes in *Solanum demissum* Lindl. and their complementary races of Phytophthora infestans (Mont.) de Bary. Euphytica 1996;15:199-203.

[135] Vossen JH, van Arkel G, Bergervoet M, Jo KR, Jacobsen E, Visser RG. The *Solanum demissum* R8 late blight resistance gene is an Sw-5 homologue that has been deployed worldwide in late blight resistant varieties. Theor Appl Genet.2016; 129:1785-1796

[136] Rodewald J, Trognitz B. Solanum resistance genes against *Phytophthora*

infestans and their corresponding avirulence genes. Mol Plant Pathol. 2013;14:740-757.

[137] Jo KR, Visser RG, Jacobsen E, Vossen JH. Characterization of the late blight resistance in potato differential MaR9 reveals a qualitative resistance gene, R9a, residing in a cluster of Tm-2 2 homologs on chromosome IX. Theor Appl Genet. 2015;128:931-941.

[138] Witek K, Jupe F, Witek AI, Baker D, Clark MD, Jones JD. Accelerated cloning of a potato late blight-resistance gene using RenSeq and SMRT sequencing. Nat Biotechnol. 2016;34:656-660.

[139] Yang L, Wang D, Xu Y, Zhao H, Wang L, Cao X, Chen Y, Chen Q. A new resistance gene against potato late blight originating from *Solanum pinnatisectum* located on potato chromosome 7. Front Plant Sci. 2017;8:1729

[140] Bethke PC, Halterman DA, Jansky S. Are we getting better at using wild potato species in light of new tools? Crop Sci. 2017;57:1241-1258.

[141] Haverkort AJ, Boonekamp PM, Hutten R, Jacobsen E, Lotz LAP, Kessel GJT, Vossen JH, Visser RGF. Durable late blight resistance in potato through dynamic varieties obtained by cisgenesis: scientific and societal advances in the DuRPh project. Potato Res. 2016; 59:35-66.

[142] Rietman H, Bijsterbosch G, Cano LM, Lee HR, Vossen JH, Jacobsen E, Visser RG, Kamoun S, Vleeshouwers VG. Qualitative and quantitative late blight resistance in the potato cultivar Sarpo Mira is determined by the perception of five distinct RXLR effectors. Mol Plant- Microbe Interact. 2012;25:910-919.

[143] Haverkort AJ, Struik PC, Visser RGF, Jacobsen E (2009) Applied biotechnology to combat late blight in potato caused by *Phytophthora infestans*. Potato Res.2009;52:249-264.

[144] Vikhe AM, Wabale HS, Wabale AS, Reddy PG. Efficacy of Plant Extracts against the Seed Borne Fungi of Zea mays L. Asian J Exper Biol Sc. 2010;5:109-111.

[145] Wani AH. An overview of the fungal rot of tomato. Mycopathologica. 2011;9: 33-38.

[146] Arya A. Recent advances in the management of plant pathogens: Botanicals in the fungal pest management. In Arya A, Perello AE, editors. In: Management of fungal plant pathogens UK: CAB International; 2010.p.1-25.

[147] Cooke L, Schepers H, Hermansen A, Bain R, Bradshaw N, Ritchie F, Shaw D, Evenhuis A, Kessel GJT, Wander JGN, Andersson B, Hansen JG, Hannukkala A, Nærstad R, Nielsen BJ. Epidemiology and integrated control of potato late blight in Europe. Potato Res. 2011;54:183-222.

[148] Agrios GN. Plant Pathology. 6th ed. New york: Elsevier Academic Press. 2005.p.922.

[149] Garrett KA, Dendy SP (2001) Cultural practices in potato late blight management. In: Complementing resistance to late blight in the Andes, 13-16 February 2001. International Potato Center, Cochabamba, Bolivia

[150] Shtienberg D, Fry WE (1990) Influence of host resistance and crop rotation on initial appearance of potato early blight. Plant Dis 74(11):849-852

[151] Venette JR, Harrison MD (1973) Factors affecting infection of potato tubers by *Alternaria solani* in Colorado. Am J Potato Res 50(8):283-292

[152] Brierley J, Sullivan L, Wale S, Hilton A, Kiezebrink D, Lees A.

Relationship between *Spongospora subterranea* f. sp. *subterranea* soil inoculum level, host resistance and powdery scab on potato tubers in the field. Plant Pathol.2013;62:413-420.

[153] Ortiz O, Nelson R, Olanya M, Thiele G, Orrego R, Pradel W, Kakuhenzire R, Woldegiorgis G, Gabriel J, Vallejo J, Xie K. Human and technical dimensions of potato integrated pest management using farmer field schools: International Potato Center and partners' experience with potato late blight management. J Integr Pest Manag. 2019;10:1-8

[154] Bisen K, Keswani C, Patel J, Singh H, Sarma B. *Trichoderma* spp.: Efficient Inducers of Systemic Resistance in Plants. In Microbial-Mediated Induced Systemic Resistance in Plants; Choudhary DK, Varma A, editors. Singapore: Springer; 2016. p.185-195. ISBN 978-981-10-0387-5.

[155] Silva HAS, Romeiro RS, Carrer Filho R, Pereira JLA, Mizubuti ESG, Mounteer, A. Induction of systemic resistance by *Bacillus cereus* against tomato foliar diseases under field conditions. J. Phytopathol. 2004;152: 371-375.

Chapter 3

The Impact of Bio-Organic and N, P, K Fertilizers on the Growth and Yield of Potato

Duraid K.A. AL-Taey and Rusul F. AL-Shmary

Abstract

Bio-organic agriculture considers the medium- and long-term impact of agricultural interferences on the agro-ecosystem. It aims to produce food while setting an ecological balance to soil fertility. Bio-organic agriculture takes a proactive design as opposed to treating problems after they emerge, so the study was conducted for studying two factors: First: the cultivars (Riviera and Arizona) class A resulting from cultivation of class E imported and cultivated in spring season 2018. The second factor: fertilizer combinations (bio-organic fertilizers compared with traditional chemical fertilizers). Arizona cultivar significantly achieved the highest values, in most of the study parameters compared to Rivera cultivar. Significant differences were observed between the treatments of fertilizer combinations, the treatment (organic fertilizer + bio-fertilizer + 25% chemical fertilizer) significantly achieved the best values compared to the control. Bi-interaction treatment (Arizona cultivar + organic fertilizer + bio-fertilizer + chemical fertilizer 25%) achieved the highest yield per hectare (43.24 $tons.ha^{-1}$).

Keywords: Sustainability, Bio fertilizers, Organic compost, Nutrients availability

1. Introduction

Potato (*Solanum tuberosum* L.) is considered one of the most important vegetable crops in the world in terms of production and cultivated area, it belongs to the Solanaceae family, which includes about 90 genera and about 2000 species [1, 2]. Cultivated areas of potato crop in Iraq are increasing, however, that the produced quantities do not meet the requirements of the Iraqi consumers. This is due to many of the problems facing the cultivation of the crop in Iraq, the most important of which is soil salinity, which plays an important role in determining productivity [3, 4], where the crop's exposure to salt stress causes a decline the production in most vegetables [5]. Therefore, research has recently tended to study raising the average of growth and production in such, improving the reality of cultivation of this crop in Iraq requires attention to the various agricultural service operations and providing plants with the necessary nutrients. Organic fertilizers are an important way to provide plants with the necessary nutrient requirements and they do not adversely affect the environment [6], where the addition of organic fertilizers to the soil improves their synthetic traits and increases the activity and numbers of microorganisms [7].

Bio-fertilizer is natural substance, which is composed of many strains of bacteria and fungus for decreasing the chemical use in fertilization applications. In addition, bio-fertilizer has a positive role in helping the plants because it contains microorganisms, which are capable of mobilizing nutrient elements from unavailable form to available form through different biological processes [8], also play a role in improving the physical, chemical and biological traits of the soil. Bio-fertilizers are one of the used materials in this field which are natural preparations containing a group of beneficial microorganisms that have an active and effective role in improving soil fertility and supplying plants with part of their nutritional needs; where it maintains the equilibrium of the elements in agricultural lands and converts the elements to the soluble and available form suitable for plant nutrition. It is also involved in the biological resistance for some pests and plant diseases [9, 10]. The concept of integrated fertilization has emerged, which is a combination of chemical, organic and bio-fertilization in order to rationalize the use of chemical fertilizers and compensating them with natural fertilizers for the purpose of increasing yield and improving quality [1].

This study aims to test the response of two potato cultivars to organic and bio-fertilization and their interaction with chemical fertilization for the traits of growth and yield in saline-affected soils.

2. Material and methods

The experiment was conducted in a private field of Babylon governorate, located on longitude 44.39 E and latitude 32.3 N during the autumn growing season (2018). Soil samples were taken from different locations and depth for the purpose of conducting some physical and chemical properties as shown in **Table 1**.

Potato tubers for the two cultivars (Arizona and Riviera) class (A) was obtained from the harvest of agricultural season (2018), which was cultivated with the class (E) and stored at 4°C in refrigerated warehouses. Tubers was cultivated on 10/9/2018 on a furrow where the length (2 m), the distance between the furrow (75 cm), 1 m was left between the experimental units and plots.

The NPK fertilizer (15:15:15) was added in two batches before culture and after 45 days of the first addition, with specified rates (25% of the recommendation of fertilizer, 50% from of the recommendation of fertilizer, 100% of the

Soil properties	Values
pH	7.5
EC ($dS.m^{-1}$)	6.23
Organic matter (%)	1.25
Nitrogen (%)	0.33
Phosphorus (%)	0.12
Potassium (%)	1.07
Sand (%)	22
Silt (%)	54
Clay (%)	24
Texture	Silty loam

Table 1.
Some of the physical and chemical properties of the soil.

recommendation of fertilizer at a rate of (600 $kg.ha^{-1}$ N: 20, P:20,K: 20), bio-fertilizer (a mixture of *Bacillus megaterium* + *Azotobacter chroococcum* + *Fluorescent Pseudomonas*) was added before cultivation according to a recommendation (200 g from bio-fertilizer to 5 L of water). Tubers were directly immersed into this solution for 30 minutes. Bio-fertilizer was obtained from the Ministry of Science and Technology, Laboratories of the Agricultural Research Department in Al-Zafaraniya.

The experiment was conducted based on Randomized Complete Block Design (RCBD), with a split-plot system with two factors, the first factor is cultivars that are symbolized by (V), which is the Main-plot, the second factor is fertilizer combinations that are symbolized by (F) with 8 treatment combinations, which is the sub-plots. Each treatment replicated with three times and the total of experimental units are (48). The significant differences between the treatments were calculated at a significant level of (0.05) for the least significant difference (LSD) using Genstat program.

2.1 Study parameters

Plant height (cm), Leaf area ($cm^2.plant^{-1}$), chlorophyll contents in leaves (SPAD unit), percentage of dry matter of leaves and total yield of tubers ($tons.ha^{-1}$).

3. Results and discussion

3.1 Plant height

Table 2 showed that a significant difference between the two cultivars, Arizona cultivar, was recorded the highest value of plant height (48.38 cm) compared to the Riviera cultivar which gave the lowest values (34.82 cm). The treatments of fertilizer combinations F6 (organic fertilizer + bio-fertilizer + 25% chemical fertilizer) achieved the highest value of plant height (45.87 cm) compared to the control treatment which amounted to 37.79 cm. As for the bi-interaction between cultivars and fertilizer combinations as shown in **Table 3**. The treatment (Arizona cultivar + 100% chemical fertilizer) has achieved the highest average amounted to 53.61 cm which did not significantly differ from the treatment (Arizona cultivar + organic fertilizer + bio-fertilizer + 25% chemical fertilizer).

3.2 Leaf area ($cm^2.plant^{-1}$)

The fertilizer combinations had a significant effect, the treatment (organic fertilizer + bio-fertilizer + 25% chemical fertilizer) had recorded the highest average amount to 9448 $cm^2.plant^{-1}$, and the lowest value at the control treatment which amounted to 5158 $cm^2.plant^{-1}$. In the bi-interaction between cultivars and fertilizer combinations **Table 3**, the treatment (Arizona cultivar + Organic Fertilizer + Bio-fertilizer + 25% Chemical Fertilizer) gave the highest value of leaf area amounted to 9454 $cm^2.plant^{-1}$.

3.3 Chlorophyll contents in leaves (SPAD unit).

Table 2 showed the highest value of chlorophyll contents in leaves amounted to 38.21 SPAD compared to the Rivera cultivar, which recorded the lowest value amounted to 35.51 SPAD. As for fertilizer combinations, the treatment (organic fertilizer + bio-fertilizer + 25% chemical fertilizer) recorded the highest value

LSD 0.05	Study factors				
	Plant height cm	**Leaf area (cm^2.$plant^{-1}$)**	**Chlorophyll content (SPAD)**	**Percent of dry weight of the leaves**	**Total yield ton.ha^{-1}**
V1	48.38	7842.88	38.21	14.98	35.49
V2	34.82	7681	35.51	14.01	31.21
LSD 0.05	**2.734**	**0.311**	**1.858**	**0.823**	**1.553**
F1	39.39	6351.50	36.77	14.10	35.62
F2	37.90	5876.50	33.19	13.50	28
F3	41.17	8079.50	37.37	14.76	33.80
F4	45.74	9218.50	37.66	15.56	36.20
F5	40.42	8911.50	37.79	14.40	37.56
F6	45.87	9448	38.70	15.09	40.58
F7	44.50	9052	36.80	14.85	33.50
F8	37.79	5158	36.62	13.67	21.55
LSD 0.05	**2.885**	**1.642**	**1.576**	**1.147**	**1.112**

V = Cultivars, V1 Arizona, V2 Rivera; F = Fertilizer combinations, F1, (corn cobs compost), F2 (corn cobs compost Organic fertilizers + bio-fertilizer), F3 (corn cobs compost + chemical fertilizer 25% of recommended fertilizer), F4 (corn cobs compost + chemical fertilizer 25% of recommended fertilizer), F5 (corn cobs compost + bio-fertilizer + chemical fertilizer 50% of recommended fertilizer), F6 (corn cobs compost + bio-fertilizer + chemical fertilizer 25% of recommended fertilizer), F7 (Chemical fertilizer 100% full recommended fertilizer), F8 (Control).

Table 2.
Effect of cultivar and fertilizer combinations on the traits of growth and yield of potato plant.

amounted 38.70 SPAD compare with other fertilizer combination treatments, while there was no significant differences among the treatments of bi interactions between cultivars and fertilizers combinations **Table 3**.

3.4 Percent of dry weight of leaves

Table 2 has shown that Arizona cultivar was achieved the highest value of percent of dry weight in leaves amounted to 14.98% compared to the Rivera cultivar which has been recorded the lowest percent of dry weight of leaves amounted to 14.01%.

The results indicated that a significant differences between the fertilizer combination treatments, the treatment (Corn cobs compost + 25% chemical fertilizer) recorded the highest average amounted to 15.56% which did not differ significantly from the treatment (corn cobs compost + bio-fertilizer + 25% chemical fertilizer) which amounted to 15.09%. Differences between bi-interaction treatments did not reach a significant level **Table 3**.

3.5 The total yield (tons.ha^{-1})

Table 2 has indicated that Arizona cultivar has been achieved the highest value of total yield amounted to 35.49 tons.ha^{-1} compared to Rivera cultivar which recorded the lowest total yield amounted to 31.21 tons.ha^{-1}.

The results in **Table 2** has indicated that Arizona cultivar has achieved the highest value of total yield amounted to 35.49 tons.ha^{-1} compared to Rivera cultivar which recorded the lowest total yield amounted to 31.21 tons.ha^{-1}. As for fertilizer

	Study factors				
LSD 0.05	**Plant height cm**	**Leaf area ($cm^2.plant^{-1}$)**	**Chlorophyll content (SPAD)**	**Percent of dry weight of the leaves**	**Total yield $ton.ha^{-1}$**
V1 F1	46.17	6543	37.63	14.27	37.50
V1 F2	45.54	5965	35.67	14.17	29.65
V1 F3	49.73	8043	38.90	15.35	37.24
V1 F4	49.89	9369	38.41	15.82	39.62
V1 F5	47.31	9056	39.33	15.14	41.35
V1 F6	50.45	9454	40.63	15.33	43.24
V1 F7	53.61	9027	37.40	15.68	33.79
V1 F8	44.31	5286	37.73	14.06	21.55
V2 F1	32.62	6160	35.90	13.94	33.75
V2 F2	30.26	5788	30.70	12.84	26.35
V2 F3	32.60	8116	35.83	14.18	30.35
V2 F4	41.59	9068	36.90	15.3	32.78
V2 F5	33.54	8767	36.25	13.66	33.76
V2 F6	41.29	9442	36.77	14.85	37.92
V2 F7	35.38	9077	36.20	14.03	33.21
V2 F8	31.27	5030	35.50	13.28	21.55
LSD 0.05	**4.024**	**2.177**	**Non-Significant**	**Non-Significant**	**1.662**

Significant at P<0.05, ANOVA; since the 2-way interaction was significant it was used to explain results.

Table 3.
Effect of interactions on the traits of yield and growth of potato plant.

combinations treatments recorded a significant difference between them, the treatment (corn cobs compost + bio-fertilizer + 25% chemical fertilizer) achieved the highest value with 40.58 $tons.ha^{-1}$, while the control treatment recorded the lowest value amounted to 21.55 $tons.ha^{-1}$. **Table 3** has shown that Arizona cultivar which was treated with (Corn cobs compost + Bio-Fertilizer + 25% Chemical Fertilizer) gave the highest total yield amounted to 43.24 $tons.ha^{-1}$ compared to other treatment.

The results above emphasized that the Arizona cultivar has been achieved the best values compared to Rivera cultivar in all study parameters, may that due to the variation of genetic traits among the cultivars as well as to the response of Arizona cultivar to the factors and conditions of the soil properties more than Riviera.

The superiority of the Arizona cultivar in the plant height, number of leaves, leaves area, led to an increase in the carbon metabolism, and accumulation of carbohydrates, amino acids and finally increased proteins, that elevated the dry matter in leaves, which reflected on total yield.

The mixing of corn cobs compost (compost) and bio-fertilizer raised the nitrogen availability in the Rhizosphere and encouraged the activity of the microorganisms, and elevation of microorganisms activity accompanied by raising the rate of organic phosphorus mineralization, then an increase of phosphorus availability, which had been effected on stimulating co-enzymes and forming chlorophyll [11, 12].

The mineral and bio-fertilizers mixing with the organic matter have a positive role in improving the vegetative traits and providing the elements necessary for plant growth and development, which contributes to increasing the photosynthesis process, thus increasing manufactured carbohydrates, and stored in tubers. These results agree with [13–15], which they found that organic fertilizers have a role in increasing the yield.

This may also due to the role of bio-fertilizers and mineral fertilizers because they contain nutrients such as where they are available to absorption after mineralizing it in the soil due to soil revitalization and this leads to improving vegetative growth, thus an increase in the yield [16, 17].

4. Conclusions

1. The results above have been confirmed the role of cultivars in obtaining an economic yield in response to the surrounding conditions, as the Arizona variety was more suitable in the conditions of the cultivated area.

2. The addition of chemical fertilizer was reduced by 75% through the combination treatment F6 (Corn cobs compost + Bio-Fertilizer + 25% Chemical Fertilizer), which realized the highest yield compare with other combination treatments.

Author details

Duraid K.A. AL-Taey[1]* and Rusul F. AL-Shmary[2]

1 Department of Horticulture, College of Agriculture, AL-Qasim Green University, Iraq

2 Agricultural Directorate of Babylon, Ministry of Agriculture, Hillah, Iraq

*Address all correspondence to: duraidaltaey@gmail.com

References

[1] Al-Juthery, H.W.A., Habeeb, K.H., Altaee, F.J.K., AL-Taey, D.K.A and Al-Tawaha., A.A.M. 2018. Effect of foliar application of different sources of nano-fertilizers on growth and yield of wheat. *Bioscience research*, 15(4): 3988-3997.

[2] Hassan, Ahmed Abdel Moneim. 2003. Potatoes. Arab Publishing House. Cairo. Egypt.

[3] Ali, N.M., D. K.A. Altaey and N. H. Altaee, 2021. The Impact of Selenium, Nano (SiO2) and Organic Fertilization on growth and yield of Potato *Solanum tuberosum* L. under Salt Stress Conditions. IOP Conf. Series: Earth and Environmental Science 735 (2021) 012042, doi:10.1088/1755-1315/735/1/012042.

[4] Al-Taey, D.K.A. and Saadoon, A.H. (2012). Effect of treatment of kinetin to reduce the salinity damage by drainage water irrigation on the growth and nitrate accumulation in the leaves of spinach, *Spenacia oleracea* L. Euphrates Journal of Agriculture Science. 4(4):11-24.

[5] AlTaey, D.K.A. (2017). Mitigation of Salt Stress by Organic Matter and GA3 on Growth and Peroxidase Activity in Pepper (*Capsicum annum*L.). Advances in Natural and Applied Sciences, 11(10): 1-11.

[6] Manea, A.I., AL-Bayati, H.J. and AL-Taey, D.K.A . 2019. Impact of yeast extract, zinc sulphate and organic fertilizers spraying on potato growth and yield. *Res. on Crops* 20 (1) : 95-100. DOI :10.31830/2348-7542.2019.013

[7] Burhan A. K. and D. K. A. AL-Taey,. (2018) Effect of Potassium humate, humic acid, and compost of rice wastes in the growth and yield of two cultivars of Dill under salt stress conditions. Advances In Natural And Applied Sciences. 12(11): pages 1-6

[8] Al-Khafajy R.A., AL-Taey D. K.A. and AL-Mohammed M.H.S. 2020. The impact of Water Quality, Bio fertilizers and Selenium Spraying on some Vegetative and Flowering Growth Parameters of *Calendula Officinalis* L. under Salinity Stress. *Int. J. Agricult. Stat. Sci*.,. 16, Supplement 1:: 1175-1180, DocID: https://connectjournals.com/03899.2020.16.1175.

[9] Al-Badawi, M. S., . 2019. The response of growth and yield traits for two potatoes cultivars, seed size, and different fertilizer treatments. Master Thesis. College of Agriculture. Al-Qasim Green University. College of Agriculture. Iraq.

[10] Al-Haddad, Zakaria Abdel Rahman. 2003. Proceedings of the Arab Conference on Organic Agriculture in order to clean the environment and strengthen the economy. Tunisia. 261-270.

[11] Al-Taey, D.K.A., M.J.H. Al-Shareefi, A.K. MijweL, A. RZ. Al-Tawaha, A. RM. Al-Tawaha.(2019). The beneficial effects of bio-fertilizers combinations and humic acid on growth, yield parameters and nitrogen content of broccoli grown under drip irrigation system. *Bulgarian Journal of Agricultural Science,* 25 (5), 959-966

[12] Al-Zogbi, M. M., Eid H. and Barhoum M., . 2007. Effect of organic and bio-fertilizers on the production of potatoes and some soil properties in Tartous. Damascus University Journal for Agricultural Sciences. 23 (2): 151-162.

[13] Al-Bayati, H. J. M., . 2010. The physiological effect of gibberellic acid (GA3) and some seaweed extracts on vegetative growth, yield, storage and consumption traits for two potato cultivars (*Solanum tuberosum* L.). College of Agriculture and Forestry. University of Al Mosul. Iraq.

[14] Al-Sharifi, M. J. H. 2015. The Effect of Organic and Chemical Fertilizer on the Growth and Yield of Potato (*Solanum tuberosum* L.) Desiree cultivar. Euphrates Journal of Agricultural Sciences 1 (2): 50-53.

[15] Bashi Q., Zaki, Z.B., .2018. Effect of imprisoning tubers and fertilizing with humic acid on the growth and yield of the autumn season for potatoes (*Solanum tuberosum* L.). Master Thesis. College of Agriculture and Forestry. University of Al Mosul. Iraq.

[16] Abed, S. M., and Fawzi M. A. and Adham A. A., . 2018. EFFECT OF BACTERIAL AND MYCORHYZA BIO-FERTILIZERS AND MINERAL RECOMMENDATION LEVELS ON GROWTH AND YIELD OF POTATO (*SOLANUM TUBEROSUM* L.). Anbar Journal of Agricultural Sciences. 16 (1): 774-788.

[17] Al-Juthery, H.W.A., Ali., N. S. , Al-Taey D.K.A and E. A. H. M. Ali. 2018. THE IMPACT OF FOLIAR APPLICATION OF NANAOFERTILIZER, SEAWEED AND HYPERTONIC ON YIELD OF POTATO. *Plant Archives* Vol. 18 No. 2, 2018 pp. 2207-2212.

Chapter 4

Solanum tuberosum Cultivation Using Nitrogen Recovered from Local Wastewater

Daniel P. Smith and Nathaniel T. Smith

Abstract

This chapter presents an approach to recover nitrogen from human waste-water at local-scale for cultivation of *Solanum tuberosum* (potato) as food crop. Nitrogen capture is by ion exchange of ammonium (NH_4^+) onto zeolite, a natural low cost mineral which is available worldwide. A coupled process is described in which waste-water ammonium is sorbed to granular zeolite, biologically extracted (desorbed), and used to support *Solanum tuberosum* growth in fill-and-drain or irrigation cultivation. The system employs separate components to optimize conditions for ammonium sorption (anaerobic ion exchange), desorption (aerobic bioextraction), and cultivation (flexible timing of water and nitrogen supply and nutrient recycle). System architecture provides a low cost and readily implemented system for highly efficient nitrogen capture and incorporation into potato tuber. The nitrogen recycle system enables sustainable local-scale intensification of *Solanum tuberosum* production and enhanced food security through use of a reliable local nutrient supply. Metrics are presented for per capita tuber production, land area, and productivity. A system design is presented with a path forward for demonstration and development.

Keywords: wastewater, nitrogen, resource recovery, ion exchange, plant nutrient, *Solanum tuberosum*

1. Introduction

The challenge of feeding the world's population requires sustainable intensification of food production– producing more food from the same amount of land with fewer external inputs and less profound negative effects on the environment [1]. Potato (*Solanum tuberosum*) has an important role to play in sustainably increasing food supply. More than a billion people worldwide eat potato (*Solanum tuberosum*) and potato consumption is steadily expanding in the face of population growth and food security needs [2]. Potato is the world's number one non-grain food, with a global crop production of 370 million metric tons in 2019 cultivated on 17.3 million ha [3]. The recent *State of the World* report by FOA stated that the cost of food is a significant factor in global food security and low levels of productivity are a significant barrier to lower costs [3]. Regional potato yields range widely, from 50 tons per hectare (t/ha) and greater in high-input agricultural systems to less than 0.6 t/ha in subsistence cropping with minimal fertilizer use [4]. Nitrogen (N) is a key nutrient required for *Solanum tuberosum* growth and is the fertilizer component that must be supplied in greatest quantity [5]. Nitrogen is essential to vegetative growth, protein synthesis, high potato yields and optimum crop quality [6]. Wastewater is a reliable local source of nutrients, and capturing nitrogen from wastewater and directing it to potato cultivation could increase productivity with low input costs. Since potato prices are often determined by local production costs [7], use of wastewater as locally available nitrogen source merits serious consideration.

2. Nitrogen for *Solanum tuberosum* production

Solanum tuberosum crop yield and quality are mostly dependent on the availability of nutrients and adequate moisture in the growth medium [8, 9]. Nitrogen is a major required macronutrient for *Solanum tuberosum* production and is essential to vegetative growth, protein synthesis, and high potato yields and quality [5, 6, 10, 11]. While potato consumption is steadily expanding in developing countries, more rapid expansion of supply is desirable to for adequate world food supply [2]. According to FOA, limited levels of productivity are a significant barrier to the lower costs needed to increase global food security [4]. However, where low input or subsistence cropping is employed, potato yields are much lower (< 5 t/ha) versus yields of 50 t/ha or more in high-input agriculture [3].

Human waste contains large quantities of nitrogen and other growth nutrients for *Solanum tuberosum* production. Where toilet systems are installed, human waste takes the form of wastewater, which is locally produced and continuously available. Wastewater treatment can supply low cost nitrogen and other nutrients at low cost. Coupling wastewater treatment with resource recovery for potato production can provide a system to realize the potential of *Solanum tuberosum* to increase global food security. Modern high-input agricultural practices contribute significantly to human alteration of the global nitrogen cycle [12]. Use of human waste for *Solanum tuberosum* production assists in the need to transform food production systems [4] and the UN sustainable development goals of higher standards of sanitation [13]. The use of controlled wastewater systems to deliberately recover nitrogen for potato production also reduces nitrogen losses to the environment and degradation of water quality [14].

The composition of major elements in potato tubers is listed in **Table 1** along with per capita generation rates for humans estimated from detailed studies of urine and fecal composition and generation rates [18]. Potato composition estimates were made from compositing multiple literature sources [11, 15–17]. Three major growth

Element		Potato tuber g/kg[1]	Human waste g/cap-day[2]
Nitrogen	N	3–14	12.80
Phosphorus	P	2.6–3.2	2.51
Potassium	K	2.9–13	2.78
Magnesium	Mg	0.21–1.3	0.35
Calcium	Ca	0.05–0.17	1.21
Iron	Fe	0.007–0.023	30.00
Sodium	Na	0.034–0.070	0.80

[1]*Millard [11], El-Latif et al. [15], Beldjilali et al. [16], Burrowes and Ramer [17].*
[2]*Estimated from Rose et al. [18].*

Table 1.
Element composition in potato tubers and human waste generation.

elements (N,P,K) are required for *Solanum tuberosum* propagation and tuber quality [19]. N, P, and K elements are present in large quantity in human waste (**Table 1**). Nitrogen (N) is the mineral nutrient most commonly deficient in agricultural soils [20] and a major determinant of tuber yield and quality [21–23]. The capture of wastewater nitrogen with zeolites and its recycle into plant protein is the major focus of this chapter. Positively charged elements in wastewater (K^+, Ca^{+2}) can also participate in ion exchange sorption and desorption cycles on zeolite [24, 25]. The provision of bulk wastewater storage in the system design described in this chapter can also be used as a source for other growth nutrients.

Wastewater is widely used for both irrigation and as nutrient source in agriculture, where the degree of treatment affects plant productivity and soil quality [26]. Wastewater agricultural uses include conventional field cropping [27], aquaponics [28] and hydroponic growth systems [29]. This chapter presents a system in which nitrogen is separated from the bulk wastewater to enable its deliberate and controlled supply for *Solanum tuberosum* cultivation.

Nitrogen is essential for conversion of solar energy into carbohydrates that are stored in the tuber. Proper nitrogen supply is needed for high yields and potato quality [30, 31]. It is desirable to match the timing of nitrogen supply with specific growth stages. Potato development generally follows sequential stages of 1. sprout development, 2. vegetative growth, 3. tuber initiation, and 4. tuber bulking. Nitrogen demand is low in the first month after planting (sprout development) and high in tuber initiation and bulking stages. The timing of growth stages is approximate and varies with environmental conditions and cultivars, and a nitrogen supply system must have a flexible nitrogen delivery rates to meet a range of plant growth needs.

For the recovery of wastewater nitrogen for potato production, it is desirable that the system reduce nitrogen losses to the environment such as occur with widely used soluble nitrogen fertilizers [32, 33]. Environmental losses can be minimized with a system that captures wastewater nitrogen on zeolite for its controlled release to match plant metabolic needs, as by a cultivation system that collects and recycles water.

Soil moisture affects the growth and yield of potato crops from both micro and seed tubers, and can soil water stress from lower irrigation rates can lead to lower tuber yields [34]. A system to capture nitrogen for *Solanum tuberosum* production must also supply adequate water throughout potato growth stages. The potato growth system described in this chapter includes storage of bulk treated wastewater that can be used for water consumptive demand and to supply nutrients other than nitrogen.

3. Capture of wastewater nitrogen with granular ion exchange media

Nitrogen in sanitation water is primarily ammonium (NH_4^+) and organic nitrogen and the organic nitrogen form is converted to NH_4^+ in anaerobic treatment [35]. Ammonium nitrogen in wastewater can be sequestered onto zeolites, natural low cost minerals with ion exchange properties which are available worldwide [36, 37]. Sorption of NH_4^+ by cation exchange zeolites is effective under anaerobic conditions [38]. Anaerobic treatment of sanitation water and NH_4^+ removal by ion exchange can comprise an integrated and low cost system to recovery of nitrogen from human sanitation water for potato production [39, 40].

Zeolites The sequestration of ammonium ion has been reported for a wide variety of natural zeolites, including salt activated Chinese (Hulaodu) zeolite [41], locally sourced zeolite in South Africa [42], natural Iranian zeolite [43, 44], Carpathian clinoptilolite [24], Malaysian zeolite [45], Doganteppe Turkish zeolite [46], natural Australian zeolite [47], natural Turkish (Yıldızeli) zeolite [48], and Serbian clinoptilolite [49].

In addition to NH_4^+, potassium ion (K^+) can also be sequestered by natural cation exchange zeolites, and desorbed [25, 45]. Potassium is a significant elemental component in potato tuber (**Table 1**). Potassium and other cations may participate along with NH_4^+ in the processes of capture and release from zeolite and cation sorption of cations is competitive [50]. Wastewater constituents other than ammonium and prominent cations can also be removed from wastewater by ion exchange though there is limited research in this area [51, 52]. Zeolite as soil amendment enhances grain crop yield and reduces nitrate leaching [53]. Zeolite-sorbed wastewater nitrogen to enhanced growth of *Arthrospira platensis* cyanobacteria [49]. Zeolite was used to separate NH_4^+ from wastewater, which was substantially recoverable and useful for slow release nitrogen fertilizer [54].

Anaerobic Baffle Reactor The Anaerobic Baffle Reactor (ABR) is suitable as for primary sanitation water treatment prior to granular zeolite media. The Anaerobic Baffle Reactor (ABR) is an anaerobic solids blanket bioreactor with multiple upflow chambers that are hydraulically linked through alternating downflow plena [55]. Flow between ABR chambers does not require pumps and is suitable for primary anaerobic treatment of sanitation water [56]. ABR has been applied in ecological sanitation systems for passive, low maintenance primary treatment of sanitation water in low-income communities [57, 58]. The solids blankets in anaerobic upflow reactors foster sedimentation, filtration and colloidal retention of sanitation water components, as well as anaerobic biological treatment [59–61].

Field IX Prototype The integration ion exchange recovery of wastewater NH_4^+ with anaerobic pre-treatment of sanitation water was verified in a field prototype study [62]. The IX reactor contained three upflow chambers, each preceded by a downward plenum and each containing granular porous zeolite (**Figure 1**). IX chambers retain NH_4^+ by ion exchange and function as anaerobic biofilters [35]. Design features of the IX prototype are listed in **Table 2**. The IX reactor had a liquid empty bed volume including down flow channels of 41.7 L. Specific construction details were presented previously [35]. Zeolite was NV-Na Ash Meadows Clinoptilolite (St. Cloud Mining Company), selected for its low cost, availability in multiple grain sizes, and its stable long-term supply. Nv-Na properties are listed in **Table 3**. Nv-Na is a hydrous sodium aluminosilicate with high specific surface area (40 m^2/g) and bulk density of ca. 800 kg/m^3. The major chemical components are 69.1% SiO_2, 11.9% Al_2O_3, 3.8% K_2O, and 3.5% Na_2O. According to the manufacturer, Nv-Na has a clean water Cation Exchange Capacity (CEC) of 1.85 meq./g (185 cmol(+)/kg). Media in Chamber 1 consisted of 100% of US 4x8 Nv-Na (2.38–4.75 mm). Media in Chambers 2 and 3 was 100% US 8x16 Nv-Na (1.18–2.38 mm).

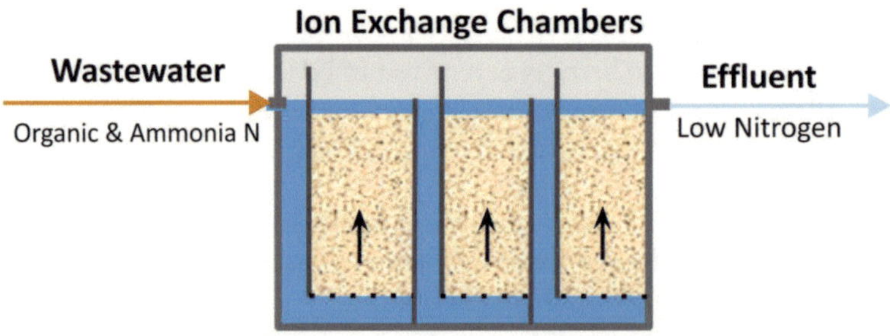

Figure 1.
IX field prototype wastewater nitrogen capture.

Chamber	Media	Empty Bed Volume (L)	Empty Bed Residence Time (hour)[2]	Zeolite Mass (kg)
	Zeolite[1]			
1	U.S. 4 × 8	15.9	37.4	6.55
2	U.S 8 × 16	13.2	31.1	7.18
3	U.S 8 × 16	12.6	29.6	5.13
Total		41.7	98.2	18.8

[1]*Clinoptilolite, 1.85 meq/g CEC.*
[2]*10.2 L/d mean flowrate.*

Table 2.
Prototype IX design.

Grain size, mm	1.5–4.5
Color	Tan - Green
Pore Volume, %	1500%
Pore Diameter, Angstrom	4.0
Specific Surface Area, m^2/g	40
Bulk density, kg/m^3	820
Solid Density, kg/m^3	1,600
Ion Exchange Capacity, meq/g	1.85

[1]*https://www.stcloudmining.com/.*

Table 3.
Properties of granular Clinoptilolite.

Field testing was conducted in Maryland at the Mayo Water Reclamation Plant in Anne Arundel County. The Mayo facility receives treats a daily flow of 1,890 m^3/day primary treated household wastewater from 3,500 residences. Influent to IX was pumped from the plant influent wet well. Zeolite was placed in the three Chambers on Day 0. The goal of initial operation was to establish the validity the IX concept and confirm its central treatment architecture. The IX prototype was dosed once per hour by peristaltic pump at 10.2 L/d from start of operation to Day 319 (4.1 day empty bed HRT. The prototype operated over an ambient temperature range of 7–24°C throughout the study. Flowrate was increased on Day 320 to accelerate the breakthrough of NH_4^+ and exhaust the sorption capacity of the IX media. Flowrate was increased to 36.5 L/d on Day 320 (factor of 3) and 71.4 L/d on Day 344 (factor of 7).

IX Prototype Performance A characteristic profile of nitrogen species through IX chambers after initial operation was established is shown in **Figure 2**. The predominant nitrogen forms in IX influent are Organic Nitrogen and ammonium, which are substantially decreased by IX Chamber 1 through Day 85. Nitrate and nitrite are not present through the IX system. Monitoring results are summarized in **Table 4** for the monitored period of Day 1–214 well before breakthrough of NH_4^+ past Chamber 1. For the Day 1–214 period, TN removal was greater than 95%. The retention of NH_4^+ by ion exchange was the major factor that determined Total Nitrogen (TN) removal by IX throughout the study. Through the entire prototype operation, Organic Nitrogen (ON) remained below 2 mg/L in IX effluent and nitrate and nitrite were below detection levels. Effective NH_4^+-N removal was calculated from effective influent NH_4^+-N and measured effluent NH_4^+-N, where effective influent NH_4^+-N is the sum of measured influent NH_4^+-N and the change in ON across the IX reactor. Effective NH_4^+-N reduction was virtually complete

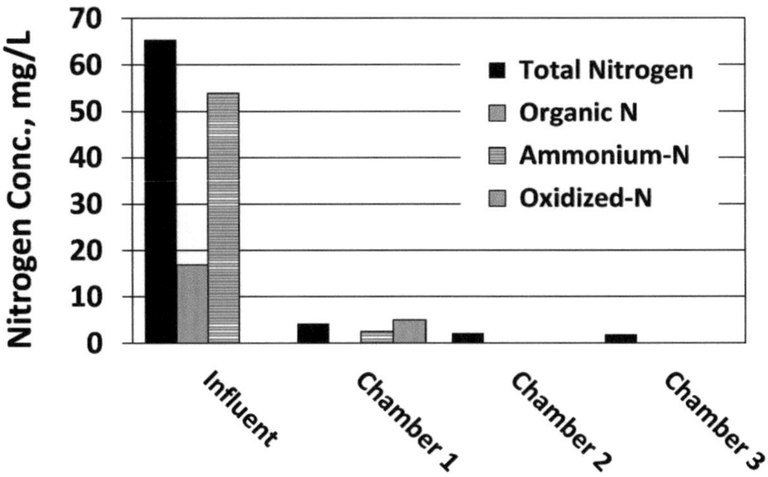

Figure 2.
IX nitrogen removal profiles (day 0–85).

Nitrogen fraction	Mean Influent	Mean Effluent	% Removal
Total	54.0	1.3	97.6
Total kjekldahl	54.0	1.9	96.5
Organic	14.3	1.3	91.2
Ammonia	42.7	0.014	100.0
Nitrate - nitrite	0.02	0	—

Table 4.
IX nitrogen removal performance (day 1–214).

from Day 1–214 (**Table 4**). For the extended operation period (Day 215–355), mean TN removal was affected by NH_4^+ breakthrough; mean removals of TN and NH_4^+-N were 29.7% and 12.1%, respectively. IX treatment substantially reduced wastewater organic matter, as indicated by a mean COD reduction of 58.7% (Day 1–214). IX effluent pH remained circumneutral throughout the study and ORP in chamber effluents remained negative.

The timecourse of ammonium nitrogen in IX chamber effluents is shown in **Figure 3**. Chambers 1, 2 and 3 showed sequential breakthrough of NH_4^+ over extended operation. IX effluent NH_4^+ (Chamber 3) remained below 0.07 mg/L through Day 214 and was ca. 2 mg/L through Day 319 after substantial breakthrough had occurred in Chambers 1 and 2 (**Figure 3**). The timecourse of NH_4^+ in chamber effluents during continuous flow showed the NH_4^+ breakthrough fronts as sorption capacity became exhausted. Flow rate was increased from Day 320 to the end of operation on Day 355. Effluent NH_4^+ increased rapidly after flow rate was increased and continued through Day 355. At the end of operation, NH_4^+ levels in the effluents of Chambers 1, 2 and 3 were at or near the influent NH_4^+ level, suggesting complete exhaustion of the ion exchange media in all IX chambers. The effective ammonium exchange capacity was 11.3 mg NH_4^+-N/g dry weight (0.81 meq/g), or 44% of the Nv-Na clean water capacity. For the Mayo wastewater matrix, effective Nv-Na capacity for NH_4^+ was 16% lower than that found in a Florida onsite wastewater IX [35]. Lower capacity in the Maryland IX was possibly due to competitive ion exchange. Calcium and sodium are prominent

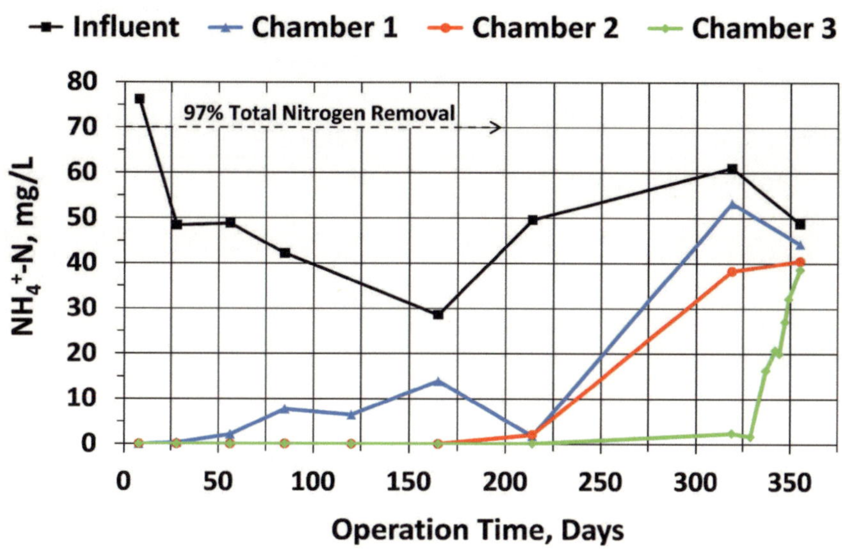

Figure 3.
Time Profiles of Ammonium Nitrogen through IX Chambers.

competing cations that could affect NH_4^+ capacity [44]. Jama and Yucel, 1989 developed forward and reverse ion-exchange isotherms for clinoptilolite and binary solutions of NH_4^+/Ca^{+2} and NH_4^+/Na^+, at a total ionic concentration of 0.10 eq/dm^3. Significant reductions in NH_4^+ capacity were observed for both competing Ca^{+2} and Na^+ ions. The conductivity of Maryland wastewater (3,650 uS/cm) suggested that NH_4^+ capacity might have been reduced by competing cations, possibly from collection system infiltration in this coastal location or Na^+ from water softener backwash.

Flow rate increases of 3 and 7 times were imposed after Day 320 (**Figure 1**) and IX Empty Bed Contact Time was decreased to as low as 0.6 d. The IX process showed no observable adverse effects on operation during this period, other than the intended acceleration of NH_4^+ breakthrough. This suggests that IX performance can be robust and resilient when challenged by the significant flow variations that are typical of local sanitation systems. IX is a highly effective system for local nitrogen recovery. It is passive, mechanically simple, has no inherent energy need, and requires little operator attention. The IX process is resilient and amenable to seasonal operation. IX a highly appropriate technology for local application and provides a new option for locations where wastewater nitrogen removal is critical. Nitrogen captured in IX can be recovered for recycling.

A field IX prototype identical to the Maryland prototype was operated in Florida. The Florida IX prototype also treated actual wastewater that had received anaerobic primary treatment. Total Nitrogen in sanitation water was reduced by over 95% by both prototypes. Nitrogen removal capacities of clinoptilolite zeolite (1.85 meq/g CEC) are shown in **Table 5** [35, 62]. Retention capacity of ammonium nitrogen was 13.5 and 11.3 g NH_4^+-N per gram clinoptilolite, or 52.1 and 43.6% of clean water CEC. The effective ammonium capacity was ostensibly reduced by competing cations (Na^+, Ca^{+2}) or other factors. Ammonium capacity reductions from competing cations would be expected to generally occur for various zeolites from different regions and sources. The operational ammonium capacity shown by the prototypes, however, is quite useful technologically for sequential sorption and bioextraction of nitrogen for plant growth. In developing countries, low per capita water usage could result in higher nitrogen concentrations in wastewater and

Site Wastewater	County Park Residence and Day Lavatory, Florida	Influent to Maryland WWTP
Days Operated	662	355
Temperature Range, °C	23–31	7–17
Mean Influent Total Nitrogen, mg/L	44.2	56.0
TN Reduction, %		97.6
NH_4^+ Capacity, mg N/g dw	13.5	11.3
% of CEC	52.1	43.6

Table 5.
IX nitrogen removal performance and capacity.

possibly increase the competiveness of NH_4^+ sorption. It is noted high effectiveness of TN recovery by IX was maintained at temperatures of 7 to 31C (**Table 5**). Ammonium recovery by IX may be suitable across many climate zones.

Summary IX is a viable means to for recover of nitrogen from wastewater over extended periods. IX treatment of primary effluent sanitation water can recover nitrogen in a passive, mechanically simple process without pumps and sophisticated controls. The system recovers a high percentage of nitrogen, is reliable, and is effective at high and low temperatures. It is effective at varying flow rates, for discontinuous operation and, suitable to local scale deployment. Final effluent of IX treatment is low in total suspended solids (TSS) and low in five-day carbonaceous biochemical oxygen demand (C-BOD_5) as a measure of bulk organic oxygen demand [35].

Solute transport model A one-dimensional solute transport model that accounted for advection, diffusion & equilibrium adsorption was used to model the transport of NH_4^+ through ion exchange chambers [62]. In the z direction:

$$\frac{dC}{dt} = \frac{1}{R}\left(D\frac{d^2C}{dz^2} - v_o\frac{dC}{dz} \right) \tag{1}$$

where C = solute concentration (mg/cm^3 NH_4^+-N), t = time (d), D = hydrodynamic dispersion coefficient (cm^2/day), z = length (cm), and v_o = pore water velocity (cm/day). The dimensionless retardation factor R encompasses instantaneous adsorption equilibria between pore water and solid phase:

$$R = 1 + \frac{p K N C^{N-1}}{\grave{e}} \tag{2}$$

where p = solid phase bulk density (g/cm^3), K = solute distribution coefficient (L/kg), N = sorption parameter (–) and θ = porosity (cm^3/cm^3). Solution of the model employed an analytical solution for fully saturated flow through porous media [63].

Simulation of NH_4^+ transport The 1-D solute transport model (Eqs. (1) and (2)) model was used to predict the NH_4^+-N concentrations in the effluents of the three ion exchange chambers. The model was applied with z axis of zero at the entrance to the first ion exchange chamber (Chamber 1) and time zero on the day of zeolite placement into Chambers 1, 2 and 3. Parameters were estimated for initial conditions and for each term in Eqs. (1) and (2). The simulation used the mean influent NH_4^+-N concentration of 52.0 mg/L that entered Chamber 1 through the study. The total mass of NH_4^+-N removed in IX operation was calculated as the difference between influent and effluent mass over 355 d of operation, which were estimated as the integrated areas under the influent and effluent time profiles of NH_4^+-N. The total NH_4^+-N mass removed divided by the dry weight of Nv-Na added to the three ion exchange chambers yielded a sorption capacity of 11.3 mg NH_4^+-N/g dw Nv-Na (0.81 meq/g) for the IX treating Mayo wastewater. The distribution coefficient of 218 L/kg was calculated from the clinoptilolite sorption capacity and the mean influent NH_4^+-N concentration. Linear sorption was assumed for the simulation (N = 1 in Eq. (2)). A media porosity of 0.45 was used based on manufacturer information and the retardation factor was 389 (dimensionless). Analytical solutions were calculated using 1-D path lengths and pore velocities in each of chamber.

Simulated breakthrough of NH_4^+-N in Chambers 1 through 3 are shown in **Figure 4** along with measured NH_4^+-N concentrations. The 1-D model provided a generally reasonable simulation of NH_4^+-N breakthrough in IX chambers. Zeolite is predicted to approach exhaustion on Days 300, 420, and 540, respectively, in Chambers 1, 2 and 3. Monitored chamber breakthroughs occurred sequentially as expected and in accord with the simulation. The 1-D model approximated measured NH_4^+-N values for Chambers 1 and 2 throughout, and for Chamber 3 up until Day 320. The monitoring data for Chambers 1 and 2 are predicted fairly well by the 1-D simulation. Model predictions are quite acceptable considering that that the 1-D solute transport solution employs a constant influent concentration versus the actual influent nitrogen level that varied significantly (**Figure 3**). The discrepancies between NH_4^+-N measured in Chamber 3 effluent versus the simulation model are due to the high increase in influent flowrate after Day 320, which invalidated the model assumption of constant flowrate and resulted in a much more rapid breakthrough of NH_4^+.

The general competence of the simulation illustrates that NH_4^+ retention by granular ion exchange media appears to be a tractable when treating actual onsite wastewater. Rational procedures for analysis, design, and monitoring can be developed for field deployments. NH_4^+ retention is the main factor affecting Total Nitrogen removal. Modeling and data suggest that operational methods can be

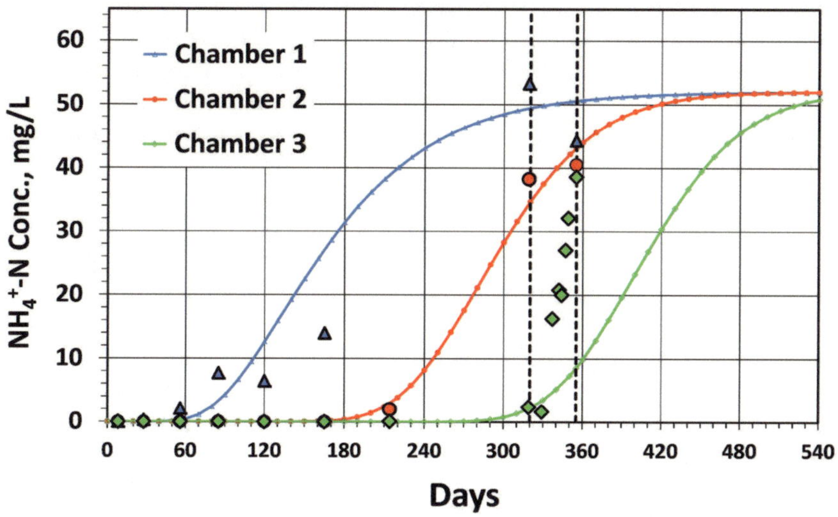

Figure 4.
Simulation model of ammonium ion breakthrough.

developed to optimize NH_4^+ retention, prevent nitrogen breakthrough and loss, cease wastewater flow to the ion exchange chambers, and initiate bioextraction of nitrogen from the spent media.

Simple field measurement of NH_4^+ in the IX chamber effluents can assess media exhaustion in each chamber, determine location of a breakthrough front, and assist on determining when to cease operation and switch to an alternate parallel IX train.

4. Incorporation of wastewater nitrogen into *Solanum tuberosum*

Bioextraction Incorporation of recovered wastewater nitrogen requires desorption of NH_4^+ from zeolite and supplying nitrogen to plant roots. Biological extraction couples biological oxidation of ammonium to nitrate (nitrification) with ammonium desorption from zeolite. The driving force for desorption is affected by the sorption density of NH_4^+ in the zeolite and the concentration of NH_4^+ in bulk water in media pores or in film water on the media surfaces. Nitrification reduces the NH_4^+ concentration and increases driving force. Nitrification rates are affected by the population of nitrifying microorganisms, temperature, oxygen supply, and pH.

Bioextraction is accomplished by circulating extraction water through zeolite(IX) to simultaneously desorb and nitrify NH_4^+. In fill and drain bioextraction, water is pumped from a bioextraction reservoir in order to fill and saturate the IX media, which then passively drains back to the reservoir when pumping is discontinued. In *fill stage* the zeolite media becomes flooded (saturated) are remains so until pumping is discontinued. In *drain stage*, passive drainage begins at high rate and gradually declines, restoring unsaturated conditions until the next *fill stage*. The frequency, duration and magnitude of pumping in the *fill stage* are important operational features that determine the quantity and timing of water supply, the temporal extents of saturated and unsaturated conditions and their relative durations, and the oxygen supply regimes.

Nitrogen bioextracted from IX accumulates in the volume of bioextraction water, generally as ammonium (NH_4^+) or nitrate (NO_3^-). Oxygen is supplied by water added during the *fill stage* and in the *drain stage* by ingress of air into the unsaturated media. Nitrification consumes alkalinity, which may depress the pH of the bioextraction solution and inhibit nitrification. Sodium carbonate and sodium bicarbonate can be amended to the bioextraction water to prevent pH decline [64]. Ammonium is inhibitory to nitrification at high concentrations. The buildup of ammonium in the bioextraction reservoir can be limited by bleed off to a separate plant growth system. Bleed off of the bioextraction reservoir can also have alkalinity preservation consequences.

The requirements of zeolite bioextraction coupled to plant growth suggests two system architectures:

- *One stage*: direct fill and drain cultivation of potato in zeolite media
- *Two stage*: fill and drain bioextraction and separate stage potato production

In the *one stage* system, potato grows directly in the zeolite media bed used for recovery of wastewater nitrogen. The *two stage* system separates fill and drain bioextraction from plant growth. The *one stage* system requires less area and has less supported volume for IX media and plant cultivation. Plant growth in the *one stage* system is obligatorily conducted in fill and drain mode. Management of the bioextraction reservoir nutrient content, chemical composition, and plant water and growth requirements are more intricately related.

In the *two stage* system, bioextraction can be optimized independently of the requirements and constraints of plant growth. Plant growth can be accomplished either in fill and drain mode with highly porous media, or with more conventional soil based systems and irrigation practices. A separate nutrient reservoir can flexibly supply nutrients and water for plant growth requirements. The bioextraction reservoir and plant growth nutrient reservoir can be linked for more flexible control of nutrient content and chemical composition. *Solanum tuberosum* cultivation produces prolific underground biomass in addition to potato tubers. An advantage of the two stage system is that *Solanum tuberosum* is not cultivated directly in the granular ion exchange media. Subsurface plant biomass products such as fine micro-roots, plant mycelium, and other constituents would not remain in the IX media after harvesting. These plant products could hamper the ability to regenerate and reuse IX media for continuous future deployments.

Transmission of pathogens and other constituents are of concern when wastewater is used for irrigation of crops, as reported for sewage farming [65, 66]. A *two stage* system incorporates inherent transmission barriers because the bioextraction and plant growth functions are separated. *Solanum tuberosum* does not grow in the zeolite media through which wastewater has passed, and transfer of IX bioextraction water to the growth system is the only communication from IX to *Solanum tuberosum*. The *two stage* system also has opportunities to create additional barriers. A *one stage* system has fewer barriers to transmission than a *two stage* system.

Direct plant growth in zeolite The integration of recovery of wastewater nitrogen with zeolite with direct growth of food crop in zeolite media has been demonstrated for *Solanum lycopersicum* [62]. The coupled system for zeolite bioextraction and plant growth is shown in **Figure 5**. Fill-and drain experiments were conducted using spent zeolite that had reached its ammonium retention capacity when treating wastewater at the Mayo Water Reclamation Plant in Anne Arundel County, Maryland (described in previous section). Spent zeolite was removed from the three IX chambers, blended, and applied in parallel treatments. Plant growth experiments were conducted in flood-and-drain regime, with a dedicated bioextraction reservoir for each planting container. A *fill cycle* was initiated on 8 hour interval (3/day) for 35 min. Establish a 2 cm standing water column above the top of the media. After

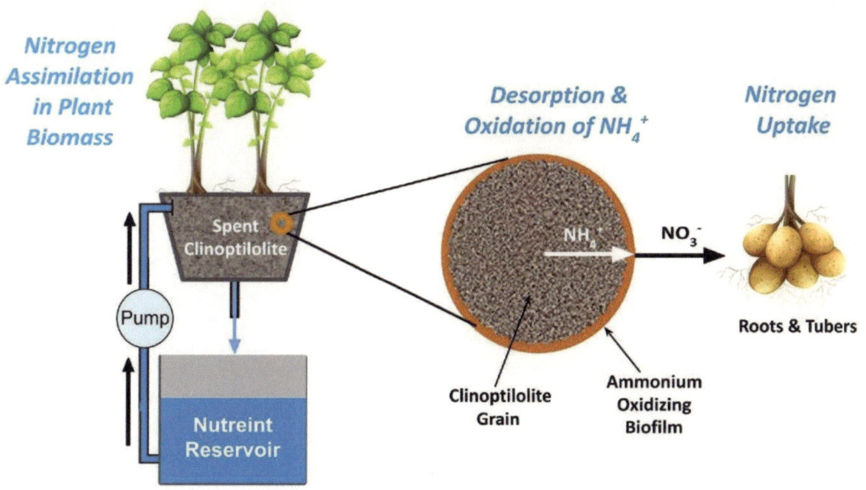

Figure 5.
Fill and drain cultivation with nitrogen from spent zeolite.

the 35 min. Fill period, water in the planting containers drained back to the bioextraction reservoir and unsaturated conditions in the media were restored.

A controlled growth chamber was used to conduct growth experiments. Parallel treatments were conducted using spent Nv-Na clinoptilolite and fresh expanded clay (**Table 6**). Each treatment consisted of a columnar planting container (21.3 cm diameter) with 12 cm media depth (3.3 L media volume). The bioextraction reservoirs served as source of external growth nutrients and enabled the nitrogen levels to be separately monitored for each treatment. Experiments were initiated by placing 15 *Solanum lycopersicum* (cherry tomato) seeds one centimeter below the media surface at the center of the planting containers. Operation of treatments was then commenced under identical conditions. Light was supplied uniformly to the growth chamber by a fluorescent 6400 K grow light fixture (Hydrofarm T-5), with a daily cycle of 12 h on/12 h off daily cycle. The Photosynthetic Photon Flux (PPF) was ~250 $\mu mol/m^2$-sec at 30.5 cm above the granular media surfaces, as measured with a quantum meter (Apogee MQ-200, Logan, Utah). The cultivation temperature varied between 13.8 to 17.7°C [62].

Bioextraction reservoir water differed in parallel treatments. The full nutrient suite contained N, P, K, Ca, Mg, and Si at the levels listed in **Tables 1**, and 10 ml/L of supernatant from an Anaerobic Baffled Reactor (ABR) treating municipal wastewater treatment plant influent. Treatment T1 had clean Nv-Na zeolite and received the full nutrient suite including synthetic nitrogen (**Table 1**). T2 received no added nutrients. T3 and T4 received no synthetic nitrogen. T4 received the full nutrient suite minus nitrogen, whereas T3 received only K and P (**Table 1**). Growth response of T4 versus T1 would ostensibly demonstrate if wastewater nitrogen on spent Nv-Na (T4) could be effectively recycled into *Solanum lycopersicum* growth. Bioextraction reservoir volumes at initial start-up were 7.57 L and the pH was adjusted to 5.9 ± 0.05. To maintain working volumes of at least 5.7 L, make-up water having the same nutrient composition as the starting solutions was added on

Treatment	T1	T2	T3	T4
Granular Media	Fresh Media	Spent Media from AN-IX Reactor		
% 4 × 8 clino	—	40	40	40
% 8 × 16 clino	60	60	60	60
% 3/8 exp. clay	40	—	—	—
Nutrient Supplementation	Full Suite	None	P & K only	Full Suite Minus N
Growth Media Ionic				
Composition, mM				
HNO_3	6.0	—	—	—
K_2HPO_4	0.5/1.5[a]	—	0.5/1.5[a]	0.5/1.5[a]
KCl	2.6	—	2.6	2.6
$CaCl_2 \cdot 2H_2O$	1.0	—	—	1.0
$MgSO_4 \cdot 7H_2O$	1.0	—	—	1.0
K_2O_3Si	1.0	—	—	1.0
$NaHCO_3$	6.0	3.0	3.0	6.0
ABR supernatant, ml/L	10	—	—	10

[a]*Before/after Day 63.*

Table 6.
Solanum Lycosperium growth on with spent Clinoptilolite.

Experiment Days 14, 28, 49, 63 and 81 listed in **Table 1**. After each make-up water addition, bioextraction reservoir pH was adjusted to pH 5.9 ± 0.05.

A comparative, non-destructive measure of plant growth for the parallel treatments was plant canopy volume. Significant differences in *Solanum lycopersicum* growth were observed (**Figure 6**). The greatest canopy volume was obtained in the T4, for which all nitrogen was provided by Nv-Na zeolite. Intermediate plant growth was obtained for treatments T1 and T3, which also received external nutrients, whereas plant canopy volume was minimal for T2, which received no external nutrients (**Figure 6**). Treatments T1 and T4 were identical with the exception of the supply synthetic nitrogen fertilizer to T1 versus growth of T4 in spent IX media without synthetic N. The greater growth of T4 versus T1 shows that nitrogen separated from human wastewater by IX can be directly recycled to production of *Solanum lycopersicum*. It also suggests that spent media may contain components other than nitrogen that are stimulatory to *Solanum lycopersicum* growth. Treatments T3 and T4 were both cultivated in spent IX media but only P and K nutrients were supplied to T3 (**Table 1**). Since both treatments would have had access to nitrogen from spent IX media, the lower canopy increase of T3 suggests that T3 growth may have been limited by trace inorganic nutrient supply or a component in ABR supernatant. The number of *Solanum lycopersicum* fruits and flowers in treatment T4 were over two times those of T1 at Day 93, which accords with the canopy volume comparison and further demonstrates that spent IX provides a favorable medium for plant propagation. The consumptive water use for crop production is a significant factor in many regions where water supplies are limited. Water use in parallel *Solanum lycopersicum* treatments was estimated by the recorded make-up volumes supplied to the bioextraction reservoirs. The increase in canopy volume on Day 93 per consumptive water use was equal to or greater for T4 (spent IX nitrogen) than synthetic nitrogen fertilizer in T1 [62].

The timecourse of NH_4^+ and NO_3^- in the dedicated hydroponic reservoirs of parallel *Solanum lycopersicum* treatments are shown in **Figure** 7. Bioextraction of ammonium ion initiated quickly and was substantial through 93 day (**Figure 6**). With spent IX media, nitrogen accumulated in the hydroponic reservoir solution as NH_4^+ and NO_3^- (NO_2^- was not detected). Spent zeolite treatments had the highest

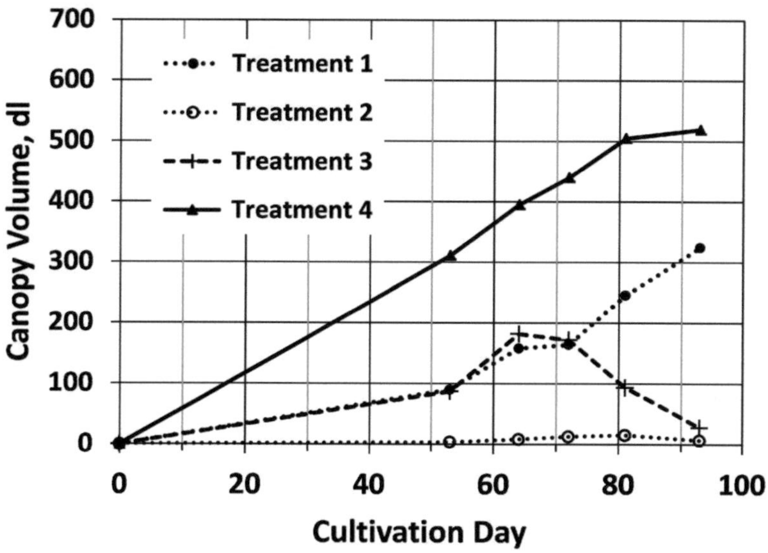

Figure 6.
Solanum Lycosperium canopy establishment.

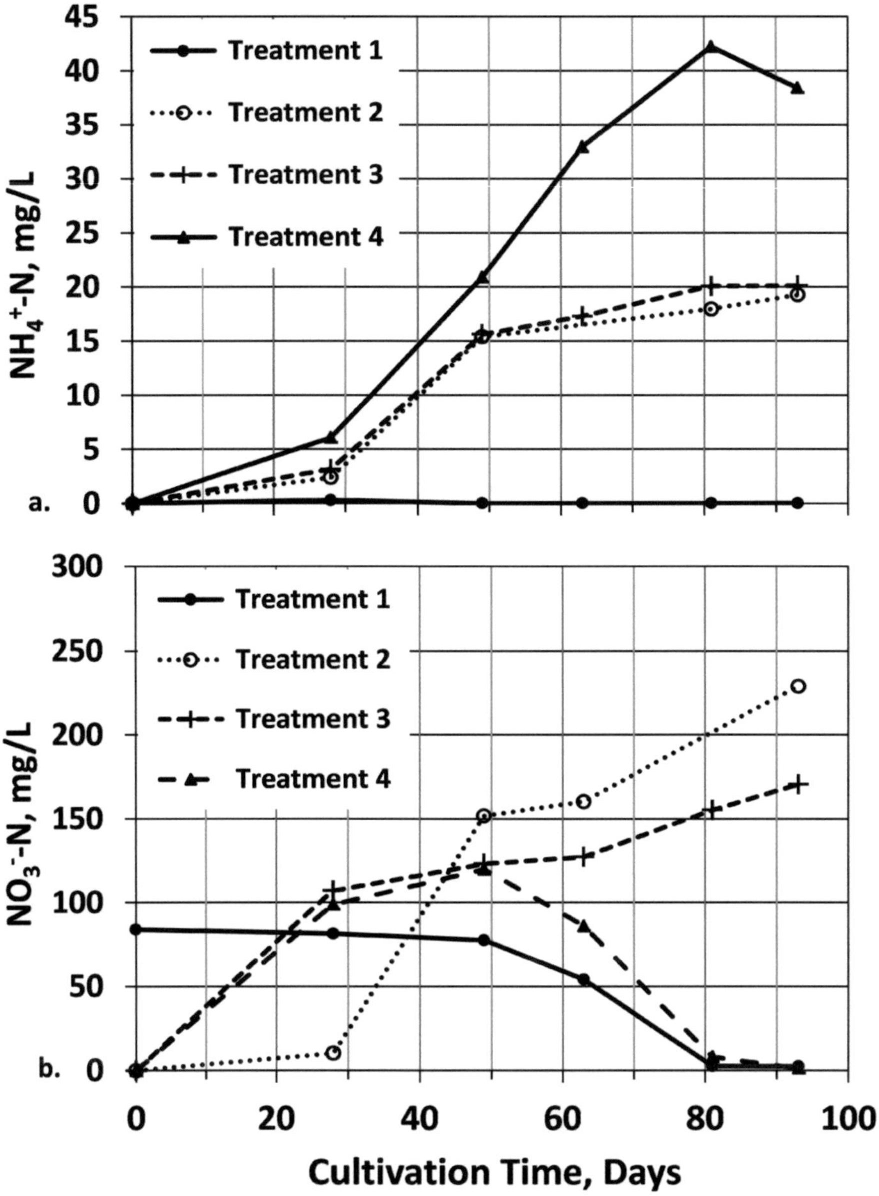

Figure 7.
Bioextraction and nitrification in reservoir nutrient solutions: a. Timecourse of ammonia and b. timecourse of nitrate in reservoir nutrient solutions.

NH_4^+-N concentrations (T4 > T3 > T2) with very low NH_4^+-N levels observed in fresh media of T1. In each of the treatments with spent IX, NO_3^--N increased to over 100 mg/L by Day 50, providing evidence that NH_4^+ in IX media was readily extracted and nitrified (**Figure 7b**). No deliberate microbial seeding was employed in T2, T3 and T4. NO_3^- levels in Treatments T2, T3, and T4 in were generally higher than NH_4^+ to Day 50, showing that nitrification rates kept up with the rates of NH_4+ extraction from the IX media.

Nitrification was slower to establish in T2, however, perhaps due to the lack of added external nutrients. After Day 50 NH_4^+-N in T4 reached substantially higher levels than T2 and T3 (**Figure 7a**), during which NO_3^- in T4 decreased substantially (**Figure 7b**). After Day 50, declining NO_3^- levels in bioextraction reservoirs

(**Figure 7b**) and highest increases in canopy volume (**Figure 5**) occurred for treatments T1 and T4, suggesting plant biomass assimilation of NO_3^-. T2 exhibited the highest NO_3^- levels, which might be explained by significant bioextraction from spent IX and limited plant growth due to the lack of external nutrients. The differences in *Solanum lycopersicum* growth and timecourse of nitrogen species in the bioextraction reservoirs illustrate the complex interactions that determine solution nitrogen levels and nitrogen availability in coupled bioextraction/plant growth systems. Nitrogen availability has important implications for plant growth in IX/hydroponic systems, as nitrogen levels and composition can affect growth rates and nitrogen allocation in leaves, stems and seeds. Further research is needed to optimize nitrogen availability in bioextraction/growth systems.

This study verified that wastewater nitrogen sorbed on zeolite IX process can be directly recycled for growth of *Solanum lycopersicum (cherry tomato)*. The bioextraction/growth system has potential for cultivation of *Solanum tuberosum (potato),* another edible plant in the nightshade family. Unlike the harvestable component of *Solanum lycopersicum*, however, *Solanum tuberosum* tubers lie below the surface of the planted medium. The significance of subsurface tuber production to potato production in a one stage bioextraction/growth is a matter that bears consideration. For *Solanum tuberosum, the* separation of the bioextraction and plant growth functions in a two stage system may be warranted.

Separate Stage Bioextraction A separate system for bioextraction of nitrogen from spent media can be optimized without the constraints of integrated plant growth. Optimization methods include the frequency, duration and magnitude of fill and drain cycles, seeding of spent IX media with nitrifying bacterial cultures, and alkalinity supplementation. Report in this arena come from wastewater treatment where zeolites are integrated into aerobic treatment processes to enhance nitrogen removal. Zeolites serve the two functions of ammonium retention through ion exchange and as solid substrate for attached growth of nitrifying microorgansms. A single reactor, two mode process for ammonium removal from secondary wastewater effluent using zeolite as the carrier for nitrifying biomass was proposed [67]. In the batch bioextraction mode, a nitrification rate of 6 g NH_4^+-N/L reactor-day (0.44 mg NH_4^+-N/g zeolite-hr.) was obtained in a fluidized bed reactor with chabazite as the carrier. Although this rate is in the high range of reported values for biofilm reactors, desorption experiments proved that nitrification will be the process's rate limiting step, rather than the desorption rate when regenerant solutions as low as 2,440 mg/L Na^+ were used. Separate mode bioextraction of chabazite zeolite with regenerant recycle and sodium bicarbonate buffer for nitrification was investigated [68]. They reported ammonium extraction rates of 0.21 g NH_4^+/L-hour which were limited by the supply of oxygen and equivalent to equivalent to 0.36 mg NH_4^+-N/g zeolite-hour in their system. Successful single stage zeolite bioextraction of zeolite has been reported at temperatures as low as 6°C, and addition of sodium carbonate and sodium bicarbonate was been used to supplement alkalinity and prevent pH decline which would be inhibitory to biological nitrification [64]. High ammonium levels may build up in the bioextraction reservoir of the fill and drain separate stage bioextraction system. High ammonium may inhibit nitrification. Coupling of bioextraction reservoir with the nutrient reservoir for *Solanum tuberosum* cultivation may be an approach to ameliorate excessive ammonium buildup in the bioextraction reservoir.

In the experiments with direct cultivation of *Solanum lycopersicum* with zeolite bioextraction, nitrogen release occurred in consort with plant uptake. Substantial nitrogen release occurred over 93 days with NO_3^- depletion in nutrient reservoirs at ~11 weeks in some cases [62]. Separate stage bioextraction enables optimizations that are free of plant growth requirements, such as seeding of spent IX media with nitrifying bacterial cultures and alkalinity supplementation. For an optimized separate stage bioextraction process, the time scale for complete ammonium ion removal from spent zeolite through oxidation desorption can be estimated as 6 to 12 weeks. A technological nitrogen capture system could employ alternating operation of two parallel IX treatment trains, with one IX train in treatment mode (i.e. receiving ABR effluent and capturing wastewater N) and the second IX train in regeneration mode (i.e. fill and drain bioextraction). An IX design with an 8 month nitrogen capture capacity (single treatment train) would enable bioextraction of spent zeolite in the second IX train well within the time to IX exhaustion.

Hydroponic Potato Cultivation There is substantial interest in potato cultivation with controlled growth including hydroponic systems. Hydroponic systems that apply controlled growth using nutrient solution feeding appear to offer significant advantages for potato production. Hydroponic concepts can draw upon to develop systems that grow potato with wastewater nitrogen recovered on zeolites, particularly for variant of fill and drain cultivation. Hydroponic systems offers higher areal yields and less space than conventional agriculture, large potential reductions in consumptive water use, high efficiency of nutrient use, faster growth and lower cultivation times [69]. Controlled growth using nutrient solution feeding appears to offer significant advantages for potato production. Greater potato productivity and high tuber quality with hydroponically grown seed tubers was reported versus those planted in porous substrate; higher efficiency of water use and greater mineral nutritional control were also advantages of hydroponic culture [70]. Hydroponic systems have the potential of discriminating nutrient control, as for example in the delineation of the interactions and effects and of nitrogen and potassium ions in nutrient growth solutions on the yield, dry matter content, and number of tubers of hydroponically grown potatoes [71]. Potato production (*Solanum tuberosum L.*) is among the most responsive of crop species to nitrogen application and controlled growth environments provide a means to optimize nitrogen supply and increase productivity [72].

The system for potato cultivation with recovered wastewater nitrogen offers some of the advantages of hydroponic systems by intensifying productivity and reducing the arable land area required [73]. Fill and drain cultivation of *Solanum tuberosum* provides some of the advantages that hydroponic systems have over field soil agriculture [74]. Additionally, the system is provides a resilient method to reduces the overreliance on rain-fed agriculture and vulnerability to climate change that are emblematic of regions in the developing world [75]. The system is intended to achieve high productivity with locally sourced nitrogen, albeit with far less critical complexity than might be found with many hydroponic growth systems.

Summary Nitrogen on granular zeolite can be incorporated into potato by bioextracting ammonium from the zeolite and supplying ammonium or nitrate for plant cultivation. One and two stage systems are envisioned with different advantages, degrees of complexity and opportunities for optimization. A system for wastewater nitrogen recovery and plant growth is an appropriate technology for sustainable intensification of *Solanum tuberosum* production at local scale. In addition to the nitrogen content, wastewater sources can provide other growth nutrients and supply consumptive water demand for *Solanum tuberosum* production. The potato growth system described in the following section includes a storage feature for bulk treated wastewater.

5. Potato production system

A formulaic design is developed for a system to recover nitrogen from wastewater by sorption on zeolite and to supply captured nitrogen for *Solanum tuberosum* production. The system extracts nitrogen from wastewater to provide a reliable and flexible nitrogen supply for potato cultivation on an as needed basis, while producing a quality treated wastewater for consumptive demand, harvesting of other constituents, and reuse.

The nitrogen recovery and recycle system:

- Uses a reliable locally generated nitrogen source
- Provide for continuous wastewater processing
- Provide high quality wastewater effluent for reuse
- Recovers and stores nitrogen
- Equalizes nitrogen supply on daily to seasonal time scales
- Provides nitrogen to *Solanum tuberosum* on as as-needed basis
- Provides transmission barriers between wastewater and *Solanum tuberosum production*
- Maximizes efficiency of nitrogen transfer from wastewater to plant biomass
- Limits nitrogen loss to groundwater as nitrate (NO_3^-)
- Limits nitrogen loss to the atmosphere as ammonia (NH_3)

Basis of System A design of the nitrogen capture and *potato* production system is based on 10 people and the nitrogen contained in their waste (**Table 7**). The per capita nitrogen excretion rate is estimated as 12.8/cap-day of which 95% can be recovered by IX. Water usage rates depend on available water sources. Higher per capita water use rates dilute waste components and reduce the nitrogen concentration (**Figure 8**). The wastewater nitrogen concentration is 128 mg/L at a design flowrate of 100 liters per capita per day, a water use rate established by the World Health Organization that ensures that most basic human needs are met and few health concerns arise [77]. The system provides the sanitation waste of 10 people to provide 44.4 kg/yr. of nitrogen for potato production.

Optimization of potato production with the controlled growth system provide advantages over non-controlled cultivation. The system design considers an aerial tuber production rate of 30 t/ha-yr. to be achievable, which is well below the reported tuber productivities of 50 t/ha-yr. and greater for high input production systems [3] and 157 t/ha-yr. in hydroponic systems [74]. If higher areal productivities can be achieved, which is quite possible, the main effect on the system architecture would be smaller area of cultivation. The nitrogen content of potato tubers is reported to increase with increasing nitrogen availability, plateauing at 1.53%, while nitrogen content of foliage was twice as high as tubers [8]. A potato tuber content of 1.53% was used in the basis of system design (**Table 7**). Tuber biomass is reported to constitute ca. 80% of total plant mass for *Ants* and *Vigri* potato varieties at growth maturity [78]. It is maintained, however, that 60% of nitrogen uptake by *Solanum tuberosum* occurs before tubularization [30]. The nitrogen use efficiency (NUE) has been defined as the tuber dry matter yield per unit of applied nitrogen [76]. For the purpose of a tuber yield calculation, a nitrogen use efficiency (NUE) of 33.3% was estimated from these reports for *Solanum tuberosum* tuber production

System Basis	
# people	10
g N/cap-day	12.8
g COD/cap-day	154
Per capita flow, L/cap-day	100
N conc., mg/L	128
COD conc., mg/L	1,540
% N capture	95
kg N/yr	44.4
tuber yield, t/ha-yr	30
tuber N content, %	1.53
NUE, %[1]	33.3
Areal N supply, g/m^2-yr	138
Cultivation Area, m^2	322
ha	0.0322
Tuber production, t/yr	0.967
Per capita yield, kg/cap-yr	96.7

[1]*Nitrogen Utiliztion Efficiency [76].*

Table 7.
Formulaic design for Solanum tuberosum growth on wastewater nitrogen: System basis.

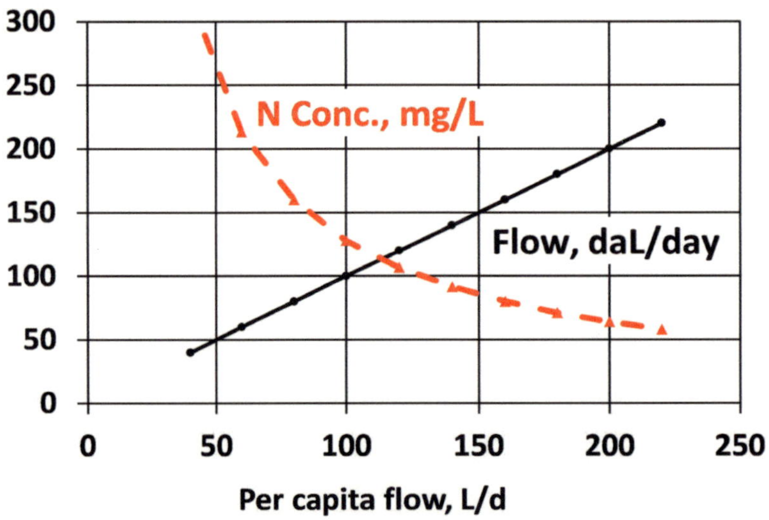

Figure 8.
Water use and wastewater nitrogen concentration.

from supplied nitrogen. An aerial nitrogen supply of 138 g/m^2-yr to 322 m^2 produces 967 kg/yr. potato tubers, a yield of 97 kg/cap-year. The system provides the high areal productivity that is central to increasing potato production in many low income areas dominated by small scale farmers [79].

System Components A schematic of system for nitrogen capture from wastewater, bioextraction of nitrogen from zeolite, and *Solanum tuberosum* production is shown in **Figure 9**. Components of the system are listed in **Table 8**. The wastewater flow

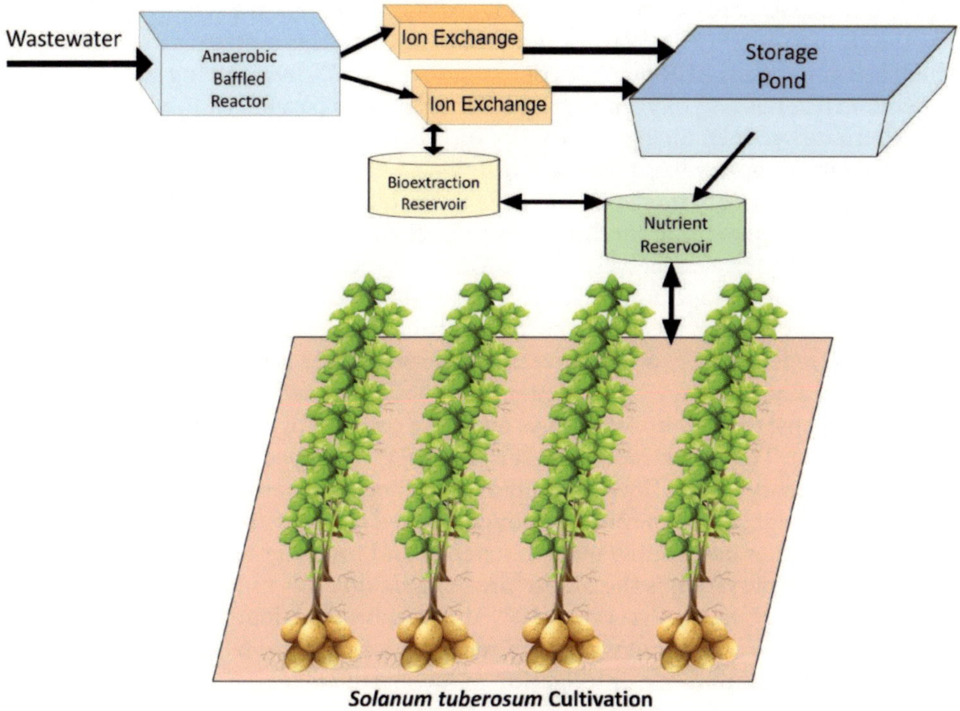

Figure 9.
System schematic.

Component	Configuration	Function	Mechanisms	Flow Regime
Anaerobic Baffled Reactor (ABR)	Series Upflow Chambers	Pre-treatment for IX Reduce suspended & colloidal solids, Ammonification	Sedimentation Filtration Hydrolysis Anaerobic treatment	Continuous flow as wastewater is generated
Ion Exchange (IX) (2 parallel modules)	Series Media Chambers	NH_4^+ sequestration NH_4^+ bioextraction	Flow through porous media Oxygenation of IX media Nitrification	Loading mode: Continuous flow Extraction mode: Fill and Drain
Storage Pond (SP)	Open Pond	Storage of IX effluent	Retention	Continuous flow
Bioextraction Reservoir (BR)	In-ground Tank	Bioextraction of NH_4^+ from IX Media	Oxygenation of IX media Nitrification Accumulation of extracted nitrogen	Fill and Drain
Solanum tuberosum **Cultivation (SC)**	Subsectioned Growth Plots	Receive nutrient solution from NR	Solarium tuberosum Growth	Fill and Drain Mode Irrigation Mode
Nutrient Reservoir (NR)	In-ground Tank	Supply nutrient solution to SC	Nutrient and water supply to SC	Fill and Dram Mode Irrigation Mode

Table 8.
System components: Solanum tuberosum growth on wastewater nitrogen.

path is into the Anaerobic Baffled Reactor (ABR), through Ion Exchange (IX), and into the storage pond (SP). Wastewater passes through this system at the rate at which it is generated. ABR provides pre-treatment for IX by reducing suspended and colloidal solids and oxygen demand. There are two parallel IX modules that each alternate between nitrogen recovery mode and regeneration mode, with each IX module in the opposite mode as the other.

In nitrogen recovery mode, IX media is saturated, preventing oxygen ingress and maintaining anaerobic conditions for which ion exchange retention of NH_4^+ is highly effective [38]. IX receives ABR effluent, extracts NH_4^+, and passes treated wastewater to the storage pond. One IX module is sized to provide an eight month NH_4^+ recovery capacity, providing sufficient time for regeneration of the other IX module. For regeneration, a Bioextraction Reservoir (BR) is placed below ground for passive drainage from IX. In regeneration mode, IX media is saturated in the *fill stage* and unsaturated in *drain stage*, enabling oxygen ingress for nitrification and desorption. The Bioextraction Reservoir (BR) is pumped to saturate the IX media with gravity return flow (fill and drain). Bioextraction results in a buildup in BR of NH_4^+ and NO_3^-. The storage pond received effluent and serves to augment consumptive water use, supply other nutrients, and for other beneficial reuses. The potato production system is the 322 m^2 area of *Solanum tuberosum* Cultivation (SC) and the coupled Nutrient Reservoir (NR) that receives nutrient solution from BR. NR supplies SC with nutrient water and receives drainage from SC. NR/SC operates in either fill and drain mode or irrigation mode. In fill and drain mode, SC media are periodically saturated and then drained; media have high porosity and hydraulic conductivity. During the drained period, oxygen ingress into unsaturated pore spaces is greater than for the finer grained soils of conventional soil-based agriculture. In irrigation mode, SC media are more conventional soils with lower porosity and hydraulic conductivity, with appropriate irrigation schedules.

Anaerobic Baffle Reactor The Anaerobic Baffle Reactor (ABR) has been used for passive, low maintenance primary treatment of sanitation water in low income communities [57]. The ABR is readily constructed and suitable for the nitrogen recycle system. The features of the ABR in the system design are listed in **Table 9**. The three chamber ABR provides a 10 day Hydraulic Residence Time (HRT) and low COD

Anaerobic Baffled Reactor	
Continuous flow through	
HRT, day	10
Liquid volume, m^3	10
COD Loading, kg/m^3-day	0.154
Chambers	
#	3
W × L × D, m	1.5 × 2.0 × 1.11
headspace height, m	0.4
total depth, m	1.51
mean upflow velocity, cm/hr	1.4
Total ABR	
W × L × D, m	1.5 × 6.0 × 1.51

Table 9.
System design: Anaerobic baffled reactor.

loading (0.15 kg/m^3-day) typical of onsite treatment of sanitation waste in anaerobic upflow reactors [60, 80]. Anaerobic treatment of organic wastes produces methane (biogas), which could be harvested from the ABR system. Biogas a local source of energy that can be used as fuel for cooking or lighting for example.

Ion Exchange and Bioextraction Reservoir The salient features of the ion exchange and bioextraction components of the system design are listed in **Table 10**. The two

Ion Exchange Module (2 parallel units)	
Continuous flow through	
Clioptilolite media	
Longevity, mouths	8
Effective CEC, meq/g	0.925
Mass, kg	2,285
Bulk density, kg/m^3	800
Volume, in^3	2.86
IX chambers	
#	4
volume, m^3	0.71
W × L × D, m	0.8 × 0.8 × 1.12
total depth, m	1.70
mean upflow velocity, cm/hr	6.5
Total IX (single module)	
W × L × D, m	0.8 × 3.2 × 1.7
Porewater volume, m^3	1.43
Bioextraction Reservoir	
Fill and draw	
volume, m^3	3.0
diameter, m	2.0
depth, m	0.96
Fill event	
events/day	6
event time increment, hr	4
pump on period, min	15
pump flowrate, L/m	150
total pumped volume, L	2,250
# porewater volumes	1.58
Storage Pond	
days storage	45
volume, m^3	45
area, m^2	49.0
depth, m	0.92

Table 10.
System design: Ion exchange modules (2 parallel units) & storage pond.

parallel IX modules are included (**Figure 9**). The IX modules are identical, each with four chambers containing 0.71 m^3 zeolite. The two IX modules operate alternately, with one in sorption mode receiving ABR effluent while the second is in regeneration mode. The zeolite in each module provides a longevity of eight months for nitrogen recovery (**Table 10**). The eight month design provides substantial storage of the nitrogen load and regeneration time in the alternate IX. When the NH_4^+ capacity of the IX module in the sorption mode approaches exhaustion, each IX module is switched to the alternate function.

When IX modules are each switched to the alternate function, regeneration of spent zeolite is initiated in the IX module that has just been switched from sorption mode. Regeneration is accomplished in fill and drain mode and is conceptually similar to one stage bioregeneration that shown in **Figure 5** without the plant growth. The Bioextraction Reservoir (BR) is pumped to the IX chambers in fill stage, which then passively drains back to BR after pumping ceases (*drain stage*). The duration and rate of pumping determines oxygen supply to nitrifiers on the zeolite surfaces and NH_4^+ and NO_3^- levels in pore water or film water on the zeolite media. The fill and drain schedule in **Table 10** shows six events per day in which IX media if fully saturated; adjustment to this schedule can readily be made during system operation. The time scale for complete ammonium ion removal from spent zeolite through oxidation desorption can be estimated as 6 to 12 weeks for an optimized bioextraction process. As bioextraction of zeolite proceeds, NH_4^+ and NO_3^- levels will build up in BR, and transfer of BR content to the *Solanum tuberosum Production* system will consume the BR nitrogen (**Figure 9**).

IX effluent is low in TSS, organic oxygen demand (COD) and carbonaceous biochemical oxygen demand (C-BOD), comparable to a well-treated wastewater effluent [35, 62]. IX effluent is directed to a storage pond for consumptive water supply, provision of other nutrients, or other reuse needs.

Solanum tuberosum Production The system for potato production includes the *Solanum tuberosum* Cultivation (SC) area and the Nutrient Reservoir (NR), which work as a coupled system (**Figure 9**). Salient design features of SC and NR are listed in **Table 11**. The cultivation area of 322 m^2 is based on the nitrogen supply from 10 people and the nitrogen requirement for tuber production at 30 t/ha-yr. (**Table 7**). NR serves as source of nutrient solution by pumping from NR to SC. NR is placed below ground for passive return drainage from SC. NR nutrient solution is managed based on the metabolic needs of *Solanum tuberosum* plants. NR receives nitrogen solution from BR on the basis of nitrogen supply needs and receives SP water on the basis of consumptive water demand and the need for other nutrients.

SC is subsectioned into twelve 26.8 m^2 plots with 42 cm media depth to accommodate potato root depth. Two manners of *Solanum tuberosum* cultivation are considered: fill and drain mode and irrigation mode (**Table 11**). Either mode enables careful control of the magnitude and timing of nitrogen and water supply to optimize *Solanum tuberosum* growth, maximize the fraction of nitrogen transferred from wastewater to nutrient reservoir to plant biomass, and limits nitrogen losses to groundwater and atmosphere. The potato has a shallow root system and significant yield response to frequent irrigation. The two growth system presented in **Table 11** provide water application rates that enable intensification of the potato yield in small areas.

The fill and drain mode of potato cultivation is similar to architecture one stage bioregeneration shown in **Figure 5**. SC media are periodically saturated and then drained, with growth media of relatively high porosity and hydraulic conductivity. The fill and drain frequency is three fill events per day at eight hour interval to each subsection (36 events/day total), where enough NR water is pumped to fully

Solanum tuberosum Cultivation System		
Cultivation		
total area, m^2	322	
# subsections	12	
area, m^2	26.8	
Nutrient Reservoir		
Sequential pumping to subsections		
volume, m^3	2.5	
diameter, m	2.0	
depth, m	0.80	
	Fill and drain mode	**Irrigation mode**
Growth media		
depth, cm	42	42
porosity	0.50	0.40
pore water volume, m^3	5.64	4.51
Subsection pumping event		
# events/day	3	1
total events/day	36	12
event time increment, hr	0.67	2.0
pump on period, min	30	15
pump flowrate, L/min	200	10
total pumped volume, L	6,000	150
water depth applied, cm	22.4	0.56
# porewater volumes	1.06	0.033
water depth applied, cm/day	67.1	0.56
nitrogen conc., mg/L	100	180
nitrogen applied, g/m^2-day	9.6	0.14

Table 11.
System design: Solanum tuberosum *cultivation.*

saturated the growth media. A nitrogen concentration of 100 mg/L in the nutrient solution provides 9.6 g/m^2-day for cultivation, with only a portion used for plant assimilation and the remainder retuning to NR. In irrigation mode, SC media are more conventional soils with lower porosity and hydraulic conductivity. The irrigation mode schedule is one application per day (12 events per day total) of 0.156 cm depth, providing a nitrogen application of 0.14 g/m^2-day from 180 mg/L nitrogen concentration in NR (**Table 11**). The irrigation schedule would be adjusted through the growth cycle to match the metabolic needs of the potato plant.

Fill and drain mode entails growth in granular, non-cohesive medium and potato prefers naturally loose soils which offer the least resistance to enlargement of the tubers [81]. It can be speculated that fill and drain cultivation of potato may be superior to irrigation mode in flushing or breaking down of potato pathogens, or in limiting their accumulation. Fill and drain and irrigation mode cultivation can both incorporate ridging (earthing up) of growth media, which is advantageous for pest control [81].

Summary A system is presented to capture nitrogen from locally generated wastewater and recycle it into potato production. Nitrogen is recovered and provided for *Solanum tuberosum* production on as as-needed basis. The system efficiently transfers nitrogen from wastewater to plant biomass and limits nitrogen losses to groundwater and atmosphere. Physical separation of wastewater treatment and *Solanum tuberosum* cultivation provides a barrier to transmission of pathogens. The nitrogen recycle system is an appropriate technology for sustainable intensification of *Solanum tuberosum* production at local scale. Projected tuber yields are 967 kg/year on a 322 m^2 plot (10 person basis). The nutritional productivity of this system can be estimated as 92.5 kg/year of crude protein [82].

Use of local wastewater nitrogen can increase *Solanum tuberosum* production and contribute to a reliable world food supply. The nitrogen recycle system meets the development goals of sustainable intensive farming, including use of local of resources to close the yield gap, reduction of footprint, and reduction of wastes [83, 84]. An alternative deployment of the nitrogen recycle system is for intensive breeding of potato seedlings to plant on adjacent areas. Potato crop is usually grown from seed potatoes, small tubers or pieces of tuber sown to a depth of 5 to 10 cm [2]. Potato seedling can be a price barrier, for example comprising 40 to 60% of the total potato production cost to smallholder farmers in African countries [79]. Dedicating the nitrogen system to seed production would focus its more intensive operation on a significant component in the price chain. Other adaptions of the nitrogen recovery and potato growth system are enclosed growth cultivation and agroforestry.

6. Summary and path forward

Local scale *Solanum tuberosum* cultivation has the potential to contribute to food security in low-income and developing countries. This chapter proposes to grow *Solanum tuberosum* using nitrogen captured from wastewater, providing a reliable and low cost nutrient supply that is available in urban, peri-urban and rural areas. A multi-element production system is envisaged that optimizes the functions of primary wastewater treatment (anaerobic upflow solids blanket), ammonium (NH_4^+) capture (anaerobic ion exchange), ammonium release (aerobic bioextraction), and *Solanum tuberosum* cultivation (fill-and-drain hydroponics and irrigation). Key to ammonium capture is the use of natural, low cost ion exchange zeolites which are available worldwide. The architecture of the system separates capture of nitrogen from nitrogen release and delivery, enabling the quantity and timing of nitrogen delivery to match the metabolic needs of *Solanum tuberosum* growth. Potential potato yields of 967 kg/year on a 322 m^2 plot (10 person basis) make the system an appropriate technology for sustainable intensification of *Solanum tuberosum* production at local scale.

This chapter provides the conceptual framework of a system focused on supplying nitrogen for *Solanum tuberosum* growth. The technique can be adapted or interfaced with other processes to provide additional growth needs including water and other nutrients. The intent of the chapter is to stimulate inventive thought and facilitate innovation, demonstration and adoption. Design and testing of field systems is needed to develop process knowledge and skill. Partnerships of environmental/sanitary engineers and agronomists would provide the most fruitful collaborative expertise. Funding from NGOs, nonprofits or governments can accelerate the path forward and bring the benefits to realization.

Author details

Daniel P. Smith* and Nathaniel T. Smith
AET Tech, LLC, Santa Cruz, CA, USA

*Address all correspondence to: daniel.smith.aet@outlook.com

References

[1] Godfray, H. 2013. The challenge of feeding 9-10 billion people equitably and sustainably. The Journal of Agricultural Science, suppl. Special Issue from the 17th International Nitrogen Workshop; Cambridge 152:2-8.

[2] International Potato Center, 2021.

[3] FAO, 2021.

[4] FAO, UNICEF, WFP and WHO, 2020. The State of Food Security and Nutrition in the World 2020. Transforming food systems for affordable healthy diets. . United Nations Food and Agricultural Organization, Rome, Italy.

[5] Koch, M., Naumann, M., Pawelzik, E., Gransee A., and Thiel, H. . 2019. The Importance of Nutrient Management for Potato Production Part I: Plant Nutrition and Yield. Potato Research 63:97-119.

[6] Goffart, J., Olivier, M., Frankinet, M. 2008. Potato crop nitrogen status assessment to improve N fertilization management and efficiency: past–present–future. Potato Research 51:355-383.

[7] Lutaladio, N. and Castaldi, L. 2009. Potato: The hidden treasure. Journal of Food Composition and Analysis 22:491-493.

[8] Ruza, A., Skrabule, I., and Vaivode, A. 2013. Influence of nitrogen on potato productivity and nutrient use efficiency. Proceedings of the Latvian Academy of Sciences: Natural, Exact, and Applied Sciences 67:247-253.

[9] da Silva, A., Zotarelli, L., Dukes, M., Agehara, S., Asseng, S., and van Santen, E. 2018. Irrigation method and application timing effect on potato nitrogen fertilizer uptake efficiency. Nutrient Cycling in Agroecosystems 112.

[10] Alva, L. 2004. Potato Nitrogen Management. Journal of Vegetable Crop Production 10:97-132.

[11] Millard, P. 1986. The nitrogen content of potato (Solanum tuberosum L.) tubers in relation to nitrogen application—the effect on amino acid composition and yields. Journal of the Science of Food and Agriculture 37:107-144.

[12] Vitousek, P., J. Aber, R. Howarth, G. Likens, P. Matson, D. Schindler, W. Schlesinger, and D. Tilman. 1997. Human Alteration of the Global Nitrogen Cycle: Causes and Consequences. Ecological Applications 7. :737-750.

[13] Corcoran, E., Nellemann, C., Baker, E., Bos, R., Osborn, D., and Savelli, H. 2010. Sick Water? The central role of Wastewater Management in Sustainable Development. A Rapid Response Assessment. United Nations Environment Programme, UN-HABITAT, GRID-Arendal.

[14] Withers, P., Jarvie, H., and State, C. 2011. Quantifying the impact of septic tank systems on eutrophication risk in rural headwaters. Environmental International 37:644-653.

[15] El-Latifa, K., Osman, E., Abdullah, R. and el Kader, N. 2011. Response of potato plants to potassium fertilizer rates and soil moisture deficit Advances in Applied Science Research 2:388-397.

[16] Beldjilali, S., Borivent, D., Mercadierm, L., Mothe, E., Clair, G. and Hermann, J. 2010. Evaluation of minor element concentrations in potatoes using laser-induced breakdown spectroscopy. Spectrochimica Act Part B 65:727-733.

[17] Burrowes, J. and Romer, N. 2008. Changes in potassium content of

different potato varieties after cooking. J Ren Nutr 18:530-534.

[18] Rose, C., Parker, A., Jefferson, B. and Cartmell, E. 2015. The Characterization of Feces and Urine: A Review of the Literature to Inform Advanced Treatment Technology. Crit Rev Environ Sci Technol. 2015 Sep 2; 45(17): 1827-1879 45:1827-1879.

[19] Leonel, M., do Carmo, E., Fernandes, A., Soratto, R., Ebúrneo, J., Garcia, E., and dos Santos, T. 2017. Chemical composition of potato tubers: the effect of cultivars and growth conditions. Journal of Food Science and Technology 54:2372-2378.

[20] Duguma, H. and Aga, M. 2019. Role of Nitrogen on Potato Production: A Review. Journal of Plant Sciences 7:36-42.

[21] Nurmanov, Y., Chernenok, V., and Kuzdanova, R. 2020. Potato in response to nitrogen nutrition regime and nitrogen fertilization. Field Crops Research 231:115-121.

[22] Tanios, S., Tegg, R., Eyles, A., Thangavel T., and Wilson, C. 2020. Potato Tuber Greening Risk is Associated with Tuber Nitrogen Content. American Journal of Potato Research 97:360-366.

[23] Milroy, S., Wang, P. and Sadras, V. 2019. Defining upper limits of nitrogen uptake and nitrogen use efficiency of potato in response to crop N supply. Field Crops Research 239:38-46.

[24] Wasielewski S., Rott, E., Minke, R., and Steinmetz, H. 2021. Application of Natural Clinoptilolite for Ammonium Removal from Sludge Water, Molecules 26:114.

[25] Guo, X., L. Zeng, and X. Jin. 2013. Advanced regeneration and fixed-bed study of ammonium and potassium removal from anaerobic digested wastewater by natural zeolite. Journal of Environmental Sciences 25:954-961.

[26] Christou, A., Maratheftis, E., Eliadou, E., Costas, M., Hapeshi, E., and Fatta-Kassinos, D. 2014. Impact assessment of the reuse of two discrete treated wastewaters for the irrigation of tomato crop on the soil geochemical properties, fruit safety and crop productivity. Agriculture, Ecosystems & Environment 192:105-114.

[27] Gatta, G., Libutti, Gagliardi, A., Beneduce, L., Brusetti, L., Borruso, L., Disciglio, G., and Tarantino, E. 2015. Treated agro-industrial wastewater irrigation of tomato crop: Effects on qualitative/quantitative characteristics of production and microbiological properties of the soil. Agricultural Water Management 149:33-43.

[28] Rana, S., Bag, S., Golder, D., Mukherjee, S., Pradhan, C. and Jana, B. 2011. Reclamation of municipal domestic wastewater by aquaponics of tomato plants. Ecological Engineering 37:981-988.

[29] Rababah, A. and Ashbolt, N. 2000. Innovative production treatment hydroponic farm for primary municipal sewage utilisation. Water Research 34:825-834.

[30] Yara America, 2021. Role of Nitrogen in Potato Production: Foliar N and potato yield. Tampa, Florida.

[31] Vos, J. 2009. Nitrogen Responses and Nitrogen Management in Potato. Potato Research 52:305-317.

[32] Hopkins, B., Rosen, C., Shiffler, A., and Taysom, T. 2008. Enhanced Efficiency Fertilizers for Improved Nutrient Management: Potato (Solanum tuberosum). Crop Management (Plant Management Network) 7:1-16.

[33] Davenport, J., Milburn,P., Rosen, C. and Thornton, R. 2005. Environmental

impacts of potato nutrient management. American Journal of Potato Research 82:321-328.

[34] Kawakami, J., Kazuto, J., and Jitsuyama, Y. 2006. Soil water stress and the growth and yield of potato plants grown from microtubers and conventional seed tubers. Field Crops Research 95:89-96.

[35] Smith, D. and Smith, N. 2015. Anaerobic-ion exchange (AN-IX) process for local-scale nitrogen recovery from wastewater. Bioresource Technology 196:324-331.

[36] Bish, D. and Ming, D., editors. 2001. Natural Zeolites: Occurrence, Properties, Applications Mineralogical Society of America, Chantilly, VA.

[37] Mumpton, F. 1999. La roca magica: Uses of natural zeolites in agriculture and industry Proc. Natl. Acad. Sci. USA 1999, 96, 3463-3470.

[38] Smith, D. 2011. Chabazite Biofilter for Enhanced Stormwater Nitrogen Removal. Water Environ. Res. 83:373-384.

[39] Smith, D. 2015. Point-of-Generation Nitrogen Recovery from Wastewater. Environmental Engineer and Scientist 51:23.

[40] Aiyuk, S., J. Amoako, L. Raskin, A. van Handel, and W. Verstraete. 2004. Removal of Carbon and Nutrients from Domestic Wastewater Using a Low Investment, Integrated Treatment Concept. Water Research 38:3031-3042.

[41] Alshameri, A., Yan, C., Al-Ani, Y., Dawood, A., Ibrahim, A., Zhou, C. and Wang, H. 2014. An investigation into the adsorption removal of ammonium by salt activated Chinese (Hulaodu) natural zeolite: Kinetics, isotherms, and thermodynamics. Journal of the Taiwan Institute of Chemical Engineers 45:554-564.

[42] Castro, C., Shyua, H., Xabab, L., Baira, R. and Yeh, D. 2021. Performance and onsite regeneration of natural zeolite for ammonium removal in a field-scale non-sewered sanitation system. Science of The Total Environment 776.

[43] Mazloomi, F. and Jalali, M. 2016. Ammonium removal from aqueous solutions by natural Iranian zeolite in the presence of organic acids, cations and anions. Journal of Environmental Chemical Engineering 4:240-249.

[44] Malekian, R., Abedi-Koupai, J., Eslamian, S., Mousavi, S. Abbaspour, K., and Afyuni, M. 2011. Ion-exchange process for ammonium removal and release using natural Iranian zeolite. Applied Clay Science, *51, 3*, 323-329.

[45] Palanivell, P., Ahmed, O., Omar, L., and Majid, N. 2021. Nitrogen, Phosphorus, and Potassium Adsorption and Desorption Improvement and Soil Buffering Capacity Using Clinoptilolite Zeolite. Agronomy 11.

[46] Sarioglu, M. 2005. Removal of ammonium from municipal wastewater using natural Turkish (Dogantepe) zeolite. Separation and Purification Technology 41:1-11.

[47] Widiastuti, N., Wu, H., Ang, H, and Zhang, D.2011. Removal of ammonium from greywater using natural zeolite. Desalination 277:15-23.

[48] Saltalı, K., Sarı, A. and Aydın, M. 2007. Removal of ammonium ion from aqueous solution by natural Turkish (Yıldızeli) zeolite for environmental quality. Journal of Hazardous Materials 141:258-263.

[49] Markou, G., Vandamme, D., and Muylaert, M. 2014. Using natural zeolite for ammonia sorption from wastewater and as nitrogen releaser for the cultivation of Arthrospira platensis. Bioresource Technology 155:373-378.

[50] Jama, M. and Yucel, H. 1989. Equilibrium Studies of Sodium-Ammonium, Potassium-Ammonium, and Calcium-Ammonium Exchanges on Clinoptilolite Zeolite. Separation Science and Technology 24:1393-1416.

[51] Ahmed, B., Zhou, L., Ngo, H., and Guo, W. 2015. Adsorptive removal of antibiotics from water and wastewater: Progress and challenges. Science of The Total Environment 532:112-126.

[52] Wang, S. and Peng, Y. 2010. Natural zeolites as effective adsorbents in water and wastewater treatment. Chemical Engineering Journal 156:11-24.

[53] Malekian, R., Abedi-Koupai, J., and Eslamian, S. 2011. Influences of clinoptilolite and surfactant-modified clinoptilolite zeolite on nitrate leaching and plant growth. Journal of Hazardous Materials 185:970-976.

[54] Beler Baykal, B., & Sari, B. (2011). An Investigation on the Recovery of Plant Nutrients from Clinoptilolite Exhausted with Domestic Wastewater. *Proceedings, Small Sustainable Solutions for Water*, International Water Association, Venice, Italy, April 18-22, (pp.470-477).

[55] Bachmann, A., Beard, V. and McCarty, P. 1985. Performance characteristics of the anaerobic baffled reactor. Water Research 19:99-106.

[56] Wang, J., Huang,Y., and Zhao, X. 2004. Performance and characteristics of an anaerobic baffled reactor. Bioresource Technology 93:205-208.

[57] Tilley, E., Ulrich, L., Lüthi, C., Reymond, P. and Zurbrügg, C. 2014. Compendium of Sanitation Systems and Technologies, 2nd Revised Edition. Eawag Department Sandec, Dübendorf, Switzerland.

[58] Foxon, K., Pillay, S., Lalbahadur, T., Rodda, N., Holder, F., and Buckley, C. 2004. The anaerobic baffled reactor (ABR): An appropriate technology for on-site sanitation http://wrcwebsite.azurewebsites.net/wp-content/uploads/mdocs/WaterSA_2004_05_69.pdf. Water SA 30:44-50.

[59] Chong, S., Sen, T., Kayaalp, A. and Ang, H. 2012. The performance enhancements of upflow anaerobic sludge blanket (UASB) reactors for domestic sludge treatment – A State-of-the-art review. Water Research 46:3434-3470.

[60] de Graff, M. 2010. Resource Recovery from Black Water. Wageningen University, Wageningen.

[61] Luostarinen, S., Sanders, W., Kujawa-Roeleveld, K. and Zeeman, G. 2007. Effect of temperature on anaerobic treatment of black water in UASB-septic tank systems. Bioresource Technology 98,:980-986.

[62] Smith, D. and Smith, N. 2017. Recovery of Wastewater Nitrogen for Solanum lycopersicum Propagation. Waste and Biomass Valorization 10:1191-1202.

[63] van Genuchten, M. and Wierenga, P. 1976. Mass transfer studies in sorbing porous media I. analytical solutions. . Soil Science Society of America Journal Division S-1- Soil Physics 40 473-480.

[64] Chen, J., Wang, X., Zhou, S. and Chen , Z. 2019. Effect of alkalinity on bio-zeolite regeneration in treating cold low-strength ammonium wastewater via adsorption and enhanced regeneration. Environmental Science and Pollution Research 26:28040-28051.

[65] Uekötter, F. 2016. City meets Country: Recycling ideas and realities on German sewage farms. Journal for the History of Environment and Society 1:89-107.

[66] Saber, M., Hoballah, E., Abouziena, H., Haggag, W., El-Ashry, S. and

Zaghloul, A. 2016. Management of Sewage Farming in Arid Region: Egyptian Experience. International Scientific Researches Journal, 72, 3, 28-56.

[67] Green, M., Mels, A., Lahav, O., and Tarre, S. 1996. Biological-ion exchange process for ammonium removal from secondary effluent. Water Science and Technology 34:449-458.

[68] Lahav, O. and Green, M. 1998. Ammonium removal using ion exchange and biological regeneration. Water Research 32:2019-2028.

[69] Pandey, R., Jain V., and Singh, K. 2009. Hydroponics Agriculture: Its Status, Scope and Limitations.

[70] Corrêaa, R., Pinto, J., Pinto, C., Faquin, V., Reis, E., Monteiro, A. and Dyer, W. 2008. A comparison of potato seed tuber yields in beds, pots and hydroponic systems. Scientia Horticulturae 116:17-20.

[71] Parra S., Rubio, W., Hernández, S., Sánchez, P. and Preciado, P. 2018. Nitrate/potassium interaction in the nutrient solution and potato yield in hydroponics. Revista Mexicana de Ciencias Agrícolas 9:993-1005.

[72] Geary, B., Clark, J., Hopkins, B., and Jolley, D. 2015. Deficient, Adequate and Excess Nitrogen Levels Established in Hydroponics for Biotic and Abiotic Stress-Interaction Studies in Potato. Journal of Plant Nutrition 38:41-50.

[73] Sardare, M. and Admane, S. 2013. A Review on Plant Without Soil-Hydroponics. International Journal of Research in Engineering and Technology 2:299-304.

[74] Khan, F.., Kurklu, A., Ghafoor, A., Ali, Q., Umair, M., and Shahzaib. 2018. A review on hydroponic greenhouse cultivation for sustainable agriculture. International Journal of Agriculture Environment and Food Sciences 2 59-66.

[75] Mugambi, M. 2020. Factors affecting the approbation of hydroponics farming as a means to mitigate and adapt to climate change amongst small scale farmers: a case study of Meru County, Kenya. MS. Norwegian University of Life Sciences, Ås.

[76] Tiemens-Hulscher, M., Lammerts, E., van Bueren, E. and Struik , P. 2014. Identifying nitrogen-efficient potato cultivars for organic farming. Euphytica 199:137-154.

[77] WHO, 2003. Domestic Water Quantity, Service Level and Health. World Health Organization, Geneva, Switzerland.

[78] Särekanno, M., Kadaja, J., Kotkas, K., Rosenberg V., and Eremeev, V. 2012. Development of field-grown potato plants derived from meristem plants multiplied with different methods. Acta Agriculturae Scandinavica, Section B - Soil & Plant Science 62:114-124.

[79] Tessema, L. and Dagne, Z. 2018. Aeroponics and Sand Hydroponics: Alternative Technologies for Pre-basic Seed Potato Production in Ethiopia. Open Agriculture 3:444-450.

[80] Krishna, G., Kumar, P. and Kumar, P. 2009. Treatment of low-strength soluble wastewater using an anaerobic baffled reactor (ABR). Journal of Environmental Management 90:166-176.

[81] FAO. 2008. International Year of the Potato 2008 New Light on a Hidden Treasure.in FAO, editor. http://www.fao.org/potato-2008/en/potato/cultivation.html. FAO, Rome.

[82] van Gelder, W. 1981. Conversion factor from nitrogen to protein for

potato tuber protein. Potato Research 24:423-425.

[83] Fróna D. and Harangi-Rákos, M. 2019. The Challenge of Feeding the World. Sustainability 11:5816.and Abiotic Stress-Interaction Studies in Potato. Journal of Plant Nutrition 38:41-50.

[84] Godfray, C., Beddington, J., Crute, I., Haddad, L., and Lawre, D. 2010. Food Security: The Challenge of Feeding 9 Billion People. Science 327:812-818.

Section 2

Origin and Wild Relatives

Chapter 5

Solanum jamesii as a Food Crop: History and Current Status of a Unique Potato

David Kinder, John Bamberg, Lisbeth Louderback, Bruce Pavlik and Alfonso Del Rio

Abstract

Solanum jamesii is a wild potato found in the US southwest. There is ample evidence that this potato was used by ancestral Puebloans as a food source, where some researchers think it was used as a starvation food while others consider it to be regular food source. Currently this potato is being grown by Native Americans, notably the Navajo, as a specialty food as well as a food crop. There are several attributes to this potato that make it especially suitable for development as our climate changes and food needs become more demanding, including its drought tolerance and ability to be crossed with other wild potato species and cultivars.

Keywords: *Solanum jamesii*, desert adapted, drought tolerant ancestral Puebloan use, starvation crop

1. Introduction

1.1 Background to *S. jamesii*

Solanum tuberosum is often regarded as a crop that originated from the US or Ireland, but in reality, only two wild potatoes are found in the United states (and none in Ireland). They are *Solanum jamesii* and *Solanum stoloniferum* which are found in the desert southwest. *S. stoloniferum* is the tetraploid relative to *S. jamesii (jam).* Jam predominates in the southwest desert regions and is found in western Texas, northern Mexico and north into southern Utah and Colorado (See **Figure 1**). Collections have been found primarily near sites of ancestral habitation which are primarily found in the high desert of the Colorado Plateau. Elevation maximum for jam can be as high as 2280 meters at Chimney Rock National Monument in SW Colorado and south at 2000 meters in Magdalena New Mexico [1]. Jam can be cultivated as far north as Salt Lake City in open areas where it survives the winter underground. The Escalante formation in central Utah is home to several stands of jam and is cultivated by some traditional Navajo farmers in that area. There is currently an effort being made to cultivate jam for sale as a specialty crop to restaurants.

Jam prefers drier climates to that of the moister environments of the east. It typically grows in sandy soils to leaf litter strewn areas in Pinyon Juniper stands as well as in open well drained silted washes. Jam is known to lie dormant for years before sprouting, which occurs generally after the monsoon rains of July and August. The

IntechOpen

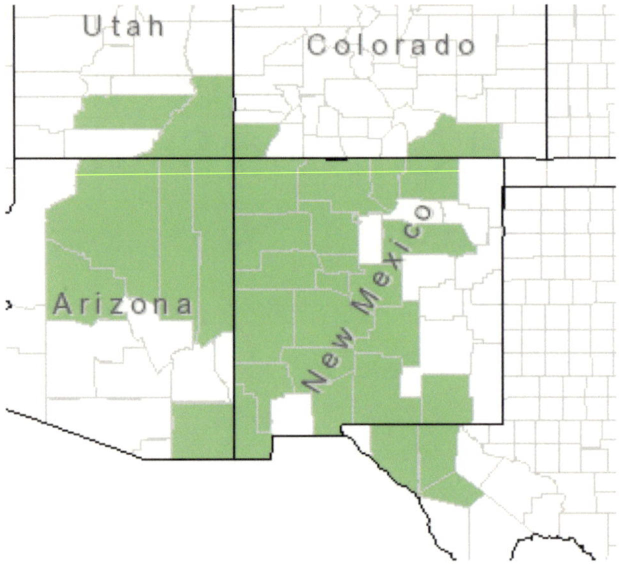

Figure 1.
Solanum jamesii Torr. *Distribution in the US SW. Data source: Plants National Database; home/profile page/ data source and documentation for* Solanum jamesii *Torr.*

plant sprouts and produces a mother tuber followed by additional tubers on stolons that depend on the length of time for growing. We have observed several tubers forming in dry years when monsoon rains did not provide more than 2–3 inches of rain in Mesa Verde's Navajo Canyon. When ample rains were received (6 inches from August to September of that year, 2014) the stolons continued to expand without producing tubers until September as fall approached. In storage Bamberg found that jam tubers that were kept for 8 years were able to sprout and produce a new crop of tubers. Additionally, in years of surveying Chaco Canyon and Mesa Verde, among other sites, there were minimally 10 years between finding sprouting jam in the canyons being examined. The drought tolerance and the ability to lie dormant for several years suggests that the genetics of jam would be well served when crossed with other species of potato. Such work is being conducted at the Wisconsin Potato Gene Bank by Bamberg and colleagues [2].

We do know that jam was used by ancestral Puebloans. At this juncture we do not know if they first found it by browsing, or if the use coincided with the beginning of agriculture. Early agricultural methods are murky where we extrapolate backwards for the methods of cultivation by native groups in the SW who practiced traditional agricultural techniques, but even those might not reflect how agriculture was carried out in the beginnings of the settling. Indeed, there is potential that the various groups or tribes were seminomadic, and left crops such as jam for their return to the area. Mesa Verde is an example where we see entering and exiting the area over the course of the year, and where grain was stored sealed in silos against animal intrusion and protected from the weather by overhangs.

Louderback [3] showed the presence of starch grains on stone tools that could be dated to 8950 BC. The starch grains are very characteristic for jam and were identified on food processing tools (metates). Other starch grains were also present suggesting a somewhat varied diet of plants that were grown in the ground. We are very confident that jam was an important food crop and provided for a varied diet to the early groups who settled the Americas first. Below is a map showing the approximate distribution of jam in the US SW. There are a few examples of stands in Northern Mexico, but research in those areas to further discover stands have been hampered by violence in the areas.

1.2 Characteristics of *S. jamesii*

Jam has the characteristic flower of the Solanum genus as shown in the two photos below (**Figure 2**). The flowers are white with yellow centers. When pollenated jam can produce fruit, but we have noted that even with pollinators present jam often will not fruit. We have rarely seen fruiting of the jam populations in MEVE and have yet to observe it in Chaco Canyon, albeit Chaco Canyon receives only 2–3 inches of moisture in a year which might speak to the non-fruiting. However, the population of Chaco is more homogeneous when compared to the population at MEVE, and Pavlik has proposed that when the genetics are similar, fruiting is limited but when more diverse genetic populations are present the fruiting becomes more robust (personal communication). We have observed this in experiments where two populations of jam are presented fruiting becomes abundant, but when single collections are grown, fruiting is rare.

Jam is also found growing in heavy grass stands where the grass has died back following sprouting in the spring. This is seen in the photo below of a stand of jam in Bandolier National Monument (**Figure 3**). This stand was adjacent to a block

Figure 2.
S. jamesii *blooms. Photo credit: David Kinder.*

Figure 3.
S. jamesii *in situ, august 2019. Photo credit: David Kinder.*

house ruin just below the cliff structures. The soil here is sandy with silt from flooding of the nearby river in the canyon.

The fruit are small and green with very small seeds (**Figure 4**). These seeds are among the smallest of the Solanaceae family [4]. They are bitter in taste, and we have not observed animal consumption of fruits; however, that may be because of limited fruiting in the wild and limited time in the field.

e tubers themselves can range from the size of a small pea up to 2 cm (**Figure 5**). Some larger tubers have been produced under cultivation. The tubers in the photo above show the variability in size. The darker colored tubers are older (and said to be stronger by native cultivators). Glycoalkaloid content is variable depending on stand. The two glycoalkaloids commonly found in jam are Chaconine and Solanine. The genin portion of the alkaloids is identical (solanidine) differing only in the sugar portion. The genin portion is not as toxic as the glycoalkaloid predominately causing liver damage. It should be noted that Chaconine possess anticancer

Figure 4.
S. jamesii *fruit. Photo credit David Kinder.*

Figure 5.
S. jamesii *tubers showing variation in size of wild harvested potatoes. Photo credit: David Kinder*

activity in cell culture. This is similar to the activity of tomatine which is primarily found in green tomatoes. To our knowledge no one has looked for tomatine in the fruit of jam.

2. Evidence for early use of jam and probability of cultivation

There is ample ethnobotanical information that indicates jam was used by multiple tribes in the American SW and northern Mexico. This ethnobotany stems from reports in the last 150 years and is assumed to have been information passed down from previous generations. The potato is ephemeral in the archaeobiology record, but there is one example of jam being found in a burial in Chaco Canyon during excavation of a grave in the 1920s [4]. This suggests that jam was at least important to those living in Chaco canyon at the time.

More recently Louderback and Pavlick found stone tools in the Escalante wash which had been used for food processing that had jam starch grains on them [3]. Jam starch grains are unique and easily distinguished from others. The tools date to over 10,000 years ago supporting the use of the potato as a food in the early first migrations. The starch grain finding supports processing of jam, but does not indicate whether it was cultivated. One assumes that cultivation began in central America and was passed northward along with corn and other food stuff, and likely with the planting of gardens the potato was also planted as well. However, the overlap between foraging and gardening is blurred by the ephemeral nature of agriculture in general with assumptions made as to what was or was not cultivated. More concrete evidence of cultivation occurs when water manipulation structures are found where those are mainly made of stone or other more permanent materials. In those same areas, wooden tools used for digging in the soils have been found (**Figure 6**). For example, in Mesa Verde there is an abundance of check dams found throughout the areas where habitation sites are found and where jam is found growing in the remnants of those structures. While this is not proof of ancestral cultivation of jam, it is compelling evidence that jam was included with other crops grown in those check dams with their fertile silty soils. This is contrasted to Chaco Canyon where the check dams were not made of stone but what remnants remain were made from soil and with time eroded. The jam found there grows in the silty washes. Chaco is unique in that it is an outlier in terms of moisture it receives, does not have a substantial water source, and was likely abandoned by the 1300s where it is thought they assimilated into the Pueblo groups located along the Rio Grande (based on Linguistic considerations by Ortman) [5].

A substantial number of jam stands that are found in areas where there is evidence of ancestral Puebloan habitation are especially large prompting one to suppose that the potatoes were cultivated along with other plants. For example, in Mesa Verde's Navajo Canyon which is adjacent to the side canyons where most of the cliff dwellings are located is a mega-population of jam that runs the length from spruce canyon's mouth to the Navajo canyon overlook where an ancient landslide blocked the valley below which then filled with silt. This canyon is thought to have been under cultivation while it was occupied more than 1000 years ago judging by the storage structures and the finding of tools for cultivating soil [4].

An interesting feature of stands of jam in major population centers such as Chaco Canyon, Mesa Verde, Bandolier Nat. Monument, and other places is that the populations are often bracketed by *Lyceum pallidum*. While this is not true of all stands, it is true of mega populations we have identified so far.

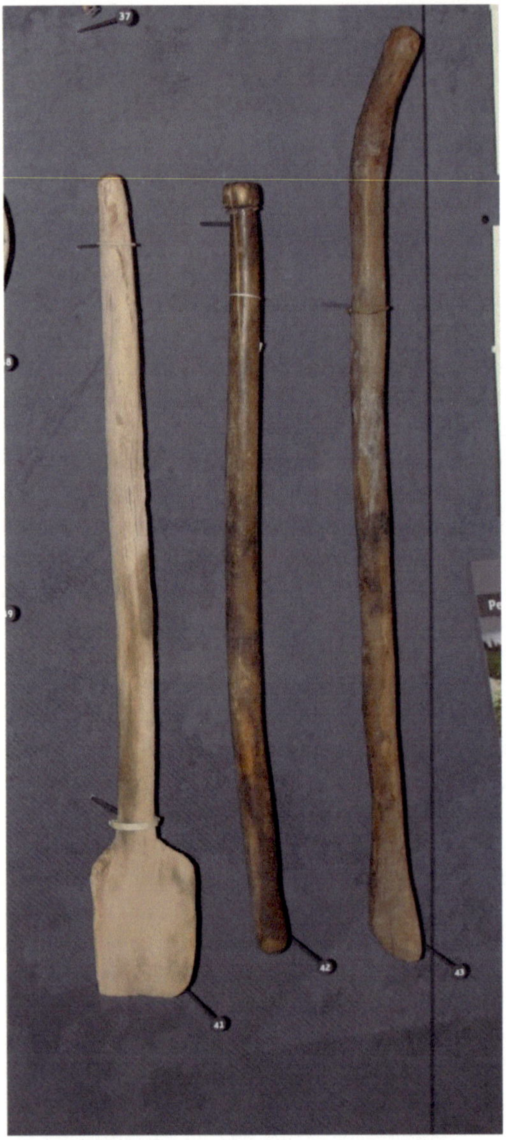

Figure 6.
Early digging tools, Museum of Natural History, University of Utah. Photo credit: David Kinder.

Genetic diversity of jam. Examination of genetic diversity can give clues to how long a population has survived in a particular area, or how the population might have drifted from surrounding populations of the same species. In the case of jam, there are a multitude of markers associated with the populations, but one remarkable finding is that the mega-population of MEVE contains 80% of the markers found in other jam populations. While one might think this is the ground zero for the beginning of jam, an alternative and more reasonable interpretation is that this populations is comprised of additions from around the southwest which were carried into MEVE and cultivated, or at least planted leading to this diversity of markers [6]. Since starch grains can give information as to sources of jam in the archeologic record, this might prove to be a useful tool applied here [7].

A further interesting finding is that the jam found in Chaco Canyon has relatively few markers that overlap with other communities without great diversity suggesting that only a single source from outside the canyon was brought in and planted/cultivated. It is known that Chaco Canyon was used for cultivation and certainly many interpretations of the use of Chaco suggest it was a trade center more so than a residential area only. No doubt ceremonies and other events were held there. The populations in Chaco are extensive with one mega-population occurring in the wash west of West Mesa beyond Peñasco Blanco. This wash contains the remnants of buildings thought to have been constructed as shelter for those engaged in tending crops, however this hypothesis is still being considered and is hampered by lack of resources to explore this canyon beyond that which occurred in the early part of the 1900s.

Jam is thought to have been consumed in several ways, with ethnobotanical information among several sources indicating that it was boiled and eaten with clay (8). The white clay is most likely kaolin which is thought to take away the bitter taste. The matter of the taste is subjective as jam is similar in taste to the russet potato. There were anecdotal reports from natives whose families had consumed jam where they roasted, sauteed the tubers in fat, flattened the tuber and roasted on a hot stone over an open fire and mashing the tubers following boiling. None of these methods are however recorded in literature and come from modern Navajo for the most part as well as Hopi.

3. Characteristics of the potato nutritionally

Potatoes are considered super foods by many, and *S. tuberosum* certainly helped maintain the Irish population during the years of English domination until the unfortunate occurrence of the potato blight. S. jamesii is no exception to the nutritional value from the perspective of minerals and other trace nutrients. Examining potatoes from Chaco Canyon and Mesa Verde for their nutritional content demonstrated that jam nutrient content is consistent between populations. It was also clear that soil content of various minerals could have some degree of influence on content but not to a great degree.

Potatoes harvested in the wild compared to *S. tuberosum* for several nutritional markers averaged 4% for protein for jam, 2% protein for *S. tuberosum*. Average amounts (in mg/100 gm wet wt. potato) for Calcium 30 mg Jam, vs. ~11 for *S. tuberosum*; for iron, ~3 mg jam, ~1 for *S. tuberosum*. For Zinc, ~0.9 vs. ~0.4 *S. tuberosum*. Calcium content was high with approximately 600 mg vs. 400 for *S. tuberosum*. In general this is twice the protein, zinc, and three times the calcium and iron [4].

Daniel Moerman's book on ethnobiology [8] of various medicinal and food plants shows among the southwestern native groups jam was an important component for the Apache, Hopi, Kawaik, Navajo and Pueblo groups along the Rio Grande river. Today many traditional Navajo grow potatoes for their own use.

The finding of jam starch grains on stone tools dating back over 10, 000 years from the present shows that the potato was an important component of the diet of ancestral Americans. That is especially important when considered in the light that corn did not reach the southwest until some 5000 years later making a slow migration onto the Colorado Plateau.

4. Potential for use in modern times – Advantages of jam for dry climate adaption

Jam has been known to exist in the American SW for some decades, and was not considered significant in the early days of plant study of indigenous people's use of this potato. More recently with interest in maintaining gene banks of native plants, and to use them for breeding purposes either by selection of plants with certain characteristic or deliberate crossing of one species with another there has been a more robust look at where these potatoes are found, and how they can be managed to the betterment of the potatoes. This work has been spearheaded by Bamberg and colleagues at the Wisconsin Potato Gene Bank (Greenbay WI) as well as others including Pavlik and Louderback at the U. of Utah. The storage potential added to various crosses with other wild or domestic potatoes holds promise for the future where potatoes can be grown in more arid climates, or can be stored for extended periods of time and maintain viable tubers for planting. In terms of third world populations where drought causes extensive starvation this small potato could be developed with potatoes with other favorable characteristics to provide a food source for those populations. Thus harnessing the potential favorable genes from jam could well produce a series of potatoes with the favorable nutritional content as well as the ability to thrive in some inhospitable climates for addressing starvation around the globe. However, it remains to be seen what this potential will mean to further production of cultivars.

5. Conclusion

Jam has a long and untold story that is just beginning to be worked out. It has characteristics that have allowed it to survive in harsh conditions, and has nutritional content that makes it even more attractive for consumption by humans. Its potential to add its genes to other potatoes is great where drought tolerance would benefit many populations greatly where more modern crops fail. Indeed, underground growing of foods would prevent browsing to adversely affect the production of potatoes in some populations unlike the attractiveness of corn for some foraging animals.

Given the findings of the use of potatoes in the American SW, as well as noting that other plant materials were used for food. The old adage of the Three Sisters – Corn, Squash and Beans – should more properly be replaced by Succotash instead to reflect the broad diet of the first Americans.

Acknowledgements

We are grateful for the help of many research workers over the course of the several years that this chapter encompasses. We would also like to thank Ohio Northern University's College of Pharmacy for providing funds for various aspects of the research conducted by DHK, and to the NSF for funding of Drs. Louderback and Pavlik for much of the work conducted. For an interesting Interview with Dr. Louderback, see: High Country News. Lisbeth Louderback, https://www.hcn.org/issues/49.20/scientific-research-escalante-valley-how-an-ancient-potato-helped-indigenous-people-survive-climate-shifts

Conflict of interest

We see no conflicts of interest for any author. The authors have collaborated on several aspects of this work, and no monetary or benefits are expected to ensue from this work.

Author details

David Kinder[1*], John Bamberg[2], Lisbeth Louderback[3], Bruce Pavlik[3] and Alfonso Del Rio[4]

1 Ohio Northern University, Ada, OH, USA

2 Wisconsin Potato Gene Bank, Green Bay, WI, USA

3 University of Utah, Salt Lake City UT, USA

4 University of Wisconsin, Madison, WI, USA

*Address all correspondence to: d-kinder@onu.edu

References

[1] Bamberg, JB, A. H. del Rio, D. Kinder, L. Louderback, B.Pavlik, C.Fernandez. (2016). Core collections of potato (solanum) species native to the USA. American Journal of Potato Research 93:564-571.

[2] Bamberg, J.B. (2010) Tuber dormancy lasting eight years in the wild potato *Solanum jamesii* . Am. J. Pot Res 87, 226-228. https://doi.org/10.1007/s12230-009-9124-9

[3] Lisbeth A. Louderback, Bruce M. Pavlik (2017) Starch granule evidence for the earliest potato use in North America PNAS 114 (29) 7606-7610 ; https://doi.org/10.1073/pnas.1705540114

[4] David H. Kinder, Karen R. Adams, Harry J. Wilson (2017) Solanum jamesii: Evidence for cultivation of wild potato tubers by ancestral Puebloan groups J. of Ethnobiology, 37(2):218-240. https://doi.org/10.2993/0278-0771-37.2.218

[5] SG Ortman (2012) Winds from the north: Tewa origins and historical anthropology. University of Utah Press. ISBN-13 : 978-1607811725

[6] John Bamberg, Kevin A Lombard, Alfonso Del Rio, Charles Fernandez, Lisbeth Louderback, Bruce Pavlik, David Kinder (2019) Field Evaluation of *Solanum jamesii* – a Native USA Wild Potato *Journal 2019 ASHS Annual Conference* 2019/7/23

[7] Nicole M.Herzogab Lisbeth A. Louderbackb Bruce M.Pavlikc (2018) Effects of cultivation on tuber and starch granule morphometrics of *Solanum jamesii* and implications for interpretation of the archaeological record. *Journal of Archaeological Science* 98, , Pages 1-6

[8] Daniel E. Moerman (1998) Native American Ethnobotany (Timber Press) ISBN13: 9780881924534

Section 3

Nutrition for Increasing Population

Chapter 6

Nutritional Composition and Biochemical Properties of *Solanum tuberosum*

Belay Dereje and Nwankwo Chibuzo

Abstract

Solanum tuberosum is the most popular vegetable in people's diets all over the world, and it's considered a staple crop in many countries. It has immense potential to reduce food insecurity and prevent malnutrition in developing and developed countries because of its productivity, nutritional composition and unique biochemical features. However, a lack of information about the nutritional composition and biochemical properties of this tuber severely limits its use. Improved awareness of the biochemical and nutritional quality, utilization, and future economic importance of the crop has important implications for human food systems, nationally and internationally. This chapter presents a brief overview of key findings that led to our current knowledge of the biochemical and nutritional composition of the *Solanum tuberosum* tuber. The wide range of *Solanum tuberosum* varieties lays a great foundation for their industrial production and applications. The biochemical and nutritional composition of the *Solanum tuberosum* is summarized briefly.

Keywords: Antioxidants, Biochemical properties, Minerals, *Solanum tuberosum*, *Solanum tuberosum* nutrition

1. Introduction

Potato is in the 4th order with respect to production and area harvested after maize, wheat and rice, as a staple crop for human nutrition with a production of more than 368 million tonnes [1, 2]. This famous tuber is grown in 80 percent of the world's countries [3, 4]. This shows that *Solanum tuberosum* is one of the most productive crops in the world. Potato can produce more nutritious food on less land and in harsher climates than most other major crops. Furthermore, this tuber can be harvested after only 8 weeks [5]. There are numerous myths about the biochemical and nutritional value of *Solanum tuberosum*. *Solanum tuberosum* is a versatile, carbohydrate-rich food that is widely consumed and prepared in a variety of ways around the world. This tuber is typically regarded as a ritual food or a garnish for other major meal components, and it is consumed as a complementary vegetable with staple foods [6]. *Solanum tuberosum* is commonly thought to contribute insignificantly to the nutritive value of a meal. Even in areas where *Solanum tuberosum* is considered staple foods, they are typically viewed solely as a source of energy, with little awareness of their vitamin or protein content [6, 7].

Solanum tuberosum contains a variety of biochemical and nutritional properties, including starch, ascorbic acid, reducing sugars, non-reducing sugars, total sugars, phenolic content, flavonoids, polyamines, and carotenoids, all of which are highly desirable in the diet due to their beneficial effects on human health [8, 9]. The nutritional value of *Solanum tuberosum* tubers is primarily defined by the presence of essential amino acids, particularly lysine, as well as high levels of starch and dietary fiber and a low concentration of fats [10]. The chemical composition of *Solanum tuberosum* determines the quality of the processing and is influenced by a series of factors including the production area, crops, soil and the climate, farming practices, storage and marketing conditions [11]. *Solanum tuberosum* tubers with no or low-fat addition have high levels of bioactive compounds and antioxidants, such as phenolic acids, primarily chlorogenic acid, ascorbic acid, and flavonoids, which are phytochemicals [10]. Increased consumption of potato tubers may increase antioxidant levels in blood and tissues and protect against oxidative stress, which is responsible for lipid, protein, and enzyme damage [3]. One of the global health goals is to increase nutrient availability to a large portion of the world's population. A sensible approach to achieving this goal would be to boost the nutritional content of commonly consumed crops like *Solanum tuberosum* [5]. Furthermore, *Solanum tuberosum* have superior biochemical and nutritional properties and are amenable to development via breeding and biotechnology methods [5, 12]. However, a paucity of information regarding the biochemical and nutritional composition of the *Solanum tuberosum* greatly limits its exploitation. Improved awareness of the biochemical and nutritional quality, utilization, and future economic importance of the crop has important implications for human food systems, nationally and internationally.

2. Nutritional composition of *Solanum tuberosum*

Solanum tuberosum have been discovered to be an especially nutritious vegetable. Freshly harvested *Solanum tuberosum* tubers contain approximately 80% water and 20% dry matter. *Solanum tuberosum* is primarily composed of starch, but they also contain trace amounts of protein and alkaline salts. They are complex carbohydrate in the form of sugars that are virtually fat and cholesterol-free. Beta-carotene, vitamin C, A, B_1, B_2, B_6, and folic acid are among many vitamins found in *Solanum tuberosum*. It also contains trace amounts of protein, amino acids, and nicotinic acid [6]. However, there have been significant variations. Because many of the nutrients in *Solanum tuberosum* are found in their skin, eating them whole rather than peeled has been linked to more health benefits [12]. *Solanum tuberosum* is not only important food security crops, but they are also excellent candidates for commercial use [13]. Processing adds value to this tuber, extends their shelf life and convenience, reduces post-harvest losses and waste, and yields a diverse range of products for various applications. *Solanum tuberosum* tubers are eaten raw or processed into products such as French fries, crisps, and canned potatoes [14].

2.1 Carbohydrate

Solanum tuberosum carbohydrates can be divided into four types: starch, non-starch, polysaccharides, and sugars. Starch is present in the form of granules, which are composed of amylopectin and amylose in a fairly constant 3:1 ratio. Amylopectin is a large, ramified molecule with approximately 10^5 glucose residues. The amylose molecule is smaller, with approximately 5000 glucose residues. There are trace amounts of phosphorus in the amylopectin fraction, which is chemically combined with starch [3]. Because of the high starch content, manufacturing *Solanum tuberosum* starch is now economically feasible in developed countries. *Solanum tuberosum* starch is used in the manufacture of adhesives, textiles, food, and the production of derived substances such as alcohol and glucose. Unlike cereal starches, these starch gels set quickly and have a high pot-paste viscosity. Non-starch polysaccharides account for only a small proportion of tuber dry matter. Because of their role as dietary fiber, non-starch polysaccharides contribute to the nutritional value of *Solanum tuberosum*. The major sugars found in white potato are sucrose, fructose, and glucose, with traces of other minor sugars [6].

2.2 Fat

Solanum tuberosum fat content is low, ranging from 0.08 to 0.13 percent fresh weight basis. This range is too low to be nutritionally significant, but it contributes to *Solanum tuberosum* flavor, tuber cellular integrity and bruising resistance, and helps to reduce enzymic darkening in tuber flesh [15]. The lipids are more important because they are susceptible to enzymatic degradation and non-enzymatic auto-oxidation, which cause off flavor and rancidity in dehydrated and instant *Solanum tuberosum* products. Polyunsaturated linoleic and linolenic acids account for 75% of total fatty acids in lipids. These factors contribute to the development of both desirable flavor characteristics in cooked tubers and undesirable off flavors in processed products. During tuber processing, lipid-degrading enzymes rapidly convert polyunsaturated acids to free fatty acids and other compounds, and they are also extremely susceptible to auto-oxidation [6].

2.3 Crude protein

Solanum tuberosum contains approximately 2 to 3% protein content on a fresh weight basis [8], and is comparable to most other root and tuber staples, except for cassava, which has half this amount. On a dry basis, it is comparable to cereals, and on a cooked basis, it is comparable to boiled rice [6]. One advantage that *Solanum tuberosum* have over cereal staples is their high lysine content. It does, however, have lower concentrations of sulfur-containing amino acids (such as methionine and cystine/cysteine) than cereals. *Solanum tuberosum*, when combined with other foods, can supplement diets that are low in lysine, such as rice accompanying *Solanum tuberosum*, which provides a higher quality protein. Meals in some developing countries are frequently served with a combination of boiled tubers of *Solanum tuberosum* and rice or pasta. However, consumers in developed countries frequently mistakenly believe that such mixtures provide nothing more than large amounts of carbohydrate energy [3]. It has been proposed that *Solanum tuberosum* comparative advantage as a food in the tropics, on a unit weight basis, stems from its ability to supply high-quality protein.

Using the most recent figures for energy and protein requirements, it can be calculated that 100 g (one small tuber) of *Solanum tuberosum* can supply 7%, 6%, and 5% of daily energy, and 12%, 11%, and 10% of daily protein needs of children aged 1–2, 2–3, and 3–5 years, respectively. Adults, depending on body weight and gender, can get from 3–6% of their daily protein needs from 100 g of tuber [15]. Bekele and Haile [16] reported that the protein contents of improved *Solanum tuberosum* varieties were 1.65% to 3.28%. The results show that these contents were variety dependent.

2.4 Nitrogen

Solanum tuberosum is rarely eaten as the sole source of nitrogen and mixing potatoes with other foods has supplementary or synergistic effects. *Solanum tuberosum* is not a high-energy food, providing only about 80 kilocalories per 100 g, but it does provide high-quality protein. This is especially important in developing countries, where energy is more readily available than protein [3]. When compared to many other vegetable crops, the nitrogenous content of the *Solanum tuberosum* tuber has a high nutritional value. The distribution of nitrogen within the tuber is not uniform, with the skin having the highest concentration, followed by the cortex, and then rising again toward the pith. *Solanum tuberosum* tubers' total nitrogen content consists of the following elements: (a) soluble, coagulable (true) protein; (b) insoluble protein; and (c) soluble non-protein nitrogen, which is composed of free amino acids, the amides asparagine and glutamine, and small amounts of nitrate nitrogen and basic nitrogen compounds including nucleic acids and alkaloids [6]. The insoluble protein fraction is mostly found in the peel. It accounts for only about 4% of total nitrogen [15].

2.5 Fiber

The dietary fiber content of raw *Solanum tuberosum* ranges between one and two g per 100 g of fresh weight. Furthermore, some of the dietary fiber may be starch that is resistant to hydrolysis by the enzymes used to remove starch before determining dietary fiber [15]. This resistant starch is created by subjecting foods to heat or dehydration, which gives the starch molecules a more ordered structure and makes them less susceptible to enzymatic digestion. In comparison to other raw items, the fresh *Solanum tuberosum* has dietary fiber content similar to sweet potatoes, but slightly lower than other roots and tubers and much lower than most cereals and dried phaseolus beans, though potatoes and cereals are similar on a dry basis [3]. Dietary fiber determinations have primarily been done on raw foods rather than cooked foods. Boiled potato flesh has fiber content comparable to cooked white rice but significantly lower than boiled green plantains or boiled phaseolus beans. Consuming the entire tuber rather than just the flesh may increase dietary fiber intake [3].

2.6 Mineral

Minerals are an important part of a healthy diet. Because of its relatively high content of certain macro and trace minerals, *Solanum tuberosum*, as a major staple food crop, could play an important role in combating mineral deficiencies. Boiling thinly sliced *Solanum tuberosum* will result in a large reduction in mineral levels while boiling whole *Solanum tuberosum* or that have been cut into large pieces increases [17].

The presence of magnesium, potassium, iron, and zinc is notable in *Solanum tuberosum* [7]. Potassium is the most abundant mineral (320 mg/100 g raw), with a higher concentration in the skin and, as a result, a lower concentration in peeled *Solanum tuberosum* products [6]. *Solanum tuberosum* is a good source of a variety of dietary minerals. *Solanum tuberosum* is listed as providing 18% of the RDA for potassium, 6% for iron, phosphorus, and magnesium, and 2% for calcium and zinc. Most minerals are well retained in boiled *Solanum tuberosum* cooked with the skin. Baking *Solanum tuberosum* with the skin on is a good way to retain minerals. There are significant differences in major and trace mineral contents between *Solanum tuberosum* genotypes. Potassium levels varied the most, while manganese levels varied the least [6].

2.6.1 Potassium

Solanum tuberosum is a valuable source of potassium in the human diet. It contains a source of dietary potassium (42.73 dried matter), which plays an important role in acid–base regulation and fluid balance and is required for optimal functioning of the heart, kidneys, muscles, nerves, and digestive systems [17, 18]. Potassium levels in *Solanum tuberosum* are comparable to those found in most fruits and vegetables per unit weight, and because potatoes are typically consumed in larger quantities, they are an important and dependable food source of this nutrient [15].

2.6.2 Phosphorus

Aside from potassium, phosphorus is the most abundant mineral in *Solanum tuberosum* (3.54 g/kg dried matter) [18]. It plays numerous roles in the human body and is essential for healthy cells, teeth, and bones. Inadequate phosphorus intake results in abnormally low serum phosphate levels, which affect appetite loss, anemia, muscle weakness, bone pain, rickets osteomalacia, susceptibility to infection, numbness and tingling of the extremities, and difficulty walking [15].

2.6.3 Calcium

Solanum tuberosum is a good source of calcium, with a wide range being reported. Two studies found calcium levels as high as 130 mg/100 dry weight and 455 mg/kg. Resistance to pathogens is linked to high levels of tuber calcium. Calcium is important for bone and tooth structure, blood clotting, and nerve transmission [6].

2.6.4 Iron, zinc and copper

Iron is found in small amounts in potatoes. A study of cultivated varieties revealed 0.3–2.3 mg of Fe per 100 g. The iron content ranges between 6 and 158 μg/g dry weight [6]. Some *Solanum tuberosum* contain iron levels comparable to those found in some cereals (rice, maize, and wheat). *Solanum tuberosum* iron should be bioavailable because, unlike cereals, it contains very little phytic acid. *Solanum tuberosum* have significant differences in zinc content [19]. The zinc content varies between 1.8 and 10.2 ug/g fresh weight. *Solanum tuberosum* from different cultivars contain zinc in 0.5–4.6 ug/g fresh weight. Zinc is required for the proper functioning of the body's immune system and is involved in cell division, growth, and wound healing. The copper content of *Solanum tuberosum* ranges from 0.23 to 11.9 mg/kg fresh weight. Copper, like zinc, is abundant in yellow-fleshed *Solanum tuberosum* [19]. Copper is required for hemoglobin synthesis, iron metabolism, and blood vessel maintenance [6].

3. Biochemical properties of *Solanum tuberosum*

In addition to supplying energy, *Solanum tuberosum* contains biochemical ingredients such as phenolics, flavonoids, anthocyanins, carotenoids, folates, ascorbic acid and sugar [3, 12, 20]. Phenolics, anthocyanins, flavonoids and carotenoids are the major antioxidants found in *Solanum tuberosum* that are beneficial for human health [8].

3.1 Phenol content

Solanum tuberosum is an excellent source of these compounds. After apples and oranges, *Solanum tuberosum* was thought to be the third having the most important source of phenols [20]. Both the skin and flesh of *Solanum tuberosum* contain phenolic compounds, whereas, the concentration is greater in the skin than in the flesh. Purple and red-skinned tubers had twice the phenolic acid concentration as white-skinned tubers. The most important phenolic acids have been identified as chlorogenic acid, protocatechuic acid, vanillic acid, and p-coumaric acid. Even though, the peels of *Solanum tuberosum* tubers contain the most phenols they discarded as waste during potato processing [12]. Fresh *Solanum tuberosum* pulp and skin contain 30 to 900 mg/kg and 1000 to 4000 mg/kg, respectively. It was also reported that the concentration of phenolic acids in purple or red-fleshed cultivars was three to four times higher than in white-fleshed cultivars. White fleshed *Solanum tuberosum* varieties were found to have fewer phenolics (less than 4 mg/g dry weight) than purple-fleshed wild species (more than 5 to 6 mg/g dry weight) [3].

The total phenolics in eleven Indian *Solanum tuberosum* varieties were evaluated after 0, 30, 60 and 90 days of storage at room temperature, 15°C and 4°C. All 11 showed a variation among the varieties and were different with storage temperature; their levels fluctuated during storage but remained above the initial level until the last day of observation [21]. Cooking significantly affects the retention and availability of phenolic compounds [22]. The effect of three domestic methods of cooking (boiling, steaming, and microwaving) on total phenols, antioxidant and anticholinesterase activities were studied. All three modes of cooking cause a decrease in the total polyphenol contents, antioxidant and anticholinesterase activities [23]. Their results show that the polyphenols are lost to different degrees according to the method of cooking, the classification of the polyphenol contents places the microwave in the first position then comes the steam cooking and lastly the cooking in the water. Similar observations have been reported in which frying causes the greatest loss of total phenolic compounds, followed by baking, steaming, boiling, and microwaving [23]. Twelve *Solanum tuberosum* landrace clones collected from established cultivations on Chiloe Island and Valdivia were selected and the total phenolic content was evaluated. The total phenolic content varied in the peeled *Solanum tuberosum* samples from 191 to 1864 mg/100 g dry matter meanwhile these parameters varied from 345 to 2852 mg/100 g dry matter in unpeeled samples [24].

3.2 Flavonols and Anthocyanins

Although *Solanum tuberosum* contains flavonols such as rutin, they are not thought to be significant sources of dietary flavonols. Flavonol concentrations increased in fresh-cut tubers, reaching up to 14 mg/100 g, implying that because of the large number of potatoes consumed, they could be a valuable dietary source. Numerous studies have suggested that flavonols have a variety of health benefits, including a lower risk of heart disease and a lower risk of certain respiratory diseases such as asthma, bronchitis, and emphysema as well as a lower risk of certain cancers such as prostate and lung cancer [3, 7, 25].

Anthocyanins are a type of pigmented flavonoids. The composition of anthocyanins in pigmented *Solanum tuberosum* is complicated by acylation in the glycoside ring. The purple and red colors of *Solanum tuberosum* varieties are due to anthocyanin pigment [26]. *Solanum tuberosum* anthocyanins have recently been recognized for their health benefits, as they have been shown to have strong antioxidative activity, anti-influenza virus activity, and anti-stomach cancer activity [27]. Flavonoids such as anthocyanins were found in high concentrations in pigmented flesh *Solanum tuberosum*, ranging from 5.5 to 35 mg/100 g fresh weight in tubers. Purple or red-fleshed *Solanum tuberosum* varieties had two times the flavonoid concentration than that of white-fleshed varieties and their concentration was significantly higher in the skin, impending 900 mg in purple-fleshed and 500 mg in red-fleshed types per 100 g fresh weight [12]. Anthocyanin pigments are found in the periderm of the tuber and impart various colors to their skin, with purple being the most common. As pigmented *Solanum tuberosum* is low-cost crops that are also a good source of antioxidant micronutrients, it could be a good source of natural anthocyanin pigments. Purple fleshed *Solanum tuberosum* had higher levels of anthocyanins than red-fleshed potatoes [28]. The extracts of flavonoids and flavones had high scavenging activities against oxygen free radicals. *Solanum tuberosum* exhibited 94 percent hydroxyl radical scavenging activity and nearly complete inhibition of superoxide radicals in the presence of anions [28]. Various biotechnological and transgenic approaches have demonstrated that it is possible to significantly increase the phenolic, anthocyanin, and flavonoid content of *Solanum tuberosum* tubers [28].

3.3 Carotenoids

Carotenoids are useful as food ingredients because they can replace synthetic pigments while also benefiting human health due to their provitamin content [29]. Carotenoids and other lipophilic compounds found in *Solanum tuberosum* tuber are also beneficial to one's diet. Carotenoids are synthesized in plastids from isoprenoids, and one of their functions is to protect against photo and oxidative stress [30]. Carotenoid concentrations in *Solanum tuberosum* germplasm have been reported to vary over a 20-fold range, with much of the variation controlled at the transcriptional level [27]. The major carotenoids found in *Solanum tuberosum* are lutein, violaxanthin, zeaxanthin, and neoxanthin, with trace amounts of -carotene. Zeaxanthin and lutein are responsible for the orange and yellow flesh colors of the tuber. White fleshed *Solanum tuberosum* varieties contained fewer carotenoids than yellow or orange-fleshed ones. Total carotenoids content in white and yellow-fleshed *Solanum tuberosum* varieties was reported to be in the range of 50–350 g/100 g fresh weight and 800–2000 g/100 g fresh weight, respectively [31]. Carotenoid levels in potato tubers vary greatly, with levels in yellow-fleshed cultivars being 20 times higher than levels in white-fleshed varieties [32]. Valcarcel *et al.* [33] reported that higher levels of total carotenoids in the skin of *Solanum tuberosum* tubers, with variety 'Burren' showing maxima values of 28 and 9 mg kg/dry weight in skin and flesh, respectively. They observed that yellow-skinned or fleshed varieties had higher contents than those with paler or white tissues, with no relationship found for other colors.

Solanum tuberosum have low basal carotenoid levels when compared to most fruits and vegetables [34]. For example, the maximum total carotenoid content of a *Solanum tuberosum* tuber is 20 mg/kg fresh weight, whereas the maximum carotenoid content of brussel sprouts is 1100 mg/kg fresh weight and carrots is 14000 mg/kg fresh weight. Despite the relatively low level of carotenoids in *Solanum tuberosum* tubers, the content of carotenoids in potato tubers is of dietary significance because *Solanum tuberosum* is a staple part of the diet.

3.4 Glycoalkaloids

Solanum tuberosum produce glycoalkaloids during germination, which protect the tuber from pathogens, insects, parasites, and predators [35]. The primary glycoalkaloids found in domestic *Solanum tuberosum* are -chaconine and -solanine, which is concentrated in the outer layers of the *Solanum tuberosum* skins (i.e., the periderm, cortex, and outer phloem). Glycoalkaloid levels in different *Solanum tuberosum* varieties can vary greatly and may be influenced postharvest by environmental factors such as light mechanical injury and storage [35]. Small *Solanum tuberosum* also have higher glycoalkaloids levels (per unit weight) than larger ones.

Plant secondary metabolites known as glycoalkaloids are toxic to microorganisms, viruses, insects, animals, and humans. The saccharide moiety of these compounds differs structurally in that solanine contains the trisaccharide solatriose, whereas chaconine has the aglycone attached to chacotriose [36]. The glycoalkaloid content of *Solanum tuberosum* tubers varies greatly and is influenced by post-harvest factors such as light exposure, irradiation, mechanical injury, and storage conditions. *Solanum tuberosum* peels are a rich source of steroidal alkaloids, which are well known for their toxicity in high concentrations for human consumption (>1 mg/g dry weight sample).

3.5 Folates

Potato is a well-known significant source of folates in the diet due to its high consumption level rather than its endogenous content. The folate concentrations in mature raw *Solanum tuberosum* range from 12 to 37 g/100 g fresh weight [37].

3.6 Ascorbic acid

Ascorbic acid is a strong reducing agent in plant metabolism, it improves the absorption and internal transport of dietary iron and zinc from other plant sources. *Solanum tuberosum* tubers have been reported to contain up to 46 mg of ascorbic acid per 100 g tuber on a fresh weight basis, and their availability is dependent on the variety, maturity status, and environmental conditions under which the crop is grown [12]. The concentration of ascorbic acid in freshly harvested peeled raw tubers ranged from 22.2 to 121.4 mg/100 g on a dry weight basis and from 6.5 to 36.9 mg/100 g on a fresh weight basis and decreased with storage period in tubers of all varieties [12]. A British study measured vitamin C levels in 33 varieties grown in three different locations across Europe [38]. The vitamin C content ranged from 13 to 30.8 mg per 100 g fresh weight. Numerous studies have shown that vitamin C levels in potatoes decrease rapidly during cold storage, with losses approaching 60% [12, 38].

Valcarcel *et al.* [33] measured the L-ascorbic acid content in 60 varieties of *Solanum tuberosum* grown in Ireland and reported the highest content of 800 mg/kg on a dry basis. They observed significant differences in L-ascorbic acid content across years and sites. The vitamin C content of eleven Indian *Solanum tuberosum* varieties was observed to be in the range from 0.0828 to 0.2416 mg/g fresh weight and these concentrations were variety and storage temperature dependent [21].

3.7 Sugars

The sugar level in *Solanum tuberosum* during suberization and harvest is heavily influenced by the variety. Low sugar content is a desirable trait for processing. Sucrose content at harvest is an indicator of the tuber's chemical maturity [12]. The higher sucrose levels in *Solanum tuberosum* tubers at harvest indicate immaturity. The sucrose content at harvest is critical because invertase hydrolysis results in the accumulation of reducing sugars, rendering *Solanum tuberosum* unfit for processing [12]. An increase in total sugars or a specific sugar and dry matter is a heritable trait, but it is also influenced by a variety of environmental factors. The sugar content of *Solanum tuberosum* during tuberization and harvest is heavily influenced by the variety. The quantity and type of sugars in a specific cultivar are inherited characteristics [12]. Since *Solanum tuberosum* increased its ability to produce sucrose as the storage period increased, more than 65 percent of the maximum sucrose accumulation occurred within 5 days of storage. According to Zhitian [39], the sucrose concentration in

Solanum tuberosum increased early in storage and then remained constant. At relatively higher temperatures (25-30°C), *Solanum tuberosum* storage in ordinary rooms, traditional heaps, and so on showed very little increase in reducing sugars. Freshly harvested mature tubers of a few Indian *Solanum tuberosum* varieties have a low level of reducing sugars. The concentration of glucose increased early in storage and then remained constant [39].

4. Conclusion

Solanum tuberosum is a staple food crop providing basic nutrition to millions of people globally. It provides numerous compounds of high nutritional value including protein, carbohydrates, minerals, carotenoids, dietary fiber, vitamins, very little fat, and sodium and other bioactive compounds. The nutrient composition of potato tubers varies greatly according to genetic and environmental factors. As a result, the nutrient content of *Solanum tuberosum* should be considered during variety screening, demonstration, and growing *Solanum tuberosum* with the climate. Phenolics, anthocyanins, flavonoids and carotenoids are the major antioxidants found in *Solanum tuberosum* that are beneficial for human health.

Acknowledgements

This chapter contains information gathered from numerous published resources, and thus, we would like to acknowledge all authors of the references used in this chapter.

Conflict of interest

The authors declare no conflict of interest.

Author details

Belay Dereje[1*] and Nwankwo Chibuzo[2]

1 Department of Food Process Engineering, Wolkite University, Ethiopia

2 Department of Food Science and Technology, Federal University of Agriculture Makurdi, Nigeria

*Address all correspondence to: belay.dereje@wku.edu.et

References

[1] M. Sulli *et al.*, "Molecular and biochemical characterization of a potato collection with contrasting tuber carotenoid content," PLoS One, vol. 12, no. 9, pp. 1-22, 2017.

[2] FAOSTAT, "Statistical database," 2019.

[3] A. S. R *et al.*, "*Solanum tuberosum* L: Botanical, Phytochemical, Pharmacological and Nutritional Significance," *Int. J. Phytomedicine*, vol. 10, no. 3, p. 115, 2018, doi: 10.5138/09750185.2256.

[4] M. Kaplan *et al.*, "Nutritional composition of potato (Solanum tuberosum L.) Haulms," Prog. Nutr., vol. 20, pp. 90-95, 2018, doi: 10.23751/pn.v20i1-S.5541.

[5] U. S. P. Genebank and S. Bay, "Effect of Propagule Type and Growing Environment on Antioxidant Activity and Total Phenolic Content in Potato Germplasm," Amer J Potato Res, vol. 84, pp. 323-330, 2007.

[6] D. A. Navarre, A. Goyer, and R. Shakya, *Nutritional Value of Potatoes*, First Edit. Elsevier Ltd., 2009.

[7] C. M. Andre *et al.*, "Andean potato cultivars (Solarium tuberosum L.) as a source of antioxidant and mineral micronutrients," J. Agric. Food Chem., vol. 55, no. 2, pp. 366-378, 2007, doi: 10.1021/jf062740i.

[8] T. Mishra, P. Raigond, N. Thakur, S. Dutt, and B. Singh, "Recent Updates on Healthy Phytoconstituents in Potato: a Nutritional Depository," Potato Res., vol. 63, no. 3, pp. 323-343, 2020, doi: 10.1007/s11540-019-09442-z.

[9] M. B. Katan and N. M. De Roos, "Promises and problems of functional foods," Crit. Rev. Food Sci. Nutr., vol. 44, no. 5, pp. 369-377, 2004, doi: 10.1080/10408690490509609.

[10] A. J. Keutgen *et al.*, "Antioxidant properties of potato tubers (Solanum tuberosum L.) as a consequence of genetic potential and growing conditions," PLoS One, vol. 14, no. 9, pp. 1-14, 2019, doi: 10.1371/journal.pone.0222976.

[11] M. Leonel *et al.*, "Chemical composition of potato tubers: the effect of cultivars and growth conditions," J. Food Sci. Technol., vol. 54, no. 8, pp. 2372-2378, 2017, doi: 10.1007/s13197-017-2677-6.

[12] A. Brar, A. K. Bhatia, V. Pandey, and P. Kumari, "Chemical Science Review and Letters Biochemical and Phytochemical Properties of Potato: A Review," Chem Sci Rev Lett, vol. 6, no. 21, pp. 117-129, 2017.

[13] K. H. Sabeena Farvin, H. D. Grejsen, and C. Jacobsen, "Potato peel extract as a natural antioxidant in chilled storage of minced horse mackerel (Trachurus trachurus): Effect on lipid and protein oxidation," Food Chem., vol. 131, no. 3, pp. 843-851, 2012, doi: 10.1016/j.foodchem.2011.09.056.

[14] T. Hussain, "Potatoes: Ensuring Food for the Future," Adv. Plants Agric. Res., vol. 3, no. 6, 2016, doi: 10.15406/apar.2016.03.00117.

[15] FAO, "Food and Agriculture Organization of the United Nations. International Year of the Potato, 2008. www.potato2008.org," 2008.

[16] B. Tilahun and H. Bewuketu, "Evaluation of improved potato (Solanum tuberosum L.) varieties for some quality attributes at Shebench Woreda of Bench-Maji Zone, Southwestern Ethiopia," African J.

Agric. Res., vol. 14, no. 7, pp. 389-394, 2019, doi: 10.5897/ajar2018.13482.

[17] P. C. B. A. S. H. JANSKY, "The Effects of Boiling and Leaching on the Content of Potassium Introduction," J. Food Sci., vol. 75, no. 5, pp. 80-85, 2008, doi: 10.1111/j.1750-3841.2008.00782.x.

[18] J. Wierzbowska, B. Cwalina-Ambroziak, M. Dosek-Sobieraj, and S. Sienkiewicz, "Contnent of minerals in tubers of potato plants treated with bioregulators," Rom. Agric. Res., vol. 2016, no. 33, 2016.

[19] G. Dugo, L. La Pera, V. Lo Turco, D. Giuffrida, and S. Restuccia, "Determination of copper, zinc, selenium, lead and cadmium in potatoes (Solanum tuberosum L.) using potentiometric stripping methods," Food Addit. Contam., vol. 21, no. 7, pp. 649-657, 2004, doi: 10.1080/026520 30410001698698.

[20] O. K. Chun, D. O. Kim, N. Smith, D. Schroeder, J. T. Han, and Y. L. Chang, "Daily consumption of phenolics and total antioxidant capacity from fruit and vegetables in the American diet," J. Sci. Food Agric., vol. 85, no. 10, pp. 1715-1724, 2005, doi: 10.1002/jsfa.2176.

[21] P. J. Nilesh, A. R. Rajeshkumar, and T. G. Jayant, "Effect of Storage Temperature on Vitamin C, Total Phenolics, UPLC Phenolic Acids Profile and Antioxidant Capacity of Eleven Potato (Solanum Tuberosum) Varieties," Hortic. Plant J., 2017, doi: 10.1016/j.hpj.2017.07.004.

[22] S. S. Jayanty, K. Diganta, and B. Raven, "Effects of Cooking Methods on Nutritional Content in Potato Tubers," Am. J. Potato Res., vol. 1, 2018.

[23] I. Laib and M. Barkat, "Optimization of Conditions for Extraction of Polyphenols and the Determination of the Impact of Cooking on Total Polyphenolic, Antioxidant," *Foods 2018,* vol. 7, no. 36, 2018, doi: 10.3390/foods7030036.

[24] K. Ah-hen, C. Fuenzalida, S. Hess, A. Contreras, and A. Vega-gálvez, "ANTIOXIDANT CAPACITY AND TOTAL PHENOLIC COMPOUNDS OF TWELVE SELECTED POTATO LANDRACE CLONES GROWN IN SOUTHERNCHILE," *Chil. J. Agric. Res.*, vol. 72, no. March, pp. 3-9, 2012.

[25] C. Del Mar Verde Méndez, M. Á. Rodríguez Delgado, E. M. Rodríguez Rodríguez, and C. Díaz Romero, "Content of Free Phenolic Compounds in Cultivars of Potatoes Harvested in Tenerife (Canary Islands)," J. Agric. Food Chem., vol. 52, no. 5, pp. 1323-1327, 2004, doi: 10.1021/jf0345595.

[26] C. R. Brown, D. Culley, C. P. Yang, R. Durst, and R. Wrolstad, "Variation of anthocyanin and carotenoid contents and associated antioxidant values in potato breeding lines," J. Am. Soc. Hortic. Sci., vol. 130, no. 2, pp. 174-180, 2005, doi: 10.21273/jashs.130.2.174.

[27] W. L. Morris, L. Ducreux, D. W. Griffiths, D. Stewart, H. V. Davies, and M. A. Taylor, "Carotenogenesis during tuber development and storage in potato," J. Exp. Bot., vol. 55, no. 399, pp. 975-982, 2004, doi: 10.1093/jxb/erh121.

[28] G. Jansen and W. Flamme, "Coloured potatoes (Solanum tuberosum L.) - Anthocyanin content and tuber quality," Genet. Resour. Crop Evol., vol. 53, no. 7, pp. 1321-1331, 2006, doi: 10.1007/s10722-005-3880-2.

[29] M. A. Beltrán-Penagos, A. del P. Sánchez-Camargo, and C. E. Narváez-Cuenca, "Proximal composition, bioactive compounds and biorefinery approach in potato tubers of Solanum tuberosum Group Phureja: a review," Int. J. Food Sci. Technol., vol. 55, no. 6, pp. 2282-2295, 2020, doi: 10.1111/ijfs.14461.

[30] D. DellaPenna and B. J. Pogson, "Vitamin synthesis in plants: Tocopherols and carotenoids," Annu. Rev. Plant Biol., vol. 57, pp. 711-738, 2006, doi: 10.1146/annurev.arplant.56.032604.144301.

[31] V. Ravindran and G. Ravindran, "Biochemical and Nutritional Assessment of Tubers from 16 Cultivars of Sweetpotato (Ipomoeu bututus L.)," J. Agric. Food Chem, vol. 43, pp. 2646-2651, 1995.

[32] J. Lachman, K. Hamouz, M. Orsák, and Z. Kotíková, "Carotenoids in potatoes - A short overview," Plant, Soil Environ., vol. 62, no. 10, pp. 474-481, 2016, doi: 10.17221/459/2016-PSE.

[33] J. Valcarcel, K. Reilly, M. Gaffney, and N. O. Brien, "Total Carotenoids and L -Ascorbic Acid Content in 60 Varieties of Potato (Solanum tuberosum L.) Grown in Ireland," Potato Res., 2014, doi: 10.1007/s11540-014-9270-4.

[34] P. D. Fraser and P. M. Bramley, "The biosynthesis and nutritional uses of carotenoids," Prog. Lipid Res., vol. 43, no. 3, pp. 228-265, 2004, doi: 10.1016/j.plipres.2003.10.002.

[35] M. Friedman, "Potato glycoalkaloids and metabolites: Roles in the plant and in the diet," J. Agric. Food Chem., vol. 54, no. 23, pp. 8655-8681, 2006, doi: 10.1021/jf061471t.

[36] A. Schieber and M. Saldaña, "Potato Peels : A Source of Nutritionally and Pharmacologically Interesting Compounds – A Review," *Food*, vol. 3, pp. 23-29, 2009, [Online]. Available: http://www.globalsciencebooks.info/JournalsSup/images/0906/FOOD_3(SI2)23-29o.pdf.

[37] E. J. M. Konings, H. H. S. Roomans, E. Dorant, R. A. Goldbohm, W. H. M. Saris, and P. A. Van den Brandt, "Folate intake of the Dutch population according to newly established liquid chromatography data for foods," Am. J. Clin. Nutr., vol. 73, no. 4, pp. 765-776, 2001, doi: 10.1093/ajcn/73.4.765.

[38] M. Finlay, B. Dale, D. W. Griffiths, and D. T. Todd, "Effects of genotype, environment, and postharvest storage on the total ascorbate content of potato (Solanum tuberosum) tubers," J. Agric. Food Chem., vol. 51, no. 1, pp. 244-248, 2003, doi: 10.1021/jf020547s.

[39] Z. Zhang, C. C. Wheatley, and H. Corke, "Biochemical changes during storage of sweet potato roots differing in dry matter content," Postharvest Biol. Technol., vol. 24, no. 3, pp. 317-325, 2002, doi: 10.1016/S0925-5214(01)00149-1.

Chapter 7

Red and Purple Flesh Potatoes a Healthy and Attractive Alternative Associated with New Market Trends

María-Teresa Pino and Cristina Vergara

Abstract

The potato is the fourth most important crop in the world in terms of human food, after maize, wheat and rice (FAOSTAT, 2019). The cultivated potato is a vital food-security crop considering its worldwide growth, from latitudes 65° Lat N to 53° Lat S, high yield, and great nutritive value. The potato is a good source of dietary energy and micronutrients, and its protein content is high in comparison with other roots and tubers. The cultivated potato is also a concentrated source of vitamin C and some minerals such as potassium and magnesium. Tuber flesh color generally ranges from white to dark yellow in cultivated potato; however, the high potato diversity shows tuber flesh color varies from white to dark purple. Red and purple-flesh potatoes are an interesting alternative for consumers due to phenolic compounds and antioxidant capacity. The goal of this publication is to show the advances in red and purple flesh potato, in terms of anthocyanin profile, color extraction and stability in simulated *in vitro* digestion.

Keywords: Antioxidant activity, Anthocyanins, *in vitro* digestion, Red and purple flesh potato, *Solanum tuberosum*

1. Introduction

The cultivated potato (*Solanum tuberosum* L.) is the fourth most important crop in the world, after maize, wheat, and rice. The cultivated potato is a vital food-security crop considering its worldwide growth and nutritional value. This crop is cultivated from latitudes 65° Lat N to 53° Lat S. However, the major potato-producing regions are in the relatively temperate zones, but it is also cultivated in Andean tropical highlands and in tropical and sub-tropical environments as a winter crop [1]. In the last decade, the developing world's potato production exceeded that of the developed world, showing a significant increase and demand in Asia, Africa, and Latin America, however in these areas it is often cultivated in marginal areas with limited access to farm inputs [2]. Potato is a very efficient food crop and produces more dry matter and proteins per unit area in comparison to cereals. In addition, potato is an efficient water user, however drought susceptible crop, because under rainfed conditions, it yields more food per unit of water than other major crops. For every m^3 of water applied to the crop, potato produces 5600 kcal of dietary energy, compared to 3860 in maize, 2300 in wheat and 2000 in rice [3]. Because of its high nutritional value and yield, cultivated potatoes constitute the bulk of the economically and agronomically important crop production. It accounts for large quantities of dietary daily energy intake compared to other crops and contributes to hunger reduction and improved nutrition. In addition, potato is also a good source of protein content, micronutrients, a concentrated source of vitamin C and potassium in comparison with other roots and tubers [4]. However, depending on potato flesh color, the nutritional value may be higher or different, because the color is associated to unique metabolite profile on phenolics, flavonoids, and carotenoids. These compounds are directly associated with antioxidant activity, and highly desirable in diet because of their beneficial effects on human health [5]. The present chapter will be focused on red and purple flesh potatoes as a healthy and attractive alternative associated with new market trends.

2. Color fleshed potatoes high in anthocyanins and antioxidant activity is promising food

2.1 Potato diversity in Chile

Chile is one of the countries with the largest potato diversity in the world and is also recognized as a center of origin (or center of diversity). Potato migration from the Andes to coastal Chile caused its adaptation to long-day conditions, this process contributed to the development of commercial cultivars worldwide [6]. Chile is the origin of the *Solanum tuberosum* group *Chilotanum* corresponding to lowland tetraploid landraces. Several Chilean potato genetic diversity and population structure studies have shown the close genetic distance between Chiloe Island landraces and the modern potato group. This germplasm appears to represent an interesting gene pool that could be exploited in potato breeding programs or also used for niche markets, by the specific needs and preferences. A collection of *S. tuberosum* consisting of 30 accessions of native landraces originating from the island of Chiloe, nine commercial cultivars commonly used in Chile and one accession of *S. fernandezianum* from Robinson Crusoe Island, located at 257 m altitude (33°39′9.03" S, 78°50′45.9" W) was evaluated; the results showed that commercial cultivars do not present the same genetic variability as native potatoes, and the allelic richness of commercial cultivars is lower than that of native *S. tuberosum ssp. tuberosum*. Most of the native potato were clustered in accordance with their geographical location, while commercial cultivars, were clustered in accordance with their breeding programs in Chile and Europe [7, 8]. The most complete morphological description of the Chilean germplasm was published in 2008 in the Catalog of Native Potatoes from Chile. Two institutions of the Chilean government, INIA, and SAG (Agricultural and Livestock Service of Chile), among 589 native accessions analyzed, 320 different allelic phenotypes were found indicating that there are at least 320 different genotypes in the collections. Of these, 158 belonging to the INIA collection were not found in the SAG collection. These 158 new genotypes should increase the number of known Chilean potatoes. As expected, different genotypes were known under the same popular name [9]. The genetic diversity and heterozygosity contain invaluable genetic, physiological, and biochemical attributes, that can guarantee new healthy food and safe global food productivity. The INIA (Agriculture Research Institute of Chile) is working to preserve that biodiversity, identifying the attributes of each landrace for further crop improvement, in terms of nutrition, flesh color, disease resistance, and other attributes.

2.2 Cultivated potato and red-purple fleshed potatoes

Cultivated potato is a high valued crop because of its nutritional properties and biochemical composition, rich in starch, reducing sugars, non-reducing sugars, proteins, and carotenoids. Other important nutrients in potatoes include minerals and vitamins such as potassium, magnesium, vitamin C as well as vitamin B6, among others [10, 11]. Potatoes are a reliable source of ascorbic acid – ranged from to 5.8 to 21 mg of vitamin C per 100 g tuber on a fresh weight (FW) basis–, however several studies have reported changes in the content of vitamin C in potato tubers depending on variety [12, 13]. Potato flesh color ranged from white to dark yellow cultivars are the most common, a recent review showed that the total carotenoids content of tubers is influenced by location, season, genotype, and their interactions, with values between 5 and 10 mg kg^{-1} FW of total carotenoids, for white-fleshed potatoes, to over 100 mg kg^{-1} FW of total carotenoids for dark-yellow potatoes [14]. Total carotenoids expression was observed in the mid of the tuber maturation process rather than in ready-to-harvest tubers. The predominant carotenoid forms found in cultivated potato were lutein, violaxanthin, zeaxanthin, and neoxanthin [15].

Today, with a major market shift for antioxidant-rich foods, the traders are also seeing an increase in the demand for red and purple fleshed potatoes, because these contain an important group of secondary plant metabolites associated with positive health benefits: phenolics, flavonoids, and anthocyanins [16]. Red and purple fleshed potatoes provide a natural source of anthocyanins and antioxidant activity [17]. Anthocyanins are recognized as natural flavonoid colorants ranged from orange-red (pelargonidin), reddish to blue-violet (malvidin), for use in food industry and pharmaceutical ingredients, because of their potential health benefits. The six predominant anthocyanidins found in higher plants (including root and tubers) are cyanidin, delphinidin, pelargonidin, peonidin, petunidin, and malvidin [18]. The phenolics compounds, flavonoids, and anthocyanins are potent antioxidants which contribute to the physiological defense against oxidative and free-radical-reactions. Food containing anthocyanins have been associated with a reduction in inflammation markers and a lower risk of chronic diseases, including obesity, diabetes, cardiovascular disease, and cancer [19, 20]. In addition, a recent study showed that anthocyanins ameliorate neurodegeneration at a molecular and clinical level and dietary anthocyanin's supplement prevents neurodegenerative diseases [21]. Colored fleshed potatoes contain relatively low amount of total phenolic acids, but its flavonoids and flavones extracts showed high scavenging activities toward oxygen compared to other fruits and vegetables [16].

In relation to the predominant anthocyanidins, a study in four potato cultivars ("Hermanns Blaue", "Highland Burgundy Red", "Shetland Black", and "Vitelotte") identified Petunidin derivatives in all of them except in "Highland Burgundy Red". Malvidin was the predominant on the "Vitelotte" cultivars. "Shetland Black" was the only one containing minor peonidin [22]. The evaluation of anthocyanin phenolic compounds of potato peels from ten colored potato cultivars (red and purple) the most prominent were pelargonidin, peonidin, and malvidin aglycones. All samples revealed antioxidant and antitumor activities, and no toxic effect [23]. Another recent study on colored potato (three red-fleshed, three-purple fleshed, and one marble-fleshed) showed that red and purple-fleshed potatoes are rich sources of anthocyanins. Pelargonidin and petunidin were the main anthocyanidin forms, and all aqueous extracts presented *in vitro* antioxidant, antibacterial and antifungal activities, and no toxic effects [24].

Because most native color fleshed potatoes have low yield, wide phenotypic variations and uneven flesh color, the INIA Chile's Potato breeding program has developed new putative color flesh potato cultivars as raw material to food coloring and ingredient extraction, with high anthocyanins flesh concentration and high yield. **Table 1** shows significant differences in color intensity (E1%), total anthocyanins (CAT), total polyphenol content (TPC), and antioxidant activity (FRAP), for selected red flesh potato (INIA RS58-3), purple flesh potato (INIA RQ12-521), blue-violet flesh potato (INIA RÑ98-9). Principal component analysis and matrix of correlation coefficients showed a good fit between color intensity (E1%) and total anthocyanins (CAT) with values from 0.63 between E1% and CAT-based in tuber dry weight between 0.90 for E1% and CAT-based in tuber fresh weight. Both red (INIA RS58-3) and blue-violet (INIA RÑ98-9) fleshed potatoes showed higher color intensity and higher total anthocyanins (CAT), also these two potato lines showed higher values in total polyphenol content (TPC), and antioxidant activity (FRAP). Conversely, the light purple flesh potato (INIA RQ12-521) showed lowest color intensity and consequently lower CAT, TPC, and antioxidant activity. Thus, selected red flesh potato (INIA RS58-3) and blue-violet flesh potato (INIA RÑ98-9) are promising raw material for natural color extraction and food coloring ingredients.

In term of Anthocyanin profile (**Table 2**), in these color fleshed potatoes, the predominant anthocyanins identified were Pelargonidin-3-glucoside, Peonidin-3-glucóside, Peonidin-3-arabinósido, Delphinidin 3-glucoside, Delphinidin 3-galactoside, Delphinidin 3-rutinoside, Delfinidina-3,5-diglucósido, Delphinidin 3-galactoside, Delphinidin 3-glucoside, Delphinidin 3-rutinoside, Malvidin-3-glucóside, and Malvidin-3,5-diglucóside. The major picks in red flesh potato (INIA RS58-3) were in Peonidin and Delphinidin derivatives, while in blue-violet flesh potato (INIA RÑ98-9) the picks were in Delphinidin and Malvidin.

2.3 Stability and Bioaccesibility: potato anthocyanins

The concentration and stability of these anthocyanins are affected by several parameters such as agronomic factors and postharvest storage. However, the stability of acylated anthocyanins is still not well addressed, and few studies in anthocyanins contents (CAT) in colored-flesh potato tubers during processing and digestion have been published [25, 26]. The stability of anthocyanins is affected by pH, temperature, and light. During the digestion process, anthocyanins stability is affected because undergo variation in pH and in digestive enzymatic activity. Therefore, the anthocyanins stabilization is needed to maintain their health effects in the human body and increase its positive effects. The anthocyanins stability could be improved by using micro-encapsulation technology such as spray-drying [26–28]. Micro-encapsulation is a technique wherein a bioactive compound is encapsulated by a biopolymer, to protect the compound from oxygen, water, or other conditions, thereby improving its stability and release in the desired stage [26, 28]. In order to know bio stabilization of anthocyanins extract from purple flesh cultivated potato, a study was addressed on the encapsulation anthocyanins' efficiency and bioaccesibility. The anthocyanin extract from INIA purple flesh potato (PPE) was micro-encapsulated by spray-drying [29] (**Figure 1**). Maltodextrin (MD) was used as the encapsulating agent, due to its high solubility in water, low viscosity, bland flavor, and colorlessness. Briefly, the mixture (extract PPE-maltodextrin) was fed into spray dryer at 130°C. The encapsulation efficiency (EE) was 86%, due the high anthocyanins-MD interactions caused by hydrogen bonding and/or electrostatic interactions. The total anthocyanins were 1.34 ± 0.02 mg cy-3-glug^{-1} and antioxidant activity (FRAP) was 10.1 ± 0.6 mg trolox equivalentg^{-1}. The moisture (5.6 ± 0.4%), water activity (aw = 0.225 ± 0.001), and particle size (6.51 ± 0.1 um)

DOI: http://dx.doi.org/10.5772/intechopen.99181

Color flesh Potato selected lines	Skin Color	Flesh Color	Tuber Shape	S.S. BRIX	Color (E1%)	CAT (mg C3G kg FW^{-1})	TPC (mg EAG kg FW^{-1})	FRAP (μmol TroloxKg DW^{-1})
INIA RS58-3	RF	RF	Rd	7.75b	0.149b	927.77b	1866.28b	12804.3b
INIARQ12-521	PF	PL	O	6.70a	0.073a	221.80a	758.10a	3687.26a
INIARÑ98-9.	VB	VB	Rd	8.7oc	0.170c	931.63b	1706.53b	12595.2b
CV				*4.28*	*9.18*	*6.12*	*6.77*	*19.84*
p-value line				<0.0001	<0.0001	<0.0001	<0.0001	<0.0001
p-value season				<0.0001	0.8241	0.5231	<0.0001	0.0001
p-value season x line				<0.0001	0.0767	0.4097	<0.0001	0.0017

Where, Red = RF, PF = Strong purple, PS = light purple, VB = blue-violet; Tuber shape Rd = Round, O = Oval. **Color intensity** *(E1%) is the optical density of a 1% juice solution at the wavelength of maximum absorbance intensity (INIA RÑ98-9: Abs = 522 nm, INIA RQ12-521: Abs = 521 y RS58-3: Abs = 506 nm) by spectrophotometer (Jasco V-700).* **Total anthocyanin content** *(TAC) by pH differential method expressed as mg of cyanidin-3-glucoside equivalents per fresh weight (FW).* **Total polyphenol content** *(TPC) by the Folin–Ciocalteu method expressed as mg Gallic acid equivalent (GAE) per fresh weight (FW).* **Antioxidant activity (FRAP)** *measured by the FRAP method as described as Trolox equivalent by spectrophotometer (Jasco V-700).*
Different letters in the same column indicate statistical difference ($p \leq 0.05$) among potato lines. Three replicates were analyzed (with 3 instrumental measures per sample). Data were analyzed by one-way ANOVA followed by the Tukey test ($p < 0.05$) by InfoStat version 2020. http://www.infostat.com.ar).

Table 1.
Color intensity (E1%), total anthocyanins (CAT), total polyphenol content (TPC) and antioxidant activity (FRAP) in tubers, for selected red flesh potato (INIA RS58-3), purple flesh potato (INIA RQ12-521), blue-violet flesh potato (INIA RÑ98-9) cultivated in Osorno (40°34′26.22"S, 73°8′0.53"W.), Chile during two seasons (2019-2020 and 2020-2021).

Color flesh Potato selected lines	Skin Color	Flesh Color	Tuber Shape	Anthocyanin profile (predominants)
INIA RS58-3	RF	RF	Rd	Pelargonidin-3-glucoside, Peonidin-3-glucoside, Delphinidin 3-galactoside
INIARQ12-521	PF	PL	O	Peonidin-3-arabinósido, Delphinidin 3-glucoside, Delphinidin 3-galactoside, Delphinidin 3-rutinoside, Malvidin-3-glucóside,
INIARÑ98-9.	VB	VB	Rd	Peonidin-3-arabinósido, Delfinidina-3,5-diglucósido, Delphinidin 3-galactoside, Delphinidin 3-glucoside, Delphinidin 3-rutinoside, Malvidin-3,5-diglucóside,

Where, Red = RF, PF = Strong purple, PS = Light purple, VB = blue-violet; Tuber shape Rd = Round, O = Oval. Anthocyanin profile by HPLC analysis was carried out in Jasco Intelligent Quaternary Gradient PumpPU-2089 Plus, UV/VIS detector e interface LC-NetII/ADC, C18 Kromasil 100-5 de 150 mm at 30C°. Detection: UV @ 520 nm.

Table 2.
Anthocyanin profile of tubers, for selected red flesh potato (INIA RS58-3), purple fleshed potato (INIA RQ12-521), blue-violet flesh potato (INIA RÑ98-9) cultivated in Osorno (40°34′26.22"S, 73°8′0.53"W.) Chile, during two seasons (2019-2020 and 2020-2021).

were within the range described for anthocyanin encapsulated obtained by spray-drying [27, 30, 31].

PPE-MD encapsulation improved its anthocyanins stability due to anthocyanin extract and encapsulating agent interaction that may occur by hydrogen bonding and/or electrostatic interactions. Reduced damage of active anthocyanins was observed under adverse storage conditions. The time-course of the storage stability assay during 140 days at 60°C showed that encapsulated extract (PPE-MD) showed significantly higher anthocyanins retention than non-encapsulated PPE (**Figure 2**), thereby extending shelf life, color, and antioxidant capacity [29]. These results agree with earlier reports on use of spray drying technique on black-carrot, black berry, maqui and plum [27, 32–34].

Other important aspect is the *in vitro* bio accessibility (BA) of anthocyanins. Bio accessibility is defined as the amount of bioactive compound (anthocyanins) that was released from the food matrix after digestion [35]. The BA of encapsulated anthocyanins extracts of purple flesh potato (PPE-MD) was significantly higher than non-encapsulated extract (**Figure 3**). Micro-encapsulation protects PPE-MD during *in vitro* digestion, against environmental conditions, especially pH. The BA of the PPE and PPE-MD was higher than previous similar studies in maqui extract [27] and blueberry extract [28]. The encapsulation technology is a useful strategy to protect anthocyanins from purple flesh cultivated potato, during storage and *in vitro* gastrointestinal digestion model, as well. The anthocyanins micro-encapsulation contributes to the development of new purple potato products in powder formulation, potentially useful as colorants for the food industry or health ingredients (antioxidant and anti-inflammatory properties).

2.4 Red and purple flesh potato-based food and natural ingredients responding to new global food market trends

The global consumer trend preferences and the health and wellness market in the next coming years, show a promising future for non-traditional color fleshed potato, as red, purple, and blue fleshed potato, because their antioxidant activity

(a)

(b)

Figure 1.
(a) Microencapsulated powders (anthocyanin extract from INIA purple flesh potato (PPE + maltodextrin) and (b) scanning electron microscopic (SEM) for anthocyanins microencapsulated powders.

and health benefit are capturing the consumers' attention. Most studies about market trends have projected that "the global health and wellness food market" would grow at a CAGR of over 6% (6–8%) during the next years. This forecast is explained in part, because the world, upon COVID-19 pandemic impact, will face the growing incidences of chronic diseases, stress, obesity, aging and other adverse health conditions, see more detail in [36–38] reports. In potato, some reports about its market under the COVID-19 pandemic situation shows that potatoes become popular due to their long shelf-life. In relation to global market, most potatoes are consumed as fresh vegetable, however, is shifting from fresh potatoes to processed potato-based foods. Based on application, the processed potato market is segmented into ready-to-cook, snacks, potato flour-gluten free, and other potato-based food additives for soups, gravies, bakery, and desserts driven by urbanization and changes in eating habits among many other factors. Thus, these global food market trends, raises further questions for food industry and R&D institutions, would be capable to develop new color fleshed potato-based foods and potato-based ingredients keeping its nutritional value and color.

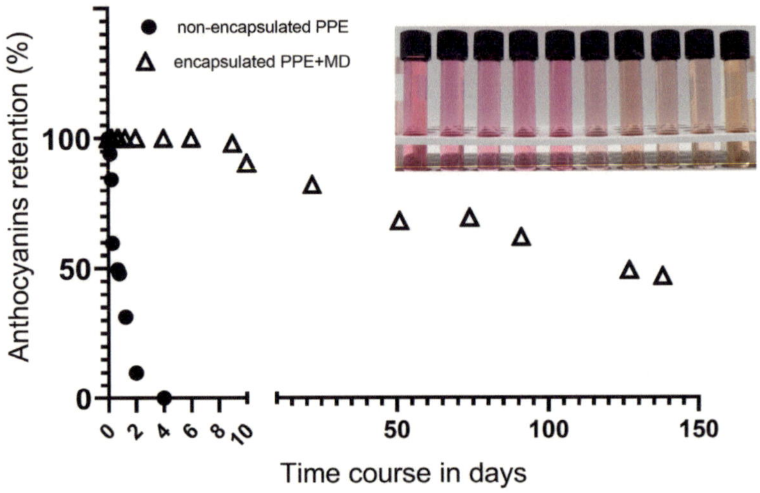

Figure 2.
Time-course storage stability assay for anthocyanins retention at 60°C for 140 days storage from non-encapsulated extract (PPE, ©) and encapsulated extract (PPE-MD, Δ), and the visual degradation of anthocyanins for liquid (analysis solution) on non-encapsulated PPE (source: Adapted Vergara et al. [29]).

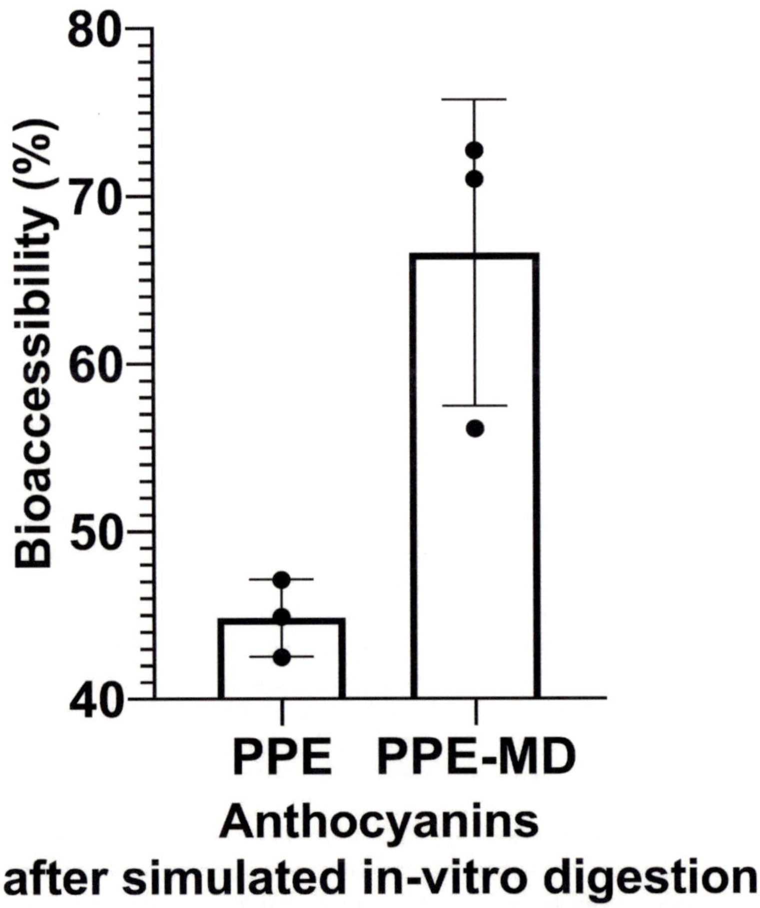

Figure 3.
Bio accessibility (%) of non-encapsulated (PPE) and encapsulated (PPE-MD) anthocyanins extract after simulated in-vitro digestion.

A recent research studied how the anthocyanin degradation and anthocyanin profile were influenced in red-fleshed potatoes (cv Herbie 26) after different methods of processing (dried cubes, French fries, chips, semi-finished products, and finished products); most evaluated processes showed losses on anthocyanin content. Chip products showed higher retention anthocyanins. Pelargonidin-3-feruloylrutinoside-5-glucoside, and pelargonidin-3-caffeoylrutinoside-5-glucoside, were most thermally stable [25]. To reduce the loss of effectiveness of plant-based compounds as anthocyanins, and polyphenols from color fleshed potatoes, micro-encapsulation arise as an alternative. This technique allows the development of novel plant-based ingredients able to keep their functionality after processing.

(a)

(b)

Figure 4.
(a) Potato-based ingredient, flakes elaborated from red flesh potato (INIA RS58-3), purple fleshed potato (INIA RQ12-521), blue-violet flesh potato (INIA RÑ98-9). (b) Flakes elaborated from light purple fleshed potato (INIA RQ12-521).

However, commercial product development depends on financial and operational viability. In the previous section of this chapter, anthocyanins' stability and bioaccesibility from color fleshed potatoes were discussed with emphasis in micro-encapsulation for INIA purple flesh potato and *in vitro* digestion. Thus, micro-encapsulated spray dried powder from purple-fleshed potato could be applied in drinks, in snacks, and in milk products because its stability and bioaccesibility [29]. The application of aqueous extracts from color fleshed potato was also tested and validated as natural colorants in a soft drink during 30-days shelf-life when compared with the commercial colorant E163 [24].

Potato flake is an ingredient with multiple applications in processed food and long shelf life. A recent study compared the convective tray drying method with a refraction-based drying method for producing potato flakes (cv. Kufri Pukhraj, a light yellow to gold flesh potato). The results showed that those flakes obtained by refraction-based drying had better nutritive value, color and acceptability. It recommended its application for the fortification of flour, baby foods, and extruded products [39]. Previously, a study in anthocyanin-rich red potato flakes showed that might improve the antioxidant system by enhancing hepatic SOD (superoxide dismutase) mRNA in mice [40]. The replacement of part of the wheat flour with purple fleshed potato powder (from freeze-dried) and albedo showed an enhancement antioxidant activity of fortified breads, and longer shelf life [41]. In addition to the previous reported health benefits, the purple fleshed potato powder (from freeze-dried) has the potential to aid in the amelioration of ulcerative colitis symptoms, a major form of inflammatory bowel disease [42].

Potato-based ingredients (flakes, spray dried powder, and freeze-dried powder) were elaborated from red flesh potato (INIA RS58-3), purple fleshed potato (INIA RQ12-521), and blue-violet flesh potato (INIA RÑ98-9) because their application in food industry (**Figure 4**). The spray dried powder shows better physical properties when compared to the freeze-dried powder. Conversely, freeze-dried powder

Potato-based ingredients	Color (E1%) Color intensity	CAT (mg C3G g^{-1}) ingredient	FRAP (μmol Trolox g-^{1}) ingredient
Flakes Red flesh INIA RS58-3	0.42 ± 0.03 ^{ab}	1.9 ± 0.2 ^{b}	45.1 ± 1.2 ^{b}
Flakes Purple flesh potato INIA RQ12-521	027 ± 0.02 ^{c}	1.2 ± 0.3 ^{c}	47.0 ± 1.7 ^{b}
Freeze-dried powder Red flesh INIA RS58-3	0.47 ± 0.01 ^{a}	2.7 ± 0.1 ^{a}	56.9 ± 4.9 ^{a}
Freeze-dried powder Purple flesh potato INIA RQ12-521	0.39 ± 0.01 ^{b}	2.2 ± 0.2 ^{ab}	46.4 ± 0.6 ^{b}

Color intensity (E1%) *is the optical density of a 1% juice solution at the wavelength of maximum absorbance intensity (INIA RÑ98-9: Abs = 522 nm, INIA RQ12-521: Abs = 521 y RS58-3: Abs = 506 nm) by spectrophotometer (Jasco V-700).* **Total anthocyanin content (TAC)** *by pH differential method expressed as mg of cyanidin-3-glucoside equivalents per fresh weight (FW).* **Antioxidant activity (FRAP)** *measured by the FRAP method as described as Trolox equivalent by spectrophotometer (Jasco V-700). Different letters in the same column indicate statistical difference ($p \leq 0.05$) among ingredient and potato lines. Three replicates were analyzed (with 3 instrumental measures per sample). Data were analyzed by one-way ANOVA followed by the Tukey test ($p < 0.05$) by InfoStat version 2020. http://www.infostat.com.ar).*

Table 3.
Color intensity (E1%), total anthocyanins (CAT) and antioxidant activity (FRAP) of potato-based ingredients (flakes and freeze-dried powder) elaborated from red flesh potato (INIA RS58-3), and purple flesh (RQ12-521).

preserves better the nutritional value such as naturally occurring. And, in spray dried powder the high temperature of heat may cause the loss of nutritional value. Red flesh potato (INIA RS58-3) and purple flesh potato (INIA RQ12-521) were selected for further evaluation because they fresh tubers show greater differences in color intensity. Potato-based ingredients as flakes and freeze-dried powder were compared (**Table 3**) for color intensity (E1%), total anthocyanins (CAT), and antioxidant activity (FRAP). As expected, freeze-dried powder preserved better the color intensity (E1%), total anthocyanins (CAT) and antioxidant activity (FRAP), however the flakes values were also attractive. These potato-based flakes and freeze-dried powder are food coloring because both ingredients provide color and bioactive compounds, with different applications.

3. Conclusion

All these antecedents, suggest that red and purple fleshed potatoes are not only a promising crop for starvation problem, also their consume promote health and may prevent chronic diseases. Anthocyanin-rich extracts from red and purple fleshed potatoes have high potential as natural colorants with multiple applications in food industry. Also, these potatoes contain an important group of secondary plant metabolites associated with antioxidant activity and positive health benefits, as phenolics, flavonoids, and anthocyanins. INIA's new putative color flesh potato cultivars (red flesh potato (INIA RS58-3), purple flesh potato (INIA RQ12-521), blue-violet flesh potato (INIA RÑ98-9)) are promising raw materials for natural color extraction and food coloring ingredients.

Acknowledgements

This chapter is a review the latest research on color fleshes potato and recent results of the authors on red to purple fleshed potato foodcoloring ingredients as well. We would like to acknowledge all authors cited in the references.

Acknowledgments to CONICYT-PAI I7817020005 and FIA PYT-2017-0488 grants.

Conflict of interest

The authors declare no conflict of interest.

Appendices and nomenclature

INIA	Instituto de Investigaciones Agropecuarias de Chile (Institute of Agricultural Research of Chile).
CAGR	Compound annual growth rate.

Author details

María-Teresa Pino* and Cristina Vergara
Food for the Future Department, Instituto de Investigaciones Agropecuarias de Chile (Institute of Agricultural Research of Chile), Chile

*Address all correspondence to: mtpino@inia.cl

References

[1] Hijmans R.J. Global distribution of the potato crop. American journal of potato research. 2001;78(6):403-412. https://doi.org/10.1007/BF02896371

[2] FAOSTAT, 2019. Available from http://www.fao.org/statistics/es/ [Accessed: 2021-06-01]

[3] Monneveux P. et al. Drought tolerance in potato (S. tuberosum L.): Can we learn from drought tolerance research in cereals? Plant Science. 2013; 205:76-86. https://doi.org/10.1016/j.plantsci.2013.01.011

[4] Beals K. A. Potatoes, nutrition and health. American Journal of Potato Research. 2019; 96(2): 102-110. https://doi.org/10.1007/s12230-018-09705-4

[5] Ezekiel R, Singh N, Sharma S, & Kaur A. Beneficial phytochemicals in potato—a review. Food Research International. 2013; 50(2):487-496. https://DOI.org/10.1016/j.foodres.2011.04.025

[6] Hardigan M.A., Laimbeer F.P.E., Newton L., Crisovan E., Hamilton J.P., Vaillancourt B., Wiegert-Rininger K., Wood J.C., Douches D.S., Farré E.M. et al. Genome diversity of tuber-bearing Solanum uncovers complex evolutionary history and targets of domestication in the cultivated potato. Proc. Natl. Acad. Sci. 2017; 114 (46): E9999-E10008. https://doi.org/10.1073/pnas.1714380114.

[7] Solano J, Mathias M., Esnault F., & Brabant P. Genetic diversity among native varieties and commercial cultivars of *Solanum tuberosum ssp. tuberosum* L. present in Chile." Electronic Journal of Biotechnology. 2013; (16) 6:8-8. http://www.ejbiotechnology.info/index.php/ejbiotechnology/article/view / v16n6-15/1791

[8] Esnault F., Solano J., Perretant M. R., Herve M., Label A. Pellé, R., ... & Chauvin, J. E. Genetic diversity analysis of a potato (*Solanum tuberosum L.*) collection including Chiloé Island landraces and a large panel of worldwide cultivars. Plant Genetic Resources. 2014; 12(1): 74. DOI:10.1017/S1479262113000300

[9] Muñoz M., Folch C., Rodriguez F., Kalazich J., Orena S., Santo, J., ... & Puga A. Genotype number and allelic diversity overview in the national collection of Chilean potatoes. Potato research. 2016; 59(3):227-240. https://doi.org/10.1007/s11540-016-9329-5

[10] Brar A., Bhatia A. K., Pandey V., & Kumari P. Biochemical and phytochemical properties of potato. Chem Sci Rev Lett. 2017; 6(21):117-129.

[11] Burgos G., Zum Felde T., Andre C., Kubow S. The Potato and Its Contribution to the Human Diet and Health. In: Campos H., Ortiz O. (eds) The Potato Crop. Springer, Cham. 2020; 37-74. https://doi.org/10.1007/978-3-030-28683-5_2 .

[12] Peña C, Palomeque L., Restrepo-Sánchez L. P., Kushalappa A., Mosquera T., & Narváez-Cuenca C.E. Variation of mineral contents with nutritional interest in a collection of Solanum tuberosum Group Phureja tubers. International Journal of Food Science & Technology. 2021; https://doi.org/10.1111/ijfs.15115

[13] Hrabovská D., Heldák J., & Volnová B. Changes in the content of vitamin C in potato tubers depending on variety. Journal of Microbiology, Biotechnology and Food Sciences. 2021; 2052-2058.

[14] Tatarowska B., Milczarek D., Wszelaczyńska E. et al. Carotenoids Variability of Potato Tubers in Relation

to Genotype, Growing Location and Year. Am. J. Potato Res. 2019; 96:493-504. https://doi.org/10.1007/s12230-019-09732-9

[15] Šulc M., Kotíková Z., Paznocht L., & Lachman J. Changes in Carotenoid Profile during Potato (Solanum tuberosum L.) Tuber Maturation. American Journal of Potato Research. 2021; 98(2):85-92. https://doi.org/10.1007/s12230-020-09805-0

[16] Ezekiel R., Singh N., Sharma S., & Kaur A. Beneficial phytochemicals in potato—a review. Food Research International, 2013; 50(2):487-496. https://doi.org/10.1016/j.foodres.2011.04.025

[17] Ercoli S., Cartes J., Cornejo P., Tereucán G., Winterhalter P., Contreras B., & Ruiz A. Stability of phenolic compounds, antioxidant activity and colour parameters of a coloured extract obtained from coloured-flesh potatoes. LWT. 2021; 36: 110370. https://doi.org/10.1016/j.lwt.2020.110370

[18] Khoo H.E., Azlan A., Tang S.T., Lim S.M. Anthocyanidins and anthocyanins: Colored pigments as food, pharmaceutical ingredients, and the potential health benefits. Food Nutr. Res. 2017; 61:1361779. DOI: 10.1080/16546628.2017.1361779

[19] Mattioli R., Francioso A.; Mosca L., Silva P. Anthocyanins: A Comprehensive Review of Their Chemical Properties and Health Effects on Cardiovascular and Neurodegenerative Diseases. Molecules. 2020; 25:3809. https://doi.org/10.3390/molecules25173809

[20] Speer H., D'Cunha N.M., Alexopoulos N.I., McKune A.J., Naumovski N. Anthocyanins and Human Health—A Focus on Oxidative Stress, Inflammation and Disease. Antioxidants. 2020; 9:366. https://doi.org/10.3390/antiox9050366

[21] Li P., Feng, D., Yang, D., Li, X., Sun, J., Wang, G., ... & Bai, W. Protective effects of anthocyanins on neurodegenerative diseases. Trends in Food Science & Technology. 2021:

[22] Eichhorn S., & Winterhalter, P. Anthocyanins from pigmented potato (Solanum tuberosum L.) varieties. Food Research International, 2005; 38(8-9):943-948.

[23] Sampaio S. L., Petropoulos, S. A., Dias, M. I., Pereira, C., Calhelha, R. C., Fernandes, Â., ... & Barros, L. Phenolic composition and cell-based biological activities of ten coloured potato peels (*Solanum tuberosum* L.). Food Chemistry. 2021; 130360.].

[24] Sampaio, S. L., Lonchamp, J., Dias, M. I., Liddle, C., Petropoulos, S. A., Glamočlija, J., ... & Barros, L. Anthocyanin-rich extracts from purple and red potatoes as natural colourants: Bioactive properties, application in a soft drink formulation and sensory analysis. Food Chemistry. 2021; 342:128526.

[25] Rytel, E., Tajner-Czopek, A., Kita, A., Tkaczyńska, A., Kucharska, A. Z., & Sokół-Łętowska, A. The Influence of the Production Process on the Anthocyanin Content and Composition in Dried Potato Cubes, Chips, and French Fries Made from Red-Fleshed Potatoes. Applied Sciences. 2021; 11(3):1104.

[26] Mahdavi S.A.; Jafari, S.M.; Ghorbani, M.; Assadpoor, E. Spray-drying microencapsulation of anthocyanins by natural biopolymers: A review. Dry technol. 2014; 32:509-518. DOI.org/10.1080/07373937.2013.839562.

[27] Fredes C.; Becerra, C.; Parada, J.; Robert, P. The microencapsulation of maqui (*Aristotelia chilensis* (Mol.) Stuntz) juice by spray-drying and freeze-drying produces powders with similar anthocyanin stability and bioaccessibility. Molecules. 2018; 23:

1227. DOI.org/10.3390/molecules23051227.

[28] Da Rosa J.R.; Nunes, G.L.; Motta, M.H.; Fortes, J.P.; Weis, G.C.C.; Hecktheuer, L.H.R.; Muller, E.; de Menezes, C.; da Rosa, C.S. Microencapsulation of anthocyanin compounds extracted from blueberry (Vaccinium spp.) by spray drying: Characterization, stability and simulated gastrointestinal conditions. Food Hydrocolloid. 2019; 89:742-748. DOI.org/10.1016/j.foodhyd.2018.11.042.

[29] Vergara, C., Pino, M. T., Zamora, O., Parada, J., Pérez, R., Uribe, M., & Kalazich, J. Microencapsulation of anthocyanin extracted from purple flesh cultivated potatoes by spray drying and its effects on in vitro gastrointestinal digestion. Molecules. 2020; 25(3):722. DOI.org/10.3390/molecules25030722

[30] Gharsallaoui, A.; Roudaut, G.; Chambin, O.; Voilley, A.; Saurel, R. Applications of spray-drying in microencapsulation of food ingredients: An overview. Food Res. Int. 2007; 40:1107-1121. DOI.org/10.1016/j.foodres.2007.07.004.

[31] Tonon R.V.; Brabet, C.; Hubinger, M.D. Anthocyanin stability and antioxidant activity of spray-dried açai (*Euterpe oleracea* Mart.) juice produced with di_erent carrier agents. Food Res Int. 2010; 43:907-914. DOI.org/10.1016/j.foodres.2009.12.013

[32] Ersus, S.; Yurdagel, U. Microencapsulation of anthocyanin pigments of black carrot (*Daucus carota* L.) by spray drier. J. Food Eng. 2007; 80:805-812. DOI.org/10.1016/j.jfoodeng.2006.07.009.

[33] Yinbin, L.; Wu, L.; Weng, M.; Tang, B.; Lai, P.; Chen, J. Effect of different encapsulating agent combinations on physicochemical properties and stability of microcapsules loaded with phenolics of plum (*Prunus salicina* lindl.). Powder Technol. 2018; 340:459-464. DOI.org/10.1016/j.powtec.2018.09.049

[34] Wang W.D.; Xu S.Y. Degradation kinetics of anthocyanins in blackberry juice and concentrate. J. Food Eng. 2007; 82:271-275. DOI.org/10.1016/j.jfoodeng.2007.01.018

[35] Lila, M.A.; Ribnicky, D.M.; Rojo, L.E.; Rojas-Silva, P.; Oren, A.; Havenaar, R.; Janle, E.M.; Raskin, I.; Yousef, G.G.; Grace, M.H. Complementary approaches to gauge the bioavailability and distribution of ingested berry polyphenolics. J Agric. Food Chem. 2011; 60:5763-5771. doi.org/10.1021/jf203526h

[36] AP NEWS. 2020. https://apnews.com/article/technology-business-lifestyle-nutrition-food-beverage-and-tobacco-products-manufacturing-2a7104b85c29462cb771aa0ee390d830 [Accessed: 2021-06-12].

[37] Technavio. 2021. https://www.prnewswire.com/news-releases/global-health-and-wellness-food-market-2020-2024--drivers-restraints-opportunities-and-threats--technavio-301287265.html [Accessed: 2021-06-12].

[38] Industry Research Biz. 2021. https://www.marketwatch.com/press-release/health-and-wellness-food-market-share-opportunities-and-strategies-2021---global-size-future-trends-segmentation-regional-analysis-company-profile-sales-revenue-market-competition-status---global-forecast-to-2026-2021-04-16 [Accessed: 2021-06-12].

[39] Zalpouri R., Kaur P., Kaur, A., & Sidhu, G. K. Comparative analysis of optimized physiochemical parameters of dried potato flakes obtained by refractive and convective drying techniques. Journal of Food Processing and Preservation. 2021; 45(1):e15077. https://doi.org/10.1111/jfpp.15077]

[40] Han, K. H., Shimada, K. I., Sekikawa, M., & Fukushima, M. Anthocyanin-rich red potato flakes affect serum lipid peroxidation and hepatic SOD mRNA level in rats. Bioscience, biotechnology, and biochemistry. 2007;0704090379-0704090379. https://doi.org/10.1271/bbb.70060]

[41] Taglieri, I., Sanmartin, C., Venturi, F., Macaluso, M., Bianchi, A., Sgherri, C., ... & Zinnai, A. (2021). Bread Fortified with Cooked Purple Potato Flour and Citrus Albedo: An Evaluation of Its Compositional and Sensorial Properties. Foods. 2021; 10:942. https://doi.org/10.3390/foods10050942]

[42] Li, S., Wang, T., Xu, B., Indukuri, V., Vanamala, J., & Reddivari, L. Anthocyanin-Containing Purple Potatoes Ameliorate DSS-Induced Colitis in Mice. Current Developments in Nutrition. (2020); 4(2):426-426. https://doi.org/10.1016/j.jnutbio.2021.108616.

Section 4

Potato Breeding

Chapter 8

Prospects for Genome Editing of Potato

Sona S. Dev, Jini Joseph and Ligi Lambert D'Rosario

Abstract

Potato (*Solanum tuberosum* L.) is a staple food crop that could play a major role in improving food security in developing nations. The sustainable production of this crop faces many challenges like pests, diseases, abiotic stresses and post-harvest problems. Transgenic technology and gene silencing strategies offered a new hope of solution to the conventional time consuming breeding programmes. However the genetically modified crops are affected by regulatory approvals and safety concerns. In this aspect, gene editing techniques like ZFNs (zinc-finger nucleases), TALENs (transcription activator-like effector nucleases), and CRISPR/Cas9 (clustered regularly interspaced short palindromic repeats/CRISPR associated Cas9), offer better choice for production of transgene and marker free disease resistant potatoes.

Keywords: Potato, ZFNs, TALENS, CRISPR/Cas9

1. Introduction

Potato (*Solanum tuberosum*) belonging to the Solanaceae family is the fourth most important staple food crop of the world consumed by more than a billion people [1]. The global total potato production exceeds 300 million metric tons [2]. Popularly called the 'poor man's friend', this crop can play a vital role to safe guard the food security and sustainability in the current scenario of surging population growth. The crop is a rich source of starch, vitamins especially C and B1 and minerals. It is also used for several industrial purposes such as for the production of starch and alcohol.

There is an urgency to increase the production and quality of potatoes to meet the demands of the rising population. However the development of new potato cultivars using traditional cross-breeding is complicated and slow due to tetrasomic inheritance and high heterozygosity of cultivated varieties [3]. Currently, research work using genome editing (GE) tools are being deployed for the precise improvement of desirable traits in crops. Genetically modified (GM) crop production faces many hurdles due to the complicated regulatory approval procedures whereas the technique of GE offers a better promise in crop improvement by making efficient and precise changes in the plant genome. This chapter describes the research advancements in potato using GE tools and the hurdles ahead due to the regulatory measures.

2. Constraints in potato production

2.1 Pests and diseases

Pests and diseases are major constraints to commercial production of potato. The major pests infesting potato include Colorado potato beetle (*Leptinotarsa decemlineata*), leafminer fly (*Liriomyza huidobrensis*), cyst nematodes (*Globodera pallida* and *G. rostochiensis*) and potato tuber moth (*Phthorimaea operculella*) during storage. The important diseases of potato include late blight (*Phytophthora infestans),* early blight (*Alternaria* ssp.), potato virus Y, potato leaf roll virus, bacterial wilt or brown rot' (*Ralstonia solanacearum*) and blackleg (*Pectobacterium carotovorum*) during storage.

2.2 Weeds

Weeds are a major problem in potato production and can reduce yields through direct competition for light, moisture, and nutrients, or by harbouring insects and diseases that attack potatoes. Weeds can have a detrimental impact on tuber yield when compared to potatoes grown in weed-free conditions [4, 5]. The weeds present at harvest can be detrimental to yield by increasing mechanical damage to the tubers and reducing harvesting efficiency by slowing the harvesting operation. Farmers mostly employ herbicides to enhance weed control.

2.3 Post-harvest shelf life

Postharvest management and storage of the potato is an important factor not only in preventing postharvest losses but also in maintaining its nutritional quality. This is because potato contains glycoalkaloids (GAs), a family of steroidal toxic secondary metabolites that occur in all parts of the potato. The levels of these toxins are significantly affected by postharvest handling stress factors with exposure to light, storage temperatures, and injuries/bruising being important stress factors. Storage is an important post-harvest activity in seed production. Storage under specific conditions is important to prevent excessive loss of weight as a result of drayage and to preserve germination quality. Prevention of diseases in storage is also important whether it be small farmer storage or commercial potato seed storage.

2.3.1 Starch composition

Potatoes are used for a variety of purposes, and not only as a vegetable for cooking at home. In fact, it is likely that less than 50 percent of potatoes grown worldwide are consumed fresh. The rest are processed into potato food products and food ingredients; fed to cattle, pigs, and chickens; processed into starch for industry; and re-used as seed tubers for growing the next season's potato crop. The commercial value of potato starch is governed by the proportion of its derivatives mainly amylose and amylopectin. There is much demand for amylose free potatoes in food and paper industries and more availability of potato cultivars with high amylopectin is warranted.

3. Genome editing and crop improvement

Crop improvement using conventional methods are often labour-intensive and time-consuming and the rarity and randomness of significant mutations to produce desirable traits hinder the development of new commercial varieties. Although genetically modified crops were introduced since 1996, concerns have been raised regarding its safety and the regulatory measures adopted by different countries has hindered its popularity. However the use of genome-editing tools for crop improvement has gained much attention because of greater accuracy and efficiency compared to conventional breeding. Genome editing has revolutionised the field of agriculture. Genome editing methods utilise sequence – specific nucleases (SSNs). The potential of genome editing using various methods like Oligonucleotide Directed Mutagenesis (ODM), Zinc-Finger Nucleases (ZFNs), bacteria-derived Transcription Activator-Like Effector Nucleases (TALENs, based on protein–DNA interactions), Meganucleases (MNs), and Clustered Regularly Interspersed Short Palindromic Repeats (CRISPRs)/CRISPR-associated 9 (Cas9) endonuclease (an RNA-guided DNA endonuclease) system are being explored by many researchers because of availability of draft sequences of various crops in public databases. These methods make precise modifications in the target genome by DNA repair mechanism to produce transgene free genetically modified desired phenotypes. It is also possible to make epigenetic changes, where the DNA sequence remains unchanged but gene expression is altered because of chromatin changes that may be heritable. Targeted mutagenesis results in double-strand breaks (dsbs) at specific genomic locations [6] and this in turn induce either of the two native DNA repair mechanisms, namely:

Non-homologous end-joining (NHEJ): an imprecise repair mechanism that introduces variable length insertions or deletions at the breaking point, rendering the target gene non- functional.

Homologous recombination (HR): that inserts homologous DNA templates at the targeted point, allowing the precise insertion or deletion of nucleotides in a specific locus [7]. This technology has proven to be an efficient mechanism for genome editing, not only for model plant organisms, such as *Arabidopsis thaliana* and tobacco, but also for economically important crop plants, including soybean, corn and rice [8–11]. This method remains more complex as it requires the simultaneous delivery of a DNA repair template that carries the desired modification to be incorporated into the repaired locus [12]. However it has wider application in site specific gene insertion, stacking of genes at a specific genome position and genome alteration to a single base level [13].

3.1 Merits of genome editing

Genome editing as already mentioned is a precise breeding method that allows for targeted single gene modifications capable of altering gene expressions throughout the entire plant genome producing desirable outcomes. Random mutagenesis breeding method using radiations or chemicals on the other hand is undirected and alters thousands of genes [14].

Genome editing or 'precision genome engineering' method offers numerous applications like [15]:

- Improvement of crop yield in varying types of soil
- Production of plants more resistant to biotic and abiotic stress
- Development of plants with better root systems for nutrient uptake and the ability to source soil moisture
- Improvement of post - harvest storage
- Increase a plant's ability to sequester carbon. – research on modifying plants to increase their CO_2 fixation ability is underway in many laboratories [16]

Hence these novel biotechnological tools offers immense scope to meet the increasing demand of food supply by increasing the productivity of crops with the same level of resources and inputs.

3.2 Major genome editing tools

3.2.1 Zinc finger nucleases

During the 1990s attempts were made by various researchers to improve the precision in genome editing with the discovery of zinc finger nucleases (ZFN). ZFNs are artificial restriction enzymes comprising of a specific zinc finger DNA-binding domain composed of 3-base pair site on DNA and a cleavage domain. The structure of ZFNs were engineered so that the DNA binding domain binds to specific DNA sequences in the genome and the cleavage domain cuts the DNA.at that specified location. The cleavage domain is a type II restriction enzyme (FokI endonuclease). Using this technique scientists can make a cut in the desired region thereby allowing to either delete the target sequence or insert a new DNA sequence via homologous recombination.

Multiple ZFNs can be combined to recognise longer sequences of nucleotides, increasing specificity and success rate of genome editing by 10 percent. The major drawbacks of ZFNs were:

- for each target a new ZFN had to be designed
- it was time consuming to engineer a successful ZFN
- poor targeting density and
- relatively high levels of off-target effects, leading to cytotoxicity

3.2.2 TALENs

With the advent of time, transcription activator-like effector nucleases (TALENs) emerged as the more powerful tool in gene editing technology. TALENs are engineered from proteins found in nature and are similar to ZFNs in that they are composed of a non-specific cleavage domain from the type II restriction endonuclease FokI, fused to DNA-binding domain sequences. The engineering of these two domains resulted in stimulating NHEJ and HR leading to precise genome editing. The main difference is that each TALE domain recognise single nucleotides rather than relying on 3-base pair sites as in ZFNs. Hence, does not affect the binding specificity of neighbouring TALEs, making the engineering of TALENs much easier than ZFNs.

Forsyth and coworkers, demonstrated that the TALEN system could be used to successfully target T-DNA incorporation into a specific pre-chosen site in the potato genome that is transcriptionally active. Importantly, these investigators designed a vector that would not allow stable integration of the *TALEN* genes into the genome. Their data indicated that TALEN-induced integration of the gene of interest at specific sites, results in co-segregation and results in predictable expression level of the integrated gene [17].

Nicolia et al., employed site-directed mutagenesis in tetraploid potato through transient TALEN expression in protoplasts. The study highlighted that the site-directed mutagenesis technology could be used as a new breeding method in potato as well as for functional analysis of important genes to promote sustainable potato production [18].

TALENs are effective genome engineering technologies but their major limitation is that tailoring the DNA binding proteins to target a sequence of interest can be costly and time-consuming [19]. Furthermore, engineering TALENs to generate targeted DSBs requires two TALEN proteins capable of binding in a tail-to-tail orientation to facilitate the dimerization of FokI nuclease domain [20]. These and other, limitations were considerably reduced in the past few years due to the advent, development, and subsequent technological advancements of the CRISPR/Cas9 system [12].

3.2.3 CRISPR

CRISPR/Cas9 system is presently the widely used genome editing technology in wide range of species ranging from the smallest microbes to the largest plants and animals. Clustered regularly interspaced short palindromic repeats (CRISPRs) are a family of DNA repeats present in most Archaea and few bacterial species that act as molecular immunity systems against invading phages and nucleic acids. These distinctive loci consist of repetitive palindromic sequences (21–47 bp), separated by hypervariable spacer sequences that exhibit homology to exogenous viral and plasmid sequences, ranging between 21 and 72 bp. These arrays are often located adjacent to helper cas (CRISPR-associated) genes that encode polymerases, nucleases and helicases. When spacer sequences are transcribed, they generate small CRISPR-RNA (crRNA) fragments that hybridise with a small non-coding transactivating crRNA (tracrRNA). This double-stranded RNA molecule is used as a guide to target invading DNA sequences as a result of complementarity, and it directs the Cas9 endonuclease to these sequences for DNA degradation by double-strand cleavage at a site preceding the protospacer associated motif (PAM) [21].

The CRISPR/Cas9 genome editing technology has been successfully employed for the genetic editing of single or multiple gene targets in several plants, such as *A. thaliana,* tobacco, rice and sweet orange [10, 22–24] and for engineering of durable resistance, even at different levels of ploidy [25].

4. Genome editing in potato

Potato (*Solanum tuberosum)* is a heterozygous polyploid crop and this makes the introgression of valuable traits from wild varieties challenging and time-consuming task. Conventional breeding therefore failed when multiple traits or novel traits not present in germplasm need to be introduced for crop improvement. Availability of genome sequence data in public database and established genetic transformation and regeneration protocols has made potato a strong candidate for genome editing. These techniques can hence be utilised to improve the production and quality traits without impacting optimal allele combinations in current varieties [26–33]. The first successful demonstration of the use of TALENs in a tetraploid potato cultivar was by knocking out all four alleles of sterol side chain reductase 2 (StSSR2) [34] involved in anti-nutritional sterol glycoalkaloid (SGA) synthesis [35, 36]. In 2015, came an important breakthrough that both TALENs [18] and CRISPR/Cas9 [37] gene-editing systems could be used to efficiently modify the potato genome. In a tetraploid plant, instead of two copies (alleles) of any particular gene present in a diploid plant, there are four copies of the same gene. Advances in gene editing techniques have shown that for several polyploid plant species, rapid and efficient modification can be achieved for most, if not all, chromosomes in the multiple chromosome sets of polyploid plants [38]. In 2015, Wang and coworkers, conducted a study in potato and demonstrated that the CRISPR/Cas9 system was highly efficient for targeted mutation of *StIAA2* gene encoding an Aux/IAA protein. They could obtain homozygous monoallelic and biallelic mutations in the first generation of transgenic plants [37].

4.1 Trait improvement in potato using genome editing

4.1.1 Disease resistance in potatoes

Plant diseases cause a major constraint in potato production and incurs huge loss to the farming community. Researchers are yet to make a major breakthrough in producing potato resistant to viruses, bacteria and fungi using the gene-editing techniques. TALEN technology has already been successfully used for engineering bacterial blight resistant rice cultivars [10]. There has also been reports on the production of virus resistant plants using CRISPR/Cas9 method either by directly targeting and cleaving the viral genome, or by modifying the host plant genome to introduce viral immunity [39].

Late blight disease, caused by fungus *Phytophthora infestans*, is the major obstacle in increasing potato production [40]. Hence a major area of focus is the production of late blight resistant potato varieties by knocking out or removing disease susceptibility genes (*S*-genes) [41]. Currently the disease is controlled by fungicide spraying and breeding for disease resistance.

R genes (Resistance genes) encode R protein that degrades the toxin produced by the pathogen and initiates defence mechanism in plant. However there are chances of losing this resistance due to high rates of evolution of effector proteins by the pathogen. Genome editing method could be applied to produce late blight resistant potatoes by editing specific amino acids in *R*-genes essential for effector recognition. Another strategy for durable broad spectrum resistance is by loss of susceptibility [42]. Silencing of multiple susceptibility genes (*S*-genes) by RNAi resulted in late blight resistance in potato [43]. The drawback of RNAi is that it does not always result in a complete knockout. Genome-editing on the other hand by the introduction of both extracellular and intracellular receptors in potato cultivars can simultaneously knockout genes belonging to the S-locus, thus aid in attaining durable broad-spectrum resistance for late blight. Du and coworkers has reported the use of an extracellular receptor protein ELR (elicitin response) from the wild potato species, *S. microdontum,* in recognising an elicitin that is highly conserved in *Phytophthora* species offering a broad spectrum durable resistance to this pathogen [44].

The team led by Aman had reported the use of Cas13 for interference against Turnip Mosaic Virus (TuMV) expressing green fluorescent protein in *Nicotiana benthamiana* both in stable and transient systems. Various potato viruses like the *Potato virus X* (PVX), *Potato virus Y* (PVY) and *Potato leafroll virus* (PLRV) account for the low production of potato. So the above study raises the hope of employing CRISPR/Cas13a system in combating the pathogenic viruses [45].

4.1.2 Herbicide resistance in potatoes

Butler et al., reported the creation of a single-stranded gemini virus-based DNA replicon (GVR) that carries *TALEN* genes targeting the potato *Acetolactate synthase1* (*ALS1*) gene and also a fragment of the *ALS1* gene that carries a mutation conferring tolerance to several classes of ALS-inhibiting herbicides. Transfection of potato cells with the gemini virus DNA replicon construct results in transient expression of TALEN genes. The double strand break created at the target site was repaired by Homologous recombination to recognise the ssDNA fragment of the *ALS1* gene carrying the desired mutation and integrate this new sequence in place of the wild-type

ALS1 sequence. The plants thus modified with GVRs did not have the presence of TALEN or gemini virus DNA sequence but held point mutations within ALS1 locus and exhibited significant tolerance to herbicide treatments [46].

4.1.3 Improving post harvest shelf life in potato

Potatoes are harvested only once annually and it therefore necessitates the cold storage of the tubers to extend its postharvest shelf life. This storage leads to the conversion of sucrose to reducing sugars (cold-induced sweetening (CIS)) that can, upon frying, lead to reactions with amino acids resulting in undesirable browning, creation of bitter tastes, and production of low amounts of toxic acrylamide. Clasen et al., targeted the vacuolar invertase (*Vinv*) genes of Ranger Russet potatoes for knockout using the TALEN gene-editing system to reduce CIS. Five out of 18 regenerated plants contained knockouts of all four *Vinv* alleles. Tubers from these plants contained no detectable reducing sugars, were light brown and after processing contained lower levels of acrylamide [30]. This *Vinv*-knockout potato was commercialised by Cellectis Plant Sciences (now Calyxt Inc.) [47].

4.1.4 Modification of starch composition of potatoes

Potato starch provides important nutrition for humans and animals besides its numerous industrial uses. The relative ratio of the two major starch types, amylose and amylopectin, determines the quality of potato starch. Hence controlling this balance has significant commercial applications. High amylopectin (amylose-free) starch has been an important common trait in staple crops due its commercial value in the food and manufacturing paper industries. In potato starchy tubers, the *GBSS* gene was successfully knocked-out to generate high-amylopectin potato using different gene editing tools.

Kusano et al. used the TALEN system to successfully disrupt copies of one key enzyme in the starch biosynthesis pathway, granule-bound starch synthase (*GBSS*) gene in potato protoplast cells [48].

In a study, Andersson et al. used transient expression of the CRISPR/Cas9 system to demonstrate complete knockout of all four GBSS alleles in PEG-treated potato protoplasts and in up to 2% of regenerated lines. The successful knockout of the *GBSS* genes completely resulted in only the amylopectin starch (amylose free) in regenerated potato microtubers [32]. In yet another study, Andersson et al. carried out a DNA-free genome editing method, using delivery of CRISPR-Cas9 ribonucleoproteins (RNP) to potato protoplasts, by targeting the gene encoding granule bound starch synthase (*GBSS*) [49].

Ma et al. used a non-viral, *Agrobacterium*-mediated infiltration method to express two TALENs with different molecular weights to target two endogenous genes -starch branching enzyme (SBE1)) and an acid invertase(INV2) into two vegetatively propagated potato cultivars, *Solanum tuberosum* Russet Burbank and Shepody. These TALENs, successfully agroinfiltrated and induced mutations at both targeted loci thus affecting the degree of branching potato cold sweetening. The agroinfiltration method was cheaper, less laborious and could save time as compared to the protoplast culture approach. The mutation was induced at the specific target site and this resulted in the production of improved plant varieties with less somaclonal variation [50]. Tuncel et al., demonstrated that Cas9-mediated mutagenesis of *SBE* genes has the potential to generate a range of new potato phenotypes with valuable starch properties without integration of foreign DNA into the genome [51].

Kusano et al. improved the gene editing system by fusing the translational enhancer dMac3 of the 5′ UTR of rice *OsMac3* mRNA to the 5′-end of Cas9 to increase its level of expression. It was found that the Granule-bound starch synthase I (*GBSSI*) gene mutant frequency induced by CRISPR/Cas9 system was greatly increased and the mutant plants produced tubers with low amylose starch [52].

In 2019, Johansen et al., reported the improvement of CRISPR/Cas9 editing efficiency in the Granule-bound starch synthase gene at the protoplast level when Arabidopsis *U6* promoter was replaced by endogenous potato *U6* promoters. This team of researchers also used the Indel Amplicon Analysis (IDAA) technique for faster and direct assessment of insertions/deletions (indels) in plants with complex genomes like potato [53].

Sevestre et al. reported the successful usage of SNP physical map of *Solanum tuberosum* L. cv. Desiree revealing the position of diverse indels for designing a specific gRNA and knocked out an isoform of starch synthase *SS6* (gene), a key enzyme of the starch biosynthetic pathway [54].

Veillet et al. used the CRISPR-Cas9 base editing, precisely in the conserved catalytic KTGGL encoding locus of the StGBSSI enzyme using a cytidine base editor (CBE). This lead to the discrete variation in the amino acid sequence and loss-of-function allele producing plants with impaired amylose biosynthesis [55].

4.1.5 Production of SGA free potatoes

Potato tubers accumulate steroidal glycoalkaloids (SGAs) α-solanine and α-chaconine that confer a bitter taste and exhibit toxicity against various organisms [56]. Commercial tuber production mandates a total glycoalkaloid content of less than 20 mg 100 g^{-1} tuber fresh weight as per industry standards, but the SGA level should be higher in the aerial parts as it can act as an allelochemical to deter insect pests like Colarado potato beetle [57, 58]. Genome editing can be utilised to target specifically the tuber expressed or aerial parts expressed genes of the SGA biosynthetic pathway leading to the development of potato cultivars with low SGA levels in tubers while maintaining higher levels in the aerial parts.

Akiyama et al., from Japan reported the successful production of potato with reduced concentrations of the toxic steroidal glycoalkaloid (SGA) compounds, α- solanine and α-chaconine that accumulate in sprouts and green tubers by genome editing. The team applied CRISPR-Cas9 system to knockout the potato *CYP88B1* gene involved in a later step of the SGA biosynthetic pathway. The *CYP88B1*-knockout potatoes showed no accumulation of SGAs. Furthermore, the corresponding amounts of steroidal saponins, important compounds in the pharmaceutical industry, accumulated in the knockout potatoes as a result of the decrease in SGA synthesis [59].

Nakayasu et al., and Yasumoto et al., used TALEN and CRISPR/Cas9 to knockout the *SSR2* gene encoding for sterol side chain reductase 2 and the *St16DOX* gene encoding for the steroid 16α-hydroxylase in the SGA biosynthetic pathway. This prevented SGA accumulation in potato tuber and hairy roots, respectively [60, 61].

4.1.6 Reduction of enzymatic browning in potato tubers

Polyphenol oxidase (PPO) catalyses the conversion of phenols to quinones resulting in browning and reducing the devaluation of the processed products from the tubers. TALEN methods was employed to knock out one of the PPO genes in potato tubers resulting in decreased browning. This technique was commercialised using different delivery techniques (PEG-mediated transfection or *Agrobacterium*-mediated transformation) by two companies (Calyxt Inc., and Simplot Plant Sciences).

Gonzalez et al., produced potatoes with reduced browning by specific editing of the polyphenol oxidase gene (*StPPO2*) in the tetraploid cultivar Desiree. CRISPR/Cas9 system using RNPs as a delivery system was employed to induce mutations in the *StPPO2* gene resulting in the production of lines with a reduction of up to 69% in tuber PPO activity and a reduction of 73% in enzymatic browning, compared to the control [62].

Khromov et al. compared *in vitro* activities of various sgRNAs designed for different regions of *phytoene desaturase* (*PDS*) from the carotenoid biosynthesis pathway and a *coilin* gene involved in plant resistance. The visual phenotype of PDS knockout makes it convenient for detection and analysis of potato genome editing due to the depigmentation in the absence of PDS. Knockout of *coilin* gene is highly desirable as deterioration of coilin is mainly involved in pathogen resistance and improving tolerance to biotic and abiotic stresses. The study revealed that the first six nucleotides located in the DNA substrate proximal to the 3'PAM site directly binded with Cas9 but did not affect the activity of Cas9-sgRNA complex. The researchers drew a conclusion that the unpaired nucleotides of target DNA with sgRNA can both stimulate or repress the activity of Cas9-sgRNA complex *in vitro* depending on the position of the mismatch [63] (**Table 1**).

4.2 Challenges in genome editing of potato

Potato is a clonally propagated highly heterozygous polyploid crop and hence complicates the use of gene editing techniques- difficulty in target designing for genome editing, obtaining homozygous mutants with all target genes mutated.

This mandates the need for screening large number of transformants to identify and propagate multiallelic mutagenic lines. Another challenge is that not all cultivars of potato are amenable to transformation and others need to be tested for transformation and regeneration in tissue culture. Protoplast transformation and regeneration of plants from leaf protoplasts also can lead to somaclonal variation, which may have negative impact(s) on plant development [67].

Attempts are being made by breeders to develop diploid potato lines inorder to understand complex agronomic traits. A major obstacle in potato breeding was the development of inbred lines due to self-incompatibility that hinders the fixing of gene edits and selection of progeny by segregating out the inserted foreign gene. Ye et al. developed self-compatible diploid potatoes by knocking out the self-incompatibility gene, Stylar ribonuclease gene (*S-RNase*) using the CRISPR-Cas9 system. This strategy opens new avenues for production of diploid inbred and self- compatible potato germplasm and pave way for studying other self-incompatible crops [66].

However many diploid, self- compatible potato germplasm were found to be recalcitrant to conventional *Agrobacterium tumefaciens* -mediated transformation [68]. Butler et al. demonstrated the utility of *A. rhizogenes* strains for rapidly generating stable mutations within hairy root clones in potato genotypes recalcitrant to *A. tumefaciens* and regenerating fertile lines capable of fixing targeted mutations, segregating out T-DNA insertions and production of additional mutants when needed. There is however a limitation to analysis of hairy root clones. CRISPR/Cas9 technology was successfully employed for targeting the potato *phytoene desaturase* (*StPDS*) gene, expressed in hairy root clones and regenerated. Targeted mutation was expressed in 64–98% of the transformed hairy root clones and this broadens the potato genotypes amenable to *Agrobacterium*-mediated transformation while reducing chimerism in primary events and accelerating the generation of edited materials [69].

Another area of concern is the occurrence of off-target mutations in non-target genes of potato during the process of GE. This results in undesired changes in plants and makes the process of mutational analysis studies more complicated. Attempts have been made to reduce or even eliminate such off-targeting by good design and test of sgRNA activity [70] and use of synthetic proofreading Cas9 variants [71].

Target Gene	Function of Target gene	Gene editing method	Gene delivery method	Trait improved	Reference
Sterol side chain reductase2 (*StSSR2*)	Steroidal glycoalkaloids reduction in tubers	TALENS	Agrobacterium	Identify key enzyme in the biosynthesis of cholesterol and related steroidal glycoalkaloids	[34]
Acetolactase synthase 1(StALS1)	Herbicide resistance	TALENS	Protoplasts	Transient expression of TALENS in potato protoplasts for targeted mutagenesis and regeneration	[18]
Acetolactase synthase 1(StALS1)	Herbicide resistance	CRISPR/ Cas9	Agrobacterium Gemini Virus Replicon (GVR)	Targeted mutagenesis and germline inheritance	[64]
Auxin/Indole 3 Acetic Acid (IAA) protein (StIAA2)	Petiole hyponasty and shoot morphogenesis	CRISPR/ Cas9	Agrobacterium	Targeted mutagenesis in the first generation of transgenic plants	[37]
Vacuolar invertase (StVInv)	Cold induced sweetening , acrylamide content in tubers	TALENS	Protoplasts	Tuber improvement for cold storage	[30]
Acetolactase synthase 1(StALS1)	Herbicide resistance	TALENS	Agrobacterium	For targeted T-DNA integration	[17]
Acetolactase synthase 1(StALS1)	Herbicide resistance	CRISPR/ Cas9 TALENS	Agrobacterium Gemini Virus Replicon (GVR)	Gene targeting via homologous recombination using donor template	[46]
Granule bound starch synthase (StGBSS)	Tuber starch quality	TALENS	Agrobacterium	Development of a Gateway system for rapid assembly of TALENS in a binary vector	[48]
1,4 alpha –glucan branching enzyme gene (SBE1), Vacuolar invertase (StVInv)	Degree of starch branching, cold induced sweetening	TALENS	Agroinfiltration	Effective delivery of TALENS and induction of mutation	[50]
Granule Bound starch synthase (StGBSS)	Tuber starch quality	CRISPR/ Cas9	Protoplasts	Targeted mutagenesis and regeneration resulting in tubers with high amylopectin starch	[32]
Transcription factor gene (StMYB44)	Phosphate transport via roots	CRISPR/ Cas9	*Agrobacterium*	Understand the molecular basis of phosphate stress responses	[65]
CYP88B1	Involved in later step of steroidal glycoalkaloid (SGA) pathway	CRISPR/ Cas9	*Agrobacterium*	Absence of steroidal glycoalkaloids	[59]

Target Gene	Function of Target gene	Gene editing method	Gene delivery method	Trait improved	Reference
Granule Bound starch synthase (StGBSS)	Tuber starch quality	CRISPR/ Cas9	Protoplasts Ribonucleoproteins (RNPs)	Regeneration of mutant lines without amylose	[49]
Steroid 16α hydroxylase (16DOX)	Encodes a steroid 16α-hydroxylase in SGA biosynthesis	CRISPR/ Cas9	*Agrobacterium rhizogenes*	Absence of steroidal glycoalkaloids-α solanine in hairy roots of potato	[60]
Stylar ribonuclease (S-Rnase)	Self incompatibility	CRISPR/ Cas9	*Agrobacterium*	Self compatibility in diploid potato lines	[66]
Granule bound starch synthase (StGBSS)	Tuber starch quality	CRISPR/ Cas9	*Agrobacterium*	Reduced amylose starch in tubers	[52]
Phytoene desaturase (PDS) and coilin gene	Carotenoid biosynthetic pathway and biotic stress resistance	CRISPR/ Cas9		Loss of colour and enhances resistance to biortic sress	[63]
Sterol side chain reductase (SSR2)	Encodes key enzyme in steroidal glycoalkaloid (SGA)synthesis	TALEN	*Agrobacterium*	Reduced steroidal glycoalkaloids	[61]
Starch branching enzymes (SBE1 and SBE2)	Introduction of α 1,6 linkages in starch	CRISPR/ Cas9	*Agrobacterium*/ Protoplasts	To generate tubers with a wide range of desirable starch content	[51]
Granule Bound starch synthase (StGBSS)	Tuber starch quality	CRISPR/ Cas9	Protoplasts with StU6 endogenoue prpmoter	Reduced amylose starch in tubers than the previous studies with foreign promoter	[53]
Granule Bound starch synthase (StGBSS)	Tuber starch quality	CRISPR/ Cas9	Agrobacterium	plants with impaired amylose biosynthesis; Base editing in conserved catalytic KTGGL encoding locus of the StGBSSI enzyme	[55]
Polyphenol oxidase (PPO2)	Conversion of phenolic substrates to quinones leading to browning	CRISPR/ Cas9	Protoplasts Ribonucleoproteins (RNPs)	Reduced browning	[62]

Table 1.
Crop improvement in potato by Gene editing techniques

A major area of focus is the generation of transgene free potato. Inorder to be accepted by the public and regulatory bodies, there should not be any trace of the exogenous DNA in the GE crops. Segregation of genetic lines is used in generation transition from T0 to T2, so that stably inherited transgene-free plants can be obtained in T2 mutant lines [72]. However, this strategy cannot easily be adopted in tetraploid potato with high allelic polymorphism. RNP delivery into protoplasts is now emerging as an excellent alternative system that avoids DNA intermediates [73].

5. Regulations on genome edited crops

The cultivation and commercialization of genetically modified crops did not attain the expected growth as it received a setback due to the strict regulations imposed by various countries. With the advent of the gene editing techniques, attempts were made to produce genome modified plants without exogenous DNA so that they do not come under the purview of the regulations.

Regulatory approaches for genome edited products is still in its infancy and different countries have issued their own legal interpretations. Different countries have adopted regulation on genome edited crops based on two types of regulatory frameworks: process-based and product based. In the case of process-based regulation, regulation is typically triggered if nucleic acids are introduced into crops or recombinant DNA technologies are deployed in the development of a crop. The European Union (EU), Argentina, Brazil and several other countries have a process-based regulatory framework [15]. EU declared that the genome edited plants can alter the natural genetic material of the plant producing adverse environmental issues and hence should be treated as transgenic plants. This stringent approach can hinder research in the development and also impact the trade of gene edited crops.

In the case of a product based regulatory framework the focus is placed on the risk inherent in the final product. The United States which has a product-based regulatory framework has no regulation for genome edited plants if no genetic elements from pathogenic species or pesticidal traits are introduced [74]. Multiple level checks are followed like FDA weighs on health benefits and the EPA weighs on the environmental impact of the edited crops. Null segregants – progeny of the transgenic, edited parent that still retain the germline edit but lack the integrated foreign DNA sequence – are exempted from regulation. Clonally propagated plants like potato normally does not produce null segregants. Japan also adopted a regulatory policy similar to the United States stating that the gene-edited plants in Japan should not be regulated (The Scientist news). Although the products of rDNA technology will still be regulated, it was stated that the genome editing technologies poses no increase in risk and therefore do not require additional regulatory oversight. No regulations were imposed by USDA on anti-browning mushrooms developed by targeting PPO using CRISPR/Cas9, indicative of the acceptance of traits created by gene editing [75].

The world's first regulation for GE crops was reported by Argentina [76]. Later on, Brazil and Chile adopted the same policies. Currently, many countries do not have a clear regulatory framework for GE crops. However, several countries like Kenya, Nigeria, and India are in the process of developing the regulatory guidelines for the application of genome editing [77].

5.1 Impact of the regulations

The commercialization of genome edited crop poses a challenge to the public sector breeders who lack funding, if they are treated equivalent to GM crops. The uncertainty in regulations will also have logistical challenges for international commodity trade. The application of genome editing can reap its benefit and ensure agricultural sustainability depending mainly on the regulatory measures adopted by each country. The potential of genome editing can be exploited fully only if it is not treated on par with genetically modified plants and not subjected to the same regulatory measures.

Another constraint in the deployment of gene editing technology is the lack of a clear implementation and effective management strategy for the sustainable development of crops produced using this tool. Do we have to adopt the practice of crop monoculture inorder to harbour durable resistance is a question under debate? From the sociological point of view also, the public acceptance of food crops engineered using genome editing technology also needs to be considered.

6. Conclusion

Genome editing could play a major role in the modification of starch content, decrease antinutrient and toxic substances and enhance the nutritive value of potatoes. This technology with high efficiency and precision raises the scope of improving other desirable plant traits. The research advancements in this field can be accelerated by the production of transgene free GE potatoes and the commercialization of the technology can be promoted only by assuring the public of its safety. Despite the challenges faced in the commercialization of GE crops and its products, intense research is being carried out in different countries. Attempts to exclude GE crops from the GMO regulations raises hope in the advancement of the editing related technology. The availability of whole genome sequence of potato, transformation and regeneration protocols of potato, and novel gene editing tools instills hope of producing elite transgene free potato plants with desirable traits in short span of time.

Conflict of interest

The authors declare no conflict of interest.

Author details

Sona S. Dev*, Jini Joseph and Ligi Lambert D'Rosario
St. Peter's College, Kochi, India

*Address all correspondence to: sona.dev@stpeterscollege.ac.in

References

[1] www.fao.org

[2] https://cipotato.org/potato

[3] Muthoni, J, Kabira, J, Shimelis, H, & Melis, R. Tetrasomic inheritance in cultivated potato and implications in conventional breeding. Australian Journal of Crop Science. 2015; 9(3): 185-190. DOI:10.3316/informit.075206802413826

[4] Love SL, Eberlein CV, Stark JC. et al: Cultivar and seed piece spacing effects on potato competitiveness with weeds. American Potato Journal.1995; 72:197-213. DOI:org/10.1007/BF02855036

[5] Nelson DC and Thoreson MC. Competition between potatoes (*Solanum tuberosum*) and weeds. Weed Science. 1981; 29(6), 672-677.

[6] Miller JC, Tan S, Qiao G, Barlow KA, Wang J, Xia DF et al. A TALE nuclease architecture for efficient genome editing. Nat. Biotechnol. 2011; 29: 143-148. DOI: 10.1038/nbt.1755

[7] Joung JK and Sander JD. TALENs: a widely applicable technology for targeted genome editing. Nat. Rev. Mol. Cell Biol. 2013; 14: 49-55. 10.1038/nrm3486.

[8] Tomas C , Erin LD, Michelle C, Li W, Yong Z, Clarice S, Joshua AB, Nikunj VS, Adam JB, Daniel FV. Efficient design and assembly of custom TALEN and other TAL effector-based constructs for DNA targeting. Nucleic Acids Res.2011; 39(12):e82.DOI: 10.1093/nar/gkr218.

[9] Haun W, Coffman A, Clasen BM, Demorest ZL, Lowy A, Ray E, Retterath A, Stoddard T, Juillerat A, Cedrone F, Mathis L, Voytas DF, Zhang F. Improved soybean oil quality by targeted mutagenesis of the fatty acid desaturase 2 gene family. Plant Biotechnology Journal. 2014; 12(7). 934-940. DOI.org/10.1111/pbi.12201

[10] Li JF, Norville JE, Aach J, McCormack M, Zhang D, Bush J, Church GM, Sheen J. Multiplex and homologous recombination-mediated genome editing in Arabidopsis and *Nicotiana benthamiana* using guide RNA and Cas9. Nat. Biotechnol. 2013; 31, 688-691.

[11] Liang Z, Zhang K, Chen K, Gao C. Targeted mutagenesis in *Zea mays* using TALENs and the CRISPR/Cas system. Journal of Genetics and Genomics. 2013; 41. 10.1016/j.jgg.2013.12.001.

[12] Stella S, Montoya G. The genome editing revolution: A CRISPR-Cas TALE off-target story. BioEssays. 2016; 38(S1). S4-S13.

[13] Nishida K, Arazoe T, Yachie N, Banno S, Kakimoto M, Tabata M, Mochizuki M, Miyabe A, Araki M, Hara KY, Shimatani Z, Kondo A. Targeted nucleotide editing using hybrid prokaryotic and vertebrate adaptive immune systems. Science. 2016; 16; 353 (6305):aaf8729. DOI: 10.1126/science.aaf8729.

[14] Hartung F, Schiemann J. Precise plant breeding using new genome editing techniques: opportunities, safety and regulation in the EU. Plant J. 2014; 78(5):742– 752. DOI.org/10.1111/tpj.12413

[15] Voytas DF and Gao C. Precision genome engineering and agriculture: opportunities and regulatory challenges. PLoS Biol. 2014;12:e1001877. DOI: 10.1371/journal.pbio.1001877

[16] Schwander T, Schada von Borzyskowski L, Burgener S, Cortina NS, Erb TJ. A synthetic pathway for the fixation of carbon dioxide in vitro. Science. 2016; 18; 354(6314):900-904. DOI: 10.1126/science.aah5237.

[17] Forsyth A, Weeks T, Richael C and Duan H. Transcription activator-like

effector nucleases (TALEN)-mediated targeted DNA insertion in potato plants. Front. Plant Sci. 2016; 7:1572. DOI: 10.3389/fpls.2016. 01572

[18] Nicolia A, Proux-Wéra E, Åhman I, Onkokesung N, Andersson M, Andreasson E, Zhu L. Targeted gene mutation in tetraploid potato through transient TALEN expression in protoplasts, Journal of Biotechnology, 2015; 204: 17-24. DOI.org/10.1016/j. jbiotec.2015.03.021.

[19] Ceasar SA, Rajan V, Prykhozhij SV, Berman JN, Ignacimuthu S. Insert, remove or replace: A highly advanced genome editing system using CRISPR/ Cas9. Biochimica et Biophysica Acta (BBA) - Molecular Cell Research, 2016;1863 (9):2333-2344., DOI. org/10.1016/j.bbamcr.2016.06.009.

[20] Sun N, Zhao H. Transcription activator-like effector nucleases (TALENs): a highly efficient and versatile tool for genome editing. Biotechnol Bioeng. 2013; 110(7):1811-1821. DOI: 10.1002/bit.24890.

[21] Horvath P and Barrangou R. CRISPR/Cas, the immune system of bacteria and Archaea. Science. 2010; 327 (5962): 167-170. DOI: 10.1126/ science.1179555

[22] Gao J, Wang G, Ma S, Xie X,Wu X, Zhang X, et al. CRISPR/Cas9-mediated targeted mutagenesis in *Nicotiana tabacum.* Plant Mol Biol. 2014;10:1007. [s11103-014- 0263-0].

[23] Jia H, Wang N. Targeted genome editing of sweet orange using Cas9/ sgRNA. PLoS One. 2014; DOI. org/10.1371/journal.pone.0093806.

[24] Shan Q, Wang Y, Li J, Zhang Y, Chen K, Liang Z, et al. Targeted genome modification of crop plants using a CRISPR-Cas system. Nat Biotechnol. 2013; 31:686-688.

[25] Wang Y, Cheng X, Shan Q, Zhang Y, Liu J, Gao C and Qiu J. Simultaneous editing of three homoeoalleles in hexaploid bread wheat confers heritable resistance to powdery mildew. Nat Biotechnol.2014; 32, 947-951. DOI. org/10.1038/nbt.2969\

[26] Diretto G, Tavazza R, Welsch R, Pizzichini D, Mourgues F, Papacchioli V et al. Metabolic engineering of potato tuber carotenoids through tuber specific silencing of lycopene epsilon cyclase. BMC Plant Biol. 6:13. 2006; DOI: 10.1186/1471-2229-6-13

[27] Diretto G, Welsch R, Tavazza R, Mourgues F, Pizzichini D, Beyer P et al. Silencing of beta-carotene hydroxylase increases total carotenoid and beta-carotene levels in potato tubers. BMC Plant Biol. 2007; 7:11. DOI: 10.1186/1471-2229-7-11

[28] Rommens CM, Ye J, Richael C and Swords K. Improving potato storage and processing characteristics through all-native DNA transformation. J. Agric. Food Chem. 2006; 54, 9882-9887. DOI: 10.1021/jf062477l

[29] Chi M, Bhagwat B, Lane W, Tang G, Su Y, Sun R et al. Reduced polyphenol oxidase gene expression and enzymatic browning in potato (*Solanum tuberosum* L.) with artificial microRNAs. BMC Plant Biol. 2014;14:62. DOI: 10.1186/1471-2229-14-62

[30] Clasen, BM, Stoddard TJ, Luo S, Demorest ZL, Li J, Cedrone F et al. Improving cold storage and processing traits in potato through targeted gene knockout. Plant Biotechnol. J. 2016; 14: 169-176. DOI: 10.1111/pbi.12370.

[31] Sun Y, Jiao G, Liu Z, Zhang X, Li J, Guo X et al. Generation of high amylose rice through CRISPR/Cas9-mediated targeted mutagenesis of starch branching enzymes. Front. Plant Sci. 2017; 8: 298. DOI: 10.3389/ fpls.2017.00298

[32] Andersson M, Turesson H, Nicolia A, Fält AS, Samuelsson M and Hofvander P. Efficient targeted multiallelic mutagenesis in tetraploid potato (*Solanum tuberosum)* by transient CRISPR-Cas9 expression in protoplasts. Plant Cell Rep. 2017; 36, 117-128. DOI: 10.1007/s00299-016-2062-3

[33] McCue KF, Breksa A, Vilches A, and Belknap WR. Modification of potato steroidal glycoalkaloids with silencing RNA constructs. Am. J. Potato Res. 2018; 95:317. DOI: 10.1007/s12230-018-9658-9

[34] Sawai S, Ohyama K, Yasumoto S, Seki H, Sakuma T, Yamamoto T et al. Sterol side chain reductase 2 is a key enzyme in the biosynthesis of cholesterol, the common precursor of toxic steroidal glycoalkaloids in potato. Plant Cell. 2014; 26: 3763-3774. doi: 10.1105/tpc.114.130096

[35] Itkin M, Rogachev I, Alkan N, Rosenberg T, Malitsky S, Masini L et al. Glycoalkaloid metabolism is required for steroidal alkaloid glycosylation and prevention of phytotoxicity in tomato. Plant Cell. 2011; 23, 4507– 4525. DOI: 0.1105/tpc.111.088732

[36] Itkin M, Heinig U, Tzfadia O, Bhide AJ, Shinde B, Cardenas PD et al. Biosynthesis of antinutritional alkaloids in solanaceous crops is mediated by clustered genes. Science. 2013; 341, 175-179. DOI: 10.1126/science. 1240230.

[37] Wang S, Zhang S, Wang W et al. Efficient targeted mutagenesis in potato by the CRISPR/Cas9 system. Plant Cell Rep. 2015; 34: 1473-1476. DOI. org/10.1007/s00299-015-1816-7

[38] Weeks DP. Gene Editing in polyploid crops: wheat, camelina, canola, potato, cotton, peanut, sugar cane, and citrus. Gene Editing in Plants. 2017; 65– 80. DOI:10.1016/ bs.pmbts.2017.05.002

[39] Zaidi SS, Tashkandi M, Mansoor S, Mahfouz M. Engineering plant immunity: using CRISPR/Cas9 to generate virus resistance. Frontiers in Plant Science. 2016; 7: 1673. DOI=10.3389/fpls.2016.01673

[40] Fisher MC, Henk DA, Briggs CJ, Brownstein JS, Madoff LC, McCraw SL et al. Emerging fungal threats to animal, plant and ecosystem health. Nature. 2012; 484: 186-194. DOI: 10.1038/ nature10947

[41] Schaart JG, van de Wiel CC, Lotz LA, Smulders MJ. Opportunities for products of new plant breeding techniques. Trends Plant Sci. 2016; 21(5):438-449. DOI.org/10.1016/j. tplants.2015.11.006.

[42] Pavan S, Jacobsen E, Visser RGF and Bai Y. Loss of susceptibility as a novel breeding strategy for durable and broad-spectrum resistance. Mol. Breed. 2009; 25:1. DOI: 10.1007/ s11032-009-9323-6

[43] Sun K, Wolters AM, Vossen JH, Rouwet ME, Loonen AE, Jacobsen E et al. Silencing of six susceptibility genes results in potato late blight resistance. Transgenic Res. 2016; 25, 731-742. DOI: 10.1007/s11248-016-9964-2

[44] Du J, Verzaux E, Chaparro-Garcia A, Bijsterbosch G, Keizer LCP, Zhou J, et al. Elicitin recognition confers enhanced resistance to *Phytophthora infestans* in potato. Nat. Plants. 2015; 1:15034. DOI: 10.1038/ NPLANTS.2015.34

[45] Aman R, Ali Z, Butt H. et al. RNA virus interference via CRISPR/ Cas13a system in plants. Genome Biol.2018; 19: 1 .DOI.org/10.1186/ s13059-017-1381-1

[46] Butler NM, Baltes NJ, Voytas DF and Douches DS. Geminivirus-mediated genome editing in potato (*Solanum tuberosum* L.) using sequence-specific nucleases. Front. Plant Sci. 2016; 7:1045. DOI: 10.3389/fpls.2016.01045.

[47] Ku HK, & Ha SH. Improving nutritional and functional quality by genome editing of crops: status and perspectives. Frontiers in Plant Science. 2020; 11: 577313. DOI.org/10.3389/fpls.2020.577313

[48] Kusano H, Onodera H, Kihira M, Aoki H, Matsuzaki H and Shimada H. A simple gateway-assisted construction system of TALEN genes for plant genome editing. Sci. Rep. 2016; 6:30234. DOI: 10.1038/srep30234

[49] Andersson M, Turesson H, Olsson N, Fält AS, Olsson P, Gonzalez MN et al. Genome editing in potato via CRISPR-Cas9 ribonucleoprotein delivery. Physiol. Plant. 2018; DOI: 10.1111/ppl.12731

[50] Ma J, Xiang H, Donnelly DJ, Meng FR, Xu H. Durnford D et al. Genome editing in potato plants by *Agrobacterium*-mediated transient expression of transcription activator-like effector nucleases. Plant Biotechnol. Rep. 2017; 11, 249-258. DOI: 10.1007/s11816-017-0448-5

[51] Tuncel A, Corbin KR, Ahn-Jarvis J, Harris S, Hawkins E, Smedley MA, Harwood W, Warren FJ, Patron NJ, Smith AM. Cas9-mediated mutagenesis of potato starch-branching enzymes generates a range of tuber starch phenotypes. Plant Biotechnol J. 2019; 17(12):2259-2271. DOI: 10.1111/pbi.13137.

[52] Kusano H, Ohnuma M, Mutsuro-Aoki, H. et al. Establishment of a modified CRISPR/Cas9 system with increased mutagenesis frequency using the translational enhancer dMac3 and multiple guide RNAs in potato. Sci Rep. 2018; 8, 13753. DOI.org/10.1038/s41598-018-32049-2

[53] Johansen IE, Liu Y, Jorgensen B, Bennett EP, Andreasson E, Nielsen KL, Blennow A and Petersen BL. High efficacy full allelic CRISPR/Cas9 gene editing in tetraploid potato. Sci Rep. 2019; 9: 17715 DOI.org/10.1038/s41598-019-54126-w

[54] Sevestre F, Facon M, Wattebled F and Szydlowski N. Facilitating gene editing in potato: a Single-Nucleotide Polymorphism (SNP) map of the *Solanum tuberosum* L. cv. Desiree genome. Sci Rep. 2020; 10: 2045. DOI.org/10.1038/s41598-020-58985-6

[55] Veillet, F, Chauvin L, Kermarrec M, Sevestre F, Merrer M Terret Z , Szydlowski N, Devaux P, Gallois J, Chauvin J. The *Solanum tuberosum* GBSSI gene: a target for assessing gene and base editing in tetraploid potato. Plant Cell Report. 2019; 38, 1065-1080.

[56] Friedman M. Potato glycoalkaloids and metabolites: roles in the plant and in the diet. J. Agric. Food Chem. 2006; 54, 8655-8681. DOI: 10.1021/jf061 471t

[57] Sanford LL, Deahl KL, Sinden SL and Ladd TL. Foliar solanidine glycoside levels in *Solanum tuberosum* populations selected for potato leafhopper resistance. Am. Potato J. 1990; 67, 461-466. DOI: 10.1007/BF0304 4513

[58] Sanford LL, Kobayashi RS, Deahl KL and Sinden SL. Diploid and tetraploid *Solanum chacoense* genotypes that synthesize leptine glycoalkaloids and deter feeding by Colorado potato beetle. Am. Potato J. 1997; 74, 15-21. DOI: 10.1007/BF02849168

[59] Akiyama R, Nakayasu M, Umemoto N, Muranaka T, Mizutani M. Molecular breeding of SGA-free potatoes accumulating pharmaceutically useful saponins. Regulation of Plant Growth & Development, 2017; 52(2):92-98. DOI.org/10.18978/jscrp.52.2_92.

[60] Nakayasu M. et al. Generation of α-solanine-free hairy roots of potato by CRISPR/Cas9 mediated genome editing of the St16DOX gene. Plant Physiol. Biochem. 2018; 131, 70-77.

[61] Yasumoto S, Umemoto N, Lee HJ, Nakayasu M, Sawai S, Sakuma T et al. Efficient genome engineering using platinum TALEN in potato. Plant Biotechnol. 2019; 36: 167-173. DOI: 10.5511/plantbiotechnology.19.0805a

[62] Gonzalez MN, Massa GA, Andersson M, Turesson H, Olsson N, Fält AS, Storani L, Oneto CAD, Hofvander P and Feingold SE..Reduced enzymatic browning in potato tubers by specific editing of a polyphenol oxidase gene via ribonucleoprotein complexes delivery of the CRISPR/Cas9 system. Front. Plant Sci. 2020; DOI.org/10.3389/fpls.2019.01649

[63] Khromov AV, Gushchin VA, Timerbaev VI et al. Guide RNA design for CRISPR/Cas9-mediated potato genome editing. Dokl Biochem Biophys. 2018; 479: 90-94. DOI.org/10.1134/S1607672918020084

[64] Butler S, Ohyama K, Yasumoto S, Seki H, Sakuma T, Yamamoto T, Takebayashi Y, Kojima M, Sakakibara H, Aoki T, Muranaka T, Saito K, Umemoto N. Sterol side chain reductase 2 is a key enzyme in the biosynthesis of cholesterol, the common precursor of toxic steroidal glycoalkaloids in potato. Plant Cell. 2015; 26: 3763-3774.

[65] Zhou X, Zha M, Huang J, Li L, Imran M, and Zhang C. StMYB44 negatively regulates phosphate transport by suppressing expression of Phosphate1 in potato. J. Exp. Bot. 2017; 68: 1265-1281. doi: 10.1093/jxb/erx026

[66] Ye M, Peng Z, Tang D, Yang Z, Li D, Xu Y, Zhang C, Huang S. Generation of self-compatible diploid potato by knockout of S-RNase. Nat Plants. 2018; 4(9):651-654. DOI: 10.1038/s41477-018-0218-6.

[67] Nadakuduti SS, Buell CR, Voytas DF, Starker CG, Douches DS. Genome Editing for Crop Improvement – Applications in clonally propagated polyploids with a focus on potato (*Solanum tuberosum* L.)Frontiers in Plant Science .2018; 9:1607. DOI=10.3389/fpls.2018.01607

[68] Leisner CP, Hamilton JP, Crisovan E, Manrique-Carpintero NC, Marand AP, Newton L, Pham GM et al. Genome sequence of M6, a diploid inbred clone of the high glycoalkaloid-producing tuber-bearing potato species *Solanum chacoense,* reveals residual heterozygosity. Plant J. 2018; 94: 562– 570.

[69] Butler NM, Jansky SH and Jiang J. First-generation genome editing in potato using hairy root transformation. Plant Biotechnology Journal.2020; 18(11):2201-2209. DOI.org/10.1111/pbi.13376

[70] Tang X, Liu G, Zhou J, Ren Q, You Q, Tian L, Xin X, Zhong Z, Liu B, Zheng X, Zhang D, Malzahn A, Gong Z, Qi Y, Zhang T and Zhang Y. A large-scale whole genome sequencing analysis reveals highly specific genome editing by both Cas9 and Cpf1 nucleases in rice. Genome Biol. 2018; 19(1):84.

[71] Chen JS, Dagdas YS, Kleinstiver BP, Welch MM, Sousa AA et al. Enhanced proofreading governs CRISPR–Cas9 targeting accuracy. Nature. 2017; 550:407-410. DOI.org/10.1038/natur e2426 8

[72] Ricroch A, Clairand P, Harwood W. Use of CRISPR systems in plant genome editing: toward new opportunities in agriculture. Emerg Top Life Sci. 2017; 1:169-182. DOI.org/10.1042/ETLS201700 85

[73] Arora L, Narula A (2017) Gene editing and crop improvement using CRISPR-Cas9 system. Front Plant Sci. 2017; 8:1932. DOI. org/10.3389/fpls.2017.01932

[74] USDA (2018) Secretary Perdue issues USDA Statement on plant

breeding innovation. Press Release No 0070.18. Available at https://www.usda.gov/media/press-releases/ 2018/03/28/secretary-perdue-issues-usda-statementplant-breeding-innovation (accessed 3 April 2018)

[75] Waltz E. Gene-edited CRISPR mushroom escapes US regulation. Nature. 2016; 532. 10.1038/nature.2016.19754.

[76] Lema MA. Regulatory aspects of gene editing in Argentina. Transgenic Res. 2019; 28: 147-150. DOI: 10.1007/s11248-019-00145-2Available at https://www.usda.gov/media/press-releases/

[77] Tripathi L, Ntui VO, Tripathi JN & Kumar PL. Application of CRISPR/Cas for diagnosis and management of viral diseases of banana. Frontiers in Microbiology. 2020; 11: 609784. DOI.org/10.3389/fmicb.2020.6097840

Chapter 9

Genetically Modified Potato for Pest Resistance: Thrift or Threat?

Martin Raspor and Aleksandar Cingel

Abstract

Significant limitations in potato production are crop loss due to the damage made by insect pests, and the cost of enormous amount of chemicals, harmful to humans and environment, extensively used in their control. As an alternative, development of genetically modified potato offered possibility for pest management in a more sustainable, environmentally friendly way. Over the past 30 years introduction of pest resistance traits progressed from a single gene to multiple stacked events and from Bt-toxin expression to expression of proteins from non-Bt sources, dsRNA and their combination, while advances in molecular biology have brought "cleaner" gene manipulation technologies. However, together with benefits any new technology also bears its risks, and there are still a range of unanswered questions and concerns about long-term impact of genetically modified crops – that with knowledge and precautionary approaches can be avoided or mitigated. Sustainability of genetically modified crops for pest control largely depends on the willingness to gain and implement such knowledge.

Keywords: potato, *Solanum tuberosum* L., genetic engineering, pest resistance, environmental safety, Bt-toxins, protease inhibitors, RNAi

1. Introduction

Almost four decades after the initial success [1], production of genetically modified plants still takes a central place in the experimental studies and biotechnology of plants. Genetic engineering has made possible introducing beneficial traits from unrelated plants, bacteria, viruses, fungi, or animal species, to overcome the major limitations of conventional plant breeding. Introduction of one or more genes into commercial crop species has helped boost crop yields due to increased resistance of transgenic lines to abiotic stress, pests and pathogens, and manipulation of metabolic pathways resulted in improving the nutritional or industrial value of genetically modified plants. Also, plant "factories" have been designed to produce high amounts of various pharmacologically important compounds, nutrients or other useful substances.

Genetically modified (GM) crops have been cultivated for more than twenty years and in 2019, the global area under GM crops was 190.4 million hectares, a 112-fold increase since their first commercialization in 1996 [2]. Gains from increased yields and cost savings brought net economic benefits amounting to more than $225 billion and added one hundred million tons to the global crop production without the need for using additional land for cultivation [3]. The development of insect resistant GM crops resulted in reduction of insecticides by 775.4 million kg (8.3%) and decreased the environmental impact of these chemicals by 18.5%. By cutting fuel usage associated with the production of chemical spray runs and tillage, this technology also reduced carbon dioxide emissions equivalent to removing more than 15 million cars from the roads [4]. However, wider adoption of GM crops remains the subject of biosafety concerns due to potential risks such as gene flow, evolution of resistance in insects and weeds, adverse effects on beneficial non-target organisms, or toxicity and allergenicity to humans.

2. Incorporating insect resistance traits

A wide range of pests and pathogens (over 50 insect and about 10 nematode species, 11 viral, 6 bacterial and over 20 fungal pathogens) [5] threaten potato (*Solanum tuberosum* L.), causing at least 40% of production losses worldwide [6]. Among them, Colorado potato beetle (CPB; *Leptinotarsa decemlineata* Say) and Potato tuber moth (PTM; *Phthorimaea operculella* Zeller) are the most widespread insect pests of potato that, if not controlled, can cause total yield or storage losses [7, 8]. CPB, particularly capable of rapid build-up of resistance to toxins, is today resistant to 56 different compounds belonging to all major insecticide classes with different modes of action [9], and tens of millions of dollars are spent annually for its management [10]. The long history of failure in chemical control of CPB and other pests, dubbed "the 125 years of mismanagement" [11], gave way to alternative means of control, including genetic engineering as more pest-specific and less risky for the environment.

Potato is one of the few crops naturally susceptible to infection by agrobacteria, so the first report on the generation of transgenic potato plants using *Agrobacterium* [12] dates from the very beginning of the "era of plant genetic engineering". Since then, many recombinant DNA delivery systems have been developed (biolistic, electroporation, PEG-mediated, etc), but, enabling high transformation frequency and efficiency, *Agrobacterium*-mediated transformation has remained the preferred method for heterologous gene integration into the potato genome, and became a routine technique in many laboratories. Over the past 30 years introduction of pest resistance traits progressed from a single gene to multiple stacked events (directed to the same or different pests), and from Bt-toxin expression to expression of proteins from non-Bt sources, dsRNA and their combination. Above all, recent advances in genome editing, with its nearly unlimited potentials, could bring about a new era in crop protection.

3. Constructing Bt-potato

Isolated in 1901 as the causative agent of silkworm disease, *Bacillus thuringiensis* (Bt) toxin became the first bioinsecticide commercially available since 1938, and remained for decades the most important microbial agent for insect control. Bt crystalline proteins (Cry toxins) appeared as an alternative to chemical insecticides, with molecular potency several hundred times greater than organophosphates and synthetic pyrethroids [13]. *Cry1Ab* was the first insecticidal gene introduced in tobacco [14], and since 1996, some Bt-plants such as maize, cotton, potato and rice, became commercialized. Now, more than 700 identified *cry* genes constitute a valuable "arsenal" with high and selective toxicity towards different insect taxa – and cloning, transfer and expression of these genes is a widely adopted strategy for incorporating resistance in commercially important crops. In the USA, for instance,

Bt-maize represents 82% of total maize production, while Bt-cotton accounted for 88% of all cotton grown in 2020 [15].

When insects feed on Bt-plants, ingested Cry protoxin is solubilized and proteolytically activated in the alkaline environment of the insect midgut to the active toxin. The activated toxin goes through complex sequential binding events with an array of receptors on the surface of midgut cells, beginning with binding to cadherin, that facilitates additional protease cleavage and assembly of oligomeric forms of the toxin. The oligomers have increased binding affinity to the secondary receptors, leading to membrane insertion and lytic pore formation [16]. Such midgut tissue disruption halts insect feeding and causes subsequent mortality.

Transgenic potato lines with introduced Cry3A delta-endotoxin from *B. thuringiensis* var. *tenebrionis*, that targets coleopteran pests, showed significantly increased resistance to the CPB. Constitutively expressed in potato, Cry3A toxin caused 100% mortality of neonate larvae within two days and 99% adult mortality within two weeks [17]. Bt-transgenic NewLeaf™ potato cultivars of Monsanto Corporation were commercialized in the USA starting in 1995, and potato became one of the first GM crops commonly used for human consumption. Next, CPB resistance was combined with virus resistance, and commercial potato cultivars NewLeafPlus™ and NewLeafY™ were launched in 1998. Additional virus resistance benefited seed producers, and commercial growers gained higher yields with reduced need for insecticides. Although commercially and agronomically successful, the NewLeaf™ varieties were withdrawn from the market in 2001, due to public concerns and competition with a new, highly effective insecticide imidacloprid [18].

Expression of several Bt-toxins of Cry1 or Cry9 classes, that target lepidopteran pests, conferred resistance to the potato tuber moth (PTM), a major potato pest in tropical and subtropical regions. Bt-lines with variable level of PTM resistance have been obtained after potato transformation with *cry1Aa* [19], *cry1Ab* [20], *cry1Ac* [21, 22], *cry1Ac9* [23, 24], *cry1Ia1* (previously known as *cry5*) [25, 26] or *cry9Aa2* [27]. Among them *cry1Ac* and *cry1Ia1* expressed in potato proved to be highly effective in PTM control, causing mortality of 80–97% of first-instar larvae fed on leaves and ~ 100% on tubers [21, 26], but none of these Bt-potato lines are available commercially. Additionally, *cry1Ac* or *cry1Ia1* expressing potato exhibit appreciable level of resistance to CPB - with up to 90% reduction of feeding, that correlates with increased first instar larvae mortality [22, 26].

Moreover *cry3A* [28], *cry1Ac9*, *cry9Aa2* [29] and *cry1Ab* [30] were independently expressed in potato under control of light-inducible promoters. Such spatial expression of *cry* genes enables high level of leaf protection against CPB or PTM, with minimal or no Cry toxin accumulation in the tubers, which represents a desirable feature for consumers.

3.1 Resistance to Bt: a CPB case

Insect resistance has become a significant problem after WWII, when intensive agriculture with reliance on chemicals and uniform cultivation practices led to about 17,000 cases of insecticide resistance among 612 insect species by 2020 [9]. Since Bt-crops also provide strong and uniform selection pressure on insect populations it is hard to believe that pest problems can be solved with Bt-approach alone. By 2017, two decades after their commercialization, reduced efficacy of Bt-plants caused by field-evolved resistance has been reported in 16 out of 33 major crop pest populations, compared to only 3 reported in 2005 [31].

In Cry3A-potato, toxin was expressed at a very high level relative to the CPB susceptibility: at least 50 times as necessary to kill first instar, and at least twofold as necessary to stop third and fourth instar development or to arrest adult egg laying [17]. Although effective in short term, this high-dose strategy represents an extremely high selection pressure for developing resistance in the insect populations, and without additional management practices, it has been predicted that CPB can develop resistance to Bt-potato within 6 generations [32]. CPB resistance potential has been demonstrated in the laboratory by repeated Cry3A toxin application, resulting in about 60-fold increase in resistance ratio after 12 generations [33], and about 300-fold increase after 35 generations [34].

Developing Bt-resistance is a complex and diverse process, and populations of the same insect species of different origins may exhibit different mechanisms of resistance to the same Cry toxin [35, 36]. Two major resistance mechanisms are: alteration of midgut proteases involved in processing of Cry proteins in the insect midgut; and modification of binding sites for Bt-toxins. Other resistance mechanisms may include retention of Bt-toxin by the midgut peritrophic membrane, aggregation of toxin proteins by the midgut esterase, elevated melanization activity of the hemolymph and midgut cells, increased rate of repair or replacement of affected epithelial cells, and increased antioxidant activity [37]. Bt-resistant CPB strains exhibit at least two levels of adaptive responses that render immunity to the Cry3A toxin: the first is lower toxin binding to the receptors, probably as a consequence of reduction of binding sites within the receptor or reduction in receptor numbers, while the second one are changes in digestive enzyme profiles and specific increase in aminopeptidase activity [38]. Although this alteration of CPB digestive profile is not connected with toxin processing or its inactivation, it can be involved in modulation and amplification of signals that activate specific innate immune responses such as melanization, coagulation and defense peptide synthesis [39] – mechanisms that have been confirmed in overcoming the exposure to Bt-toxin in other insect species [35, 40].

Moreover, plasticity of its life cycle, large pool of genetic variation in life history traits and capability to effectively cope with naturally occurring host plant toxins or almost every chemical insecticide, leave no doubt that CPB can develop resistance to Bt-potato, given sufficient time. This also brings concerns on whether CPB can be prevented from developing resistance to Bt-potato – since with only a single resistance gene expressed, the high dose/refuge strategy is the only resistance management option available [41]. Although such strategy can hinder accumulation of initially rare homozygous resistance genes in Bt-exposed insect populations by decreasing selection pressure, its effectiveness is questionable in the case of CPB. While the susceptible beetles are "arrested" on Bt-potato, in the resistant strains ingestion of Cry3A toxin significantly increased both CPB larval motility and adult flight activity, whereby more physiologically resistant individuals showed higher behavioral responsiveness. Such behavioral resistance can affect gene flow between susceptible and resistant beetles, increasing distribution of resistant homozygous CPB offsprings within and between Bt-potato fields [33, 42]. In addition, effectiveness of the refuge strategy will be compromised not only when expressed toxin genes do not kill all of the heterozygous progeny, but also if resistance is non-recessive. Evidence of both the laboratory-selected [43] and field-evolved [44] resistance to Cry toxins indicates that some populations of target pests evolve dominant resistance alleles, which can be hardly defeated with the refuge strategy.

3.2 Improving toxicity and preventing resistance

When exploring the functions of specific regions of Cry proteins, some of site-directed mutations resulted in increased binding affinity of Cry toxins to insect midgut receptors, conferring additional toxicity. For example, a triple Cry1Ab mutant protein showed up to 36-fold increase in toxicity [45], while multiple Cry3A mutations conferred 2-fold higher toxicity against CPB [46] compared to wild-type Cry toxins. Deletion of small regions of the toxin can result in increased toxicity or in toxins that could counter insect resistance to native Cry toxins. Deletion of 42 residues of the amino-terminal region resulted in an up to 6.6-fold increase in Cry2A toxicity against a lepidopteran pest [47], while Cry1AMod toxins (that due to the lack of α-helix can form oligomers in the absence of cadherin receptor) are effective against Cry1A-resistant target pests with mutations in the cadherin gene [48]. Additionally, added cadherin receptor fragment showed significant synergistic effect with Cry toxins, including 3.7-fold and 6.4-fold enhanced toxicity of Cry3Aa and Cry3Bb, respectively, to CPB [49].

The specificity of Cry proteins allows targeting a single pest or closely related insect species within the same order, but such specificity does not provide a wide range of protection. Improving or broadening the range of protection (as well as minimizing secondary pest infestations upon primary pest control) can be achieved through combining multiple resistance factors – a strategy that at the same time prevents or delays the evolution of insect resistance. The construction of hybrid Cry toxins can confer a wider target spectrum or higher toxicity than each of the parental toxins from which they are derived. Examples include hybrid Cry1Aa/Cry1Ac and Cry1Ab/Cry1C toxins, that exhibited 30- and 10-fold higher toxicity against target pests [50, 51]. Furthermore, a *cry1Ba/cry1Ia* hybrid gene (*SN19*) driven by a light- or wound-inducible promoter protects potato leaves from attacks of coleopteran (CPB) and lepidopteran (PTM, European corn borer and tomato leaf miner) pests, causing 100% mortality of first instar larvae when fed on *SN19*-transformed potato [52, 53]. However, among all these strategies gene stacking appeared as most effective, and there are numerous examples of introducing multiple resistance or other agronomic enhancement factors in commercially grown plants, including potato where pyramided *cry3A* and *SN-19* genes can provide 100% control of CPB [54]. The first stacked-traits crop that gained regulatory approval in 1995 was cotton expressing *cry1Ab* and *epsps* (conferring resistance to the herbicide glyphosate), leading to the several hundred stacked events for increased pest resistance in commercial crops, approved to date. The recently released ten-gene maize under the name SmartStax™ Pro x Enlist™, combines three herbicide tolerance genes, six Bt-genes (targeting both lepidopteran and coleopteran pests) and *dvsnf7* dsRNA [55]. However, benefits of Bt-gene pyramiding can be compromised due to inappropriate management strategies, as well as insects capable of cross-resistance.

For instance, concurrent use of one-toxin and pyramided two-toxin crops will enhance resistance to pyramided Bt-plants if the two-gene plants produce a similar toxin as the single-gene plants (for example, this is the case for marketed maize and cotton where the additional Bt-gene was "added" to an already existing Bt-line). Target pests can evolve a single gene resistance that overcomes both Bt genes used in the pyramiding, even if expressed Bt-toxins have different binding sites. A clear example are *Helicoverpa zea* populations that exhibit increased survival on cotton with stacked *cry1Ac* and *cry2Ab* genes, as result of extensive exposure to Cry1Ac before two-toxin cotton was introduced [56]. Mechanisms that could cause cross-resistance in the target insects may include alteration in digestive proteases (if the same proteases activate or degrade both Bt-toxins) or changes affecting pore formation or pore function, a general step in the action mechanism of many Cry proteins [37]. Thus, the promising strategy for stacking varieties should be combining genes with different mechanisms of actions, such as a *cry* gene with host plant resistance or other heterologous factors (including Vip toxins, protease inhibitors or dsRNA, combined in some approved events) to minimize the possibility that random mutations in a single insect gene could confer resistance to both or more introduced traits.

3.3 Bt-related concerns

In 1999, laboratory studies showed that Bt-maize pollen had deleterious effects on Monarch butterfly larvae [57], raising questions and concerns about Bt-crop impacts on non-target organisms. Additionally, since both target and non-target insect pests ingest toxin when feeding on Bt-plants, Bt-toxin may also affect beneficial predatory arthropods through consumption of target pests or by facultative feeding on transformed plants.

Riddick and Barbosa [58] showed no adverse effect on survival, fitness or predation potential of *Coleomegilla maculata*, an entomophagous and pollenophagous beetle, when fed on Cry3A-intoxicated CPB. Similarly, another beneficial carabid beetle, *Nebria brevicollis*, was not affected with Cry3A when fed with non-target potato pest *Lacanobia oleracea* larvae [59], indicating that, due to its high specificity, Cry3A toxin presents a very low risk to coleopterans other than the targeted CPB. In addition, EPA (Environmental Protection Agency) studies on impacts of Cry3A-potato found no adverse effects on non-target wildlife exposed to the crop, indicating that beneficial arthropods were generally more abundant in Bt-potato plots compared to those treated with synthetic insecticides. Natural enemies are sufficient for aphid control on Bt-potato, while high numbers of this secondary potato pest populations are present in plots where beneficial arthropods were eliminated by insecticide treatment and no chemical aphid control was applied [60]. For instance, ladybird beetles, that are abundant and valued predatory species, preferably feeding on aphids and occasionally pollen when prey is scarce, remain unaffected on Cry3A expressing potato [59]. It was shown that Bt-potato fields were inhabited with diverse populations of these aphidophagous coccinellids, whose numbers significantly decreased with application of chemical insecticides [61]. Also, Bt-potato is not a threat to other endangered coleopteran species, since their habitat does not overlap with potato fields and their larvae do not feed on potato [60]. In addition, 25 studies that assessed potential effects of Bt-toxins introduced in commercialized GM crops (lepidopteran-active Cry1, Cry2, or Cry9 and coleopteran-active Cry3 class) found no negative effect on survival of either honey bee larvae or adults [62]. However, it may be also expected that some CPB predators will be less abundant in Bt-potato fields due to low pest densities (rather than Cry3A toxicity), such as in the case of carabid *Lebia grandis* [63], or that complexity of interaction on tritrophic (plant-pest-natural enemy) level can be altered in an unexpected way. For instance, survival, weight gain and fecundity of the wasp *Aphidius nigripes*, parasitoid of the potato aphid (*Macrosiphum euphorbiae*), was negatively affected on Bt-potato, although Cry3A did not directly affect the aphid, nor should be toxic to parasitic wasps [64].

Furthermore, studies on commercialized SmartStax maize with six Bt-genes (*cry34Ab1*, *cry35Ab1*, *cry3Bb1*, *cry1F*, *cry1A.105* and *cry2Ab2*) provided evidence that the different Cry proteins do not interact in a way that poses a risk to the investigated non-target species under controlled laboratory conditions [65, 66]. However, data available in the literature regarding the impact of Bt-crops on non-target arthropods are mostly incomplete and sometimes controversial. Most studies have focused on certain but not all aspects of non-target or beneficial insect fitness and most of the field trials were conducted on a small scale, over a relatively short period of time.

Although free Bt-toxin released in root exudates and from decaying plant residues is rapidly degraded by soil microbes, it can be stabilized by binding on clays or humic substances and stay unchanged for two weeks to 6 months [67], depending on soil composition and pH, or crop species [68]. However, studies on Bt-crops have generally revealed no or minor transient effect on earthworms, nematodes, protozoans, bacteria, and fungi in soil [68].

Due to the acidic environment of the mammalian digestive tract and the absence of specific receptors, it is generally accepted that Bt-toxins do not bear substantial risk for human health. Additionally, about 60 years of history of using Bt-products as biopesticides showed that risks of toxicity or allergenic reactions to the Cry proteins are minimal. Cry3A toxin does not exhibit acute oral toxicity to mammalians in doses 10,000 times higher than its amount in potato tubers, and is rapidly digested *in vitro* [60, 69]. In simulated digestion models the protein is degraded within 30 s to polypeptides less than 2 kDa, suggesting that Cry3A will be even more efficiently degraded in robust gastrointestinal systems of humans and other mammals. Efficient degradation and lack of structural similarity to known allergenic proteins significantly minimize the potential for Cry3A to induce allergic reactions [69]. Likewise, similar findings on safety exist for other Cry toxins introduced in maize, cotton and soy, that are authorized for cultivation in one or more countries [70]. The only exception is Cry9c toxin, which due to its resistance to breakdown by digestive enzymes may be found in the bloodstream after oral feeding in the rat model, with potency to induce immunological responses [71]. In 1998, *cry9c*-expressing maize named 'Starlink' has been approved only for animal feed and industrial use, but recalled two years later in the USA, EU, Japan and South Korea, after detection of Cry protein residues through human food supply. This controversy indicated the need for a broader and properly managed assessment in monitoring and enforcement concerning potential health risks of toxicity, allergenicity and genetic hazards associated with Bt-crops, to ensure their greater acceptance. Although majority of studies indicate that Bt-crops would be as safe as parental lines – with few exceptions [72, 73] that were rather critiqued than accepted in scientific community – studies on the long-term health effects of Bt-plants will still be necessary [74]. Also, the potential of cumulative, combined or unexpected effects in the "next generations" Bt-crops with stacked *cry* genes, or combined with other resistance factors, clearly calls for revisions of "outdated" risk assessments made based on single Bt-gene expression.

4. Targeting digestive enzymes

As a reflection of more than one hundred million years of coevolutionary "arms race", plants developed numerous mechanisms to resist the attacks of pathogens and herbivores. Here, being part of the plant "chemical warfare" arsenal, secondary metabolites take an important place, with more than 200,000 known compounds with defensive activity. Among that broad repertoire, protein antimetabolites such as lectins, α-amylase inhibitors and especially plant protease inhibitors (PIs) are the most used for engineering crop resistance against various pests.

The most important role of PIs in plants is protection from both biotic and abiotic stresses. They may also have other functions: from tissue-specific regulation of endogenous proteases – especially in storage organs such as seeds and tubers [75], to the regulation of programmed cell death [76]. About 500 plant PIs were described, and according to the protease type they inhibit, PIs are classified as cysteine, serine, aspartyl and metallo protease inhibitors [77]. Generally, the inhibition is based on PIs binding to or near the enzyme active site, forming a stable complex with a low dissociation constant. This complex is often additionally "locked" by disulphide bonds, so that upon eventual hydrolysis the inhibitor remains associated to the enzyme, effectively blocking access of the substrate [78]. The mechanism of PIs antimetabolic effect on insects has not been fully elucidated and, due to its high specificity, it is assumed that different types of PIs also have different modes of action. The simplest model implies a direct antidigestive effect due to inhibition of proteolysis [79].

The second, more accepted model, is based on compensation for the loss of proteolytic activity – proteinase hyperproduction – which by redirecting amino acid utilization reduces their availability for insect growth and development [80] which, in addition to reduced performance, often increases insect mortality. PIs can also disrupt processes such as molting, neuropeptide synthesis, water balance, and enzyme regulation [81–83] or directly interfere with insect reproductive processes [84].

The early evidence on the protective role of PIs came in mid-20th century, when it was observed that soybean products negatively affect development of red flour beetle larvae [85]. In a pioneering research, Green and Ryan [86] reported on a rapid, both local and systemic, accumulation of PIs in potato and tomato leaves upon CPB attack, demonstrating the importance of PIs in plant defense against insects. Not long after, the first PI-transformed plant, tobacco expressing cowpea trypsin inhibitor, CpTI, conferred increased resistance to several lepidopteran, coleopteran and orthopteran insect pests [87]. This initial success triggered a generation of numerous transgenic plants expressing different PIs, more or less efficient in control of target pests. However, despite this promising development, none of PI-transgenic plants have been commercialized to date. One of the reasons is the conclusive "acute mortality" efficacy of Bt-plants, similar to the chemical insecticides. By contrast, PIs often cause decrease in insect fitness on a relative level, such as a reduction in growth and reproduction or extended development, that in a time scale can significantly reduce the size of pest population (for example, prolonged larval development brings longer exposition to predators, while the reduction in body mass decreases investment in reproduction). Secondly, a more important reason are adaptive capacities of insects that can compromise this approach, clearly demonstrated in some cases. These evolutionary, diet-induced strategies include overproduction of sensitive digestive enzymes that outnumber inhibitors, switching to digestive protease complements insensitive to PI or PI degradation with non-target proteases [88].

After evidence of deleterious effects of E-64, a broad spectrum thiole cysteine PI isolated from *Aspergillus japonicum,* on larval growth, survival, and adult fecundity of CPB [89], CPB cysteine proteinases (that account for most of CPB digestive proteolysis) have become target for heterologous cystatins expressed in potato plants. Two rice cystatins, oryzacystatins I and II (OCI and OCII), although exhibiting inhibition of CPB larvae cathepsin H-like proteases *in vitro* [90] proved ineffective in CPB control. With no increase in mortality, CPB larvae overcame initial digestive inhibition by hypertrophic behavior and restored cysteine proteinase activity by introducing isoforms insensitive to OCI [91] or OCII [92]. Contrary to expectations, some aspects of CPB larvae performance were actually enhanced by chronic ingestion of each of the two rice cystatins: faster growth and leaf consumption, shorter development time and even increase in body mass before pupation in case of OCI [91]. Slight reduction in insect growth rate was also observed with recombinant CDI (cathepsin D inhibitor from tomato), as a result of overproduction of inhibitor-sensitive proteases. However, after this initial response CPB larvae switched their digestion to the CDI insensitive protease complement, resuming normal growth and development despite ingestion of the inhibitor expressed in potato plants [93].

These results clearly demonstrate that, due to its exceptional adaptability to the different host plant protective compounds [88], CPB can hardly be controlled by a single, narrow spectrum PI. Thus, to achieve more efficient control and prevent compensatory insect responses, broadening the spectrum of inhibition by protein fusion, transgene stacking or using multidomain PIs appeared as a possible solution. However, only a slight reduction in CPB larvae performance was achieved in potato expressing stacked rice cystatins, OCI and OCII [94, 95] or with multidomain serine PI from locust (LIP), active against both trypsin and chymotrypsin [96].

In contrast to this, equistatin, a PI from the sea anemone, with one domain that inhibits cysteine and a second domain active against aspartic proteases, had detrimental effect on CPB larvae growth and significantly increased their mortality after ingestion of equistatin-coated potato leaves [97]. Unfortunately, with expression of this potent PI in potato very low resistance level against CPB was achieved: the amount of active inhibitor in leaves was considerably reduced due to its degradation by native potato proteinases [98]. The promising results came with a hybrid CDI-CCII inhibitor (fusion of CDI with maize cystatin II), also active against both aspartate and cysteine proteinases. When painted on potato leaves, CDI-CCII initially reduced CPB larvae growth and food consumption by about 50% [99], but its real effects still remain to be proved in long-term feeding assays. Finally, fungal cysteine PIs, macrocypin and clitocypin, emerged as more favorable. Exhibiting strong inhibition of CPB cysteine proteinases, these PIs, introduced in potato, reduced growth and increased development time of CPB larvae [100, 101]. Moreover, the most promising trait of macrocypin and clitocypin is the absence of CPB digestive compensatory responses [100, 101] observed for PIs derived from other sources. However, relatively low expression was achieved in transgenic potato and, since they act in dose dependent manner, it is necessary to improve macrocypin and clitocypin expression levels for more pronounced negative effects on CPB larvae.

Additionally, potato expressing serine PI (CpTI or Soybean Kunitz, C-II and PI-IV) exhibited enhanced resistance to the lepidopteran larvae with about 50% reduction in total insect biomass [82, 102].

Several approaches based on structure–function models have been used to improve the inhibitory potency of protease inhibitors against specific proteases, including site directed mutagenesis of specific amino acids, molecular phage display procedures involving random mutagenesis in specific regions of the inhibitor sequence, or activity-based functional proteomics approach. By single mutations at the positively selected amino acid sites of the tomato multicystatin SlCYS8, variants with improved inhibitory potency toward the CPB digestive proteases were generated [103], and functional proteomics approach was used for identifying variants that efficiently capture CPB digestive protease targets [104]. P2V10, the most potent variant of SlCYS8 PI, expressed in potato, significantly reduced growth of CPB larvae in a 72 h feeding assay [104]. Similarly, after 4 days of feeding on potato expressing a modified variant of cystatin from barley (HvCPI-1 C68 fi G), that targets the cathepsin B-like fraction of cysteine digestive proteolysis, CPB larvae had about 23% lower weight, probably due the metabolic cost associated with the hyperproduction of inhibited digestive proteases [105]. However, knowing the remarkable CPB larvae adaptability to adjusting their digestive profile to functionally distinct plant PIs, studies assessing the long-term detrimental effects of these engineered cystatins are needed.

Although the usefulness of recombinant PIs expressed alone still remains to be proved or improved, they can enhance Cry toxicity. Several serine protease inhibitors can increase the insecticidal activity of Cry toxins 2–20 fold [106] and delay the resistance evolution of the targeted pest [107]. Although it is not known how PIs enhance Bt-toxin activity, it is supposed that they may inhibit the inactivation of Bt-toxins by specific gut proteases, or prevent the degradation of membrane receptors, increasing binding ability of Cry toxins [108]. In such way, hybrid SN19 (*cry1Ba/cry1Ia*) combined with OCII in potato caused 100% mortality of all CPB larval stages within 6 days, and adults within 2 weeks [54]. However, as of today there are only three approved events with PI (all stacked with *cry1Ac*): cotton co-expressing *CpTI*, maize with *pinII* (from potato) and poplar with *API* (from *Sagittaria sagittifolia*) [55].

Due to the existence of targets in most organisms in nature, beside the toxic effect on the pest, recombinant PIs can directly affect the digestive proteolysis in pollinators, symbionts and/or indirectly, through prey feeding on transgenic plants, they can endanger the ecological function of predators. However, although artificial diet studies indicate that predatory insects may be susceptible to the PI, prey-mediated effects are usually not observed when cystatins or CpTI are expressed in transgenic potato. When *Podisus maculiventris* was fed with tomato moth (*L. oleracea*) caterpillars reared on CpTI-potato plants, no negative effects on the predator were observed [109]. Predation on neither CPB nor Egyptian cotton leafworm (*Spodoptera littoralis*) larvae reared on potato plants expressing barley cystatin had negative effects on survival and growth of the predatory bug *P. maculiventris* [105]. Also, no detrimental effects were observed on larvae and adults of the ladybird *Harmonia axyridis* upon consuming larvae of diamondback moth (*Plutella xylostella*) reared on OCI-expressing plants [110], or in *Diaeretiella rapae*, a parasitoid of potato-peach aphid (*Myzus persicae*) [111]. Stinkbug *Perillus bioculatus* feeding on CPB reared on OCI-potato compensated for the effects of this cystatin by introduction of serine-type proteases [112], while improved performance of secondary pest *Macrosiphum euphorbiae* on the same host plant also improved performance of the parasitoid wasp *Aphidius nigripes* [113].

On the other hand, although the effects of native plant PIs, such as CpTI or OCI, on non-target organisms have been well documented, there is little evidence of effects of new-generation inhibitors with stronger effects on pest proteinases, hybrid inhibitors or combined effects of several different insecticidal proteins. The challenge, of course, is to find or devise those variants of PIs that show increased activity against the target pest proteinases and decreased activity against proteinases of the host plant or of beneficial insects. Also, cystatins that occur naturally in seeds of rice and maize, present in potato tubers or in egg-white, are not novel in the human diet, and expressed in transgenic plants should not cause public concerns [114] – but the expression of strong broad-spectrum aspartate and serine PIs may raise many questions in the future.

5. Lectins

Widely distributed in nature, lectins are a heterogeneous group of sugar-binding proteins with numerous biological functions. In plants they are involved in the transport and utilization of carbohydrates, cell organization, division and signaling, embryomorphogenesis, phagocytosis or as mediators of plant-microorganism symbiosis [115]. However, their most distinctive role is in plant defense mechanisms against pathogens and pests. Binding to a variety of glycoproteins, plant lectins can inhibit absorption of nutrients by disruption of insect gut epithelium structure or, by interacting with targets in insect hemolymph, fat tissue and ovaries, interfere with a number of physiological processes, such as growth, development and detoxification [116]. Although they can exhibit protective roles against insect pests from different orders, lectins are particularly useful for controlling Hemiptera, that are generally less sensitive to Bt or PIs.

Snowdrop mannose-binding lectin (*Galanthus nivalis* agglutinin, GNA) is the first lectin known for insecticidal activity. Expressed in potato, GNA can decrease growth and fecundity of potato-peach aphid (*M. persicae*) or glasshouse-potato aphid (*Aulacorthum solani*), reducing the rate of their population growth up to four times [117, 118]. Effects of GNA on *M. persicae* vary with its expression level in potato plants: at low level GNA reduces colonization of transformed potato, without significant impact on insect performance [119], while highly expressed, GNA can reduce aphid survival and performance [120]. Besides, GNA can be effective in control of lepidopteran pests – tomato moth (*L. oleracea)* larvae exhibited about 50% reduction in biomass, prolonged development and 40% increased mortality rate when fed on transformed potato [121]. Concanavalin A (ConA), a glucose/mannose-binding lectin from jackbean (*Canavalia ensiformis*), can also be effective in control of both hemipteran and lepidopteran potato pests. Despite its relatively low expression level in potato plants, ConA decreased the fecundity of *M. persicae* (up to 45%) and reduced *L. oleracea* larval weight (about 50%) and retarded their development [122].

However, lectins can negatively impact beneficial non-target organisms, and for instance, preys that were fed on GNA potato were less favored or resulted in smaller, shorter-lived predators or parasitoids [123, 124]. Although they are present in most plants – especially abundant in cereal and legume seeds or potato tubers – lectins are generally considered toxic to animals and humans. So even though GNA did not show considerable toxicity in rat feeding studies [125], there is no doubt that food expressing such proteins requires long-term studies to evaluate its potentially harmful effects.

6. Silencing vital genes

After the Nobel prized discovery of RNA interference (RNAi) as a basic mechanism of post-transcriptional gene silencing by double-stranded RNA (dsRNA) [126] RNAi has become a powerful experimental tool for determining gene functions, had an immense impact on biomedical research and found its application in the management of insect pests. Evolutionarily conserved in all eukaryotes, the mechanism of RNAi is involved in different processes including internal gene regulation (micro RNA or miRNA pathway), genome protection against transposons (piwi-interacting RNA or piRNA pathway) and defence against viral infections (small interfering RNA or siRNA pathway) [127]. Althought the siRNA pathway in insects mostly represents the first line of defense against viral RNA, it can be exploited for introduction of specific dsRNA that, through mechanism of RNAi, can initiate degradation of complementary endogenous insect mRNA. Thus, selection of any target gene and delivery of its sequence-specific dsRNA to cells can lead to functional knockout of that gene – affecting insect growth and development or increasing their mortality. A first proof-of-concept came in 2007, when transgenic maize expressing V-ATPase-specific dsRNA showed significant reduction in feeding damage caused by western corn rootworm (WCR) [128]. Maize with dsRNA transcript containing a 240 bp fragment of the WCR *Snf7* gene (encoding a membrane-remodeling protein) stacked with several *cry* genes (*cry3Bb1* and *cry34/35Ab*) was first such crop commerically approved in 2017, and five more events expressing *Snf7* dsRNA and different Cry proteins stacked in maize were approved to date [55].

However, various studies showed that different insect orders differently respond to orally delivered dsRNA – coleopterans are mostly sensitive, while RNAi efficiency is low for most lepidopterans. Multiple mechanisms contribute to this variability, including instability of dsRNA upon ingestion, insufficient dsRNA internalization, endosomal entrapment, deficient function of the RNAi machinery and reduced systemic spreading. Once consumed, the dsRNA first has to avoid degradation by dsRNases (dsRNA-specific ribonucleases) on their way through insect digestive tract. Level of dsRNA degradation by saliva or midgut nucleases varies among different insect orders and, for instance, midgut stability of dsRNA is greater in the CPB (Coleoptera) than in *Schistocerca gregaria* (Orthoptera) or budworm *Heliothis virescens* (Lepidoptera) [129, 130]. The next barrier is the internalization of the dsRNA in the cell. Two mechanisms of cellular uptake of dsRNA have been identified in insects: SID-like (Systemic RNA Interference Deficient) transmembrane channels, and clathrin-dependent endocytosis. The latter mechanism seems to play the primary role in the uptake of dsRNA in many insect species, whereas SID-like genes have been identified in Hemiptera, Lepidoptera and Coleoptera but their additional role in mediating dsRNA uptake has only been confirmed for WCR and CPB [131, 132]. In clathrin-dependent endocytosis, after binding to the receptors and forming endosomes, the dsRNA is released into the cytoplasm before reaching the lysosomes. In CPB, such endosomal escapes occur easier than in most lepidopterans, where the dsRNA can enter the cells but remains trapped in the endosomes [130].

Once taken up in the cytoplasm, dsRNA is recognized by the core RNAi machinery and processed into 21–23 bp siRNA by the enzyme Dicer 2 (DCR-2). The siRNA are loaded onto Argonaute 2 (Ago-2) protein and incorporated into the RNA-induced Silencing Complex (RISC). Upon degradation of the passenger strand of siRNA, RNase active domain of Ago-2 cleaves the mRNA recognized by the siRNA guide strand, inducing gene silencing. One of the reasons for efficient RNAi in coleopterans is the duplication of core RNAi pathway genes, including DCR-2 and Ago-2 [133]. Additionally, in CPB, components of miRNA and piRNA pathways are also critical for effectiveness of gene silencing by the siRNA pathway, but their involvement in dsRNA-mediated RNAi needs to be further investigated in Coleoptera and other insects [134]. A particularly interesting aspect of the RNAi response in insects is its potential systemic character, whereby the silencing signal can spread from the midgut to other tissues, causing systemic RNAi. The exact nature of this signaling pathway still remains elusive, and efficient silencing of genes in midgut tissue was predominant, especially in more derived dipteran and lepidopteran species that appear to be more refractory to systemic RNAi [135].

Although there is a vast number of essential genes in insect genomes, the choice of the target gene can significantly affect the efficiency of RNAi – but the factors making one essential gene a better target than another one are not currently understood. Variation in transcriptional activity, mechanisms of expression regulation, mRNA stability and its accumulation level may play an important role in defining a particular gene susceptibility to dsRNA, and screening of a larger number of potential target genes for RNAi efficiency remains the only reliable method of choice.

6.1 Targeting CPB

The availability of the CPB transcriptome [136] allows specific targeting of CPB genes critical for normal physiological processes and numerous studies demonstrated successful knockdown of target genes in dsRNA-fed CPB. Silencing the expression of genes that are crucial for maintaining physiological functions, such as actin and V-ATPase genes, or genes coding components involved in protein transportation (*Sec23* and *COPβ*) can directly impair growth and induce mortality [137]. Knockdown of genes crucial for synthesis of 20-hydroxyecdysone and juvenile hormone, disrupts larval molting and pupal metamorphosis, decreasing the emergence of adults [138–141], while suppression of proline degradation (necessary for ATP production) reduces flight ability and increases mortality of CPB adults [142, 143]. In addition, RNAi can enhance the effectiveness of other control measures or resistance factors introduced in potato. For instance, suppression of CncC, a transcription factor regulating multiple cytochrome P450 genes, increased CPB susceptibility to insecticide imidacloprid [144], while silencing of a Cry3Aa-binding protein, prohibitin, enhanced the toxicity of Cry3Aa [145].

Reduction in CPB juvenile hormone (JH) titer, that regulates metamorphosis and reproduction in insects, was achieved by knockdown of JHAMT (JH acid methyltransferase), the last rate-limiting enzyme in JH biosynthesis. Feeding on transgenic potato plants expressing ds*JHAMT* had negative impact on CPB larvae growth and development, increased larval mortality (about 25%) and reduced pupation rate by 50%. Moreover, emerged CPB adults had lower weight and females lay fewer or no eggs, which was confirmed in field trials [146]. Additionally, feeding CPB larvae on transgenic potato expressing *EcR* (molting-associated Ecdysone receptor) gene dsRNA resulted in 15–80% mortality, reduction in body weight and disturbed metamorphosis [147]. However, the success of the RNAi gene silencing is limited by the level of dsRNA expression and dsRNA stability in transgenic plants. Since insects lack RNA-dependent RNA polymerase, the RNAi signal cannot be amplified in their cells, and efficiency of target gene knockout mostly depends on the amount of ingested dsRNA. Also, insects are more responsive to longer dsRNA – but dsRNAs produced in plant cytoplasm are usually processed into siRNAs by native plant RNAi machinery. For example, dsRNAs longer than 60 bp can trigger *DvSnf7* gene silencing in WCR, while 21 bp siRNAs were not efficient [148].

On the other hand, transformation of chloroplast DNA has potential for overcoming the constraints of nuclear transformation in dsRNA-mediated pest control. First advantage of transplastomic plants are markedly high gene expression levels, that due to tissue specificity, occur predominantly where functional plastids are present. An example is expression of Cry2Aa2 protoxin in tobacco chloroplasts in 20- to 30-fold higher levels than current commercial nuclear transgenic plants, which is lethal for both susceptible and Bt-resistant target insects [149]. Secondly, a great advantage of plastid transformation is the stability of dsRNA in plastids, as chloroplasts do not have the RNAi machinery. Among about 130 genes encoded by the chloroplast genome, none is Dicer-like or Argonaute-like, and there is no evidence of import of these nuclear-encoded proteins in chloroplasts [150]. Three recent studies demonstrated that when expressed from chloroplast genome, hp/dsRNA can confer a high level of protection against either lepidopteran (*Helicoverpa armigera*) [151, 152] or coleopteran (CPB) pests [153], compared to their nuclear transgenic counterparts [152, 153]. Transplastomic potato expressing *β-actin* (*ACT*) or *SHRUB* (analog to *Snf7*) dsRNA, or both, produced large amounts of unprocessed dsRNA in leaves (but not in tubers) with detrimental effect on CPB growth and development. All first-instar larvae fed on transplastomic *ACT* dsRNA-expressing plants died within 5 days, while 40% of larvae survived on *SHRUB* dsRNA-expressing leaves. Nuclear-transformed plants produced much less dsRNA but more siRNAs, exhibiting a weaker suppression of target mRNA and almost no mortality was observed in CPB fed with leaves from nuclear transgenic potato [153].

Furthermore, chloroplast genome transformation also offers other advantages over nuclear transformation, including introduction of multiple genes in a single transformation event and lack of gene silencing, position or pleiotropic effects. Additionally, maternal inheritance excludes plastid genes and therefore reduces dispersion of the transgene by pollen transmission, increasing the biosafety of transgenic plants. However, plastid transformation is still much more challenging than nuclear transformation and limited by the methods of DNA delivery, homologous recombination efficiency and the methods for efficient selection and regeneration of transformants [154].

6.2 RNAi-related concerns

Numerous studies have shown that under long-term pressure of control strategies such as chemical insecticides or Bt-toxin, insects can rapidly evolve resistance, and there is no reason to believe that it would be differently with RNAi. Theoretically, there are three possible sources of resistance: mutations in the sequence of the target gene, mutations inactivating the RNAi machinery and mutations that affect the stability and/or uptake of ingested dsRNAs in the insect digestive tract. First two mechanisms are unlikely to become source of resistance. For instance, in CPB the mismatch rate of *β-Actin* dsRNA and a target mRNA lower than 3% does not reduce the RNAi efficiency [155], while drastic sequence changes in target (essential) genes or those that inactivate the highly conserved genes of the RNAi machinery can easily jeopardize insect fitness and survival. However, the third scenario is quite possible and a first insect population, WCR with developed resistance to RNAi was reported in a transgenic maize field. Moreover, *DvSnf7*-dsRNA resistance in WCR is not sequence-specific, and cross-resistance to other dsRNAs is connected with dsRNA uptake rather than degradation [156]. Similarly, cross resistance to dsRNAs was achieved in a laboratory population of CPB, where foliar application of V-ATPaseA dsRNA resulted in >11,100-fold resistance after nine generations of selection [157]. Again, reduced uptake of dsRNA in midgut cells was responsible for the evolution of RNAi resistance.

With perfect sequence homology between dsRNA and mRNA only target gene suppression is expected, but it appears that siRNAs operate within cells with a certain level of "freedom" among targets. Mutation analyses showed that RNAi can be efficiently triggered with >80% sequence identity between siRNA and mRNA [158] but this mismatching tolerance can vary with insect species, target gene and dsRNA concentration [159, 160]. Moreover, dsRNA can provoke responses independently of its sequence, affecting insect antiviral immunity, gene expression and performance [158, 160]. Although not fully understood, these effects are particularly pronounced for dsRNA administered at high concentrations, supposing that high levels of siRNA may saturate the core RNAi machinery [161]. Given the small sizes of siRNAs, off-target effects that can appear in RNAi are probably quite common [162] and not considered as a concern in target organisms, but off-target binding in non-target organisms can represent a hazard if they are sufficiently exposed to the RNAi. To date, question how dsRNAs affect target and off-target genes in non-target organisms has received little attention, and existing studies indicate that the insecticidal effects of *V-ATPase*, *DvSnf7* or *NUC* (nuclease) dsRNAs are narrow, presuming adverse effects on non-target arthropods to be very low [163–165]. Additionally, in crops expressing dsRNA non-target insects can be only affected by feeding on plant. In the case of transplastomic potato expressing *β-actin* dsRNA [153], non-target insects had to consume potato leaves to be affected by RNAi – but by doing so they were considered pests. At the same time, pollinators and pollen-eating insects are exposed to minimal amounts of dsRNA, since chloroplasts are excluded from pollen due to maternal inheritance. Thus, careful design of the dsRNA and bioinformatic analyses can minimize non-target or off-target effects, but they cannot be completely excluded, since siRNAs do not need to share perfect sequence identity with target mRNAs to inhibit their translation in both predictive and unpredictive ways.

dsRNAs exhibit low persistence in environment and microbial degradation of nucleic acids has been shown to be a key driver for such lack of stability. Biological activity of *DvSnf7* dsRNA expressed in maize was undetectable within approximately 2 days after application to soil [166], and within 7 days in the aquatic environment [167]. In addition, biodegradation kinetics of dsRNA were independent of the dsRNA concentration, sequence length and secondary structure (hairpin or linear) [166].

Vertebrates are exposed to dietary intake of a number of various dsRNAs from animal, plant or microbial origin. Some are completely complementary to human or animal genes [168] and capable of initiating the RNAi pathway if they reach a target cell. However, there are numerous biological barriers, including nucleases along the digestive tract, and in bloodstream, series of cellular membrane barriers and endosomes significantly reduce dsRNAs to the levels insufficient for mediating RNAi. In 28-day repeat oral toxicity study in mice with *DvSnf7* dsRNA or with siRNAs and a long *V-ATPase* dsRNA (effective in WCR control and with 100% sequence complementarity to mouse vacuolar *ATPase*) no adverse effect was observed, even with doses billions time higher than anticipated human dietary exposure [169]. Thus, according to available data, it is likely that consumption of plants expressing dsRNA will not present a safety issue. However, whether ingestion of dsRNA can affect the immune systems of humans and animals, both directly or through impacting the gut microbiota, is currently unclear [170].

7. Transgene flow

Gene flow is the transfer of genetic material from one organism to another, including inheritance (vertical gene transfer) or transfer between unrelated species (horizontal gene transfer). Althought horizontal gene transfer can contribute in "shaping" genomes of both prokaryotes and eukaryotes, there are almost no evolutionary examples of gene transfer from eukaryotes to bacteria [171]. Transfer of plant DNA to bacteria has been demonstrated at a very low frequency under artificial conditions, and the only genes from GM plants that are likely to be successfully transferred are other bacterial genes, commonly used for selection in transformation [172]. More than 90% of transgenic plants that have been generated in different laboratories carry one of the three genes used for selection (resistance to antibiotics kanamycin or hygromycin, or herbicide phosphinothricin) [173], all of bacterial origin. Antibiotics are the most effective selection system for potato transformation, increasing its efficiency from 0.2%–4.5% under non-selective conditions to over 80% [174, 175]. However, they generally have no use after the selection phase of transformation, and can be completely removed or excised by different approaches, including segregation from the gene of interest after co-transformation, and different site-specific or homologous recombination systems [176]. In this way, using self-crossing segregation or inducible self-excision by the Cre-loxP system, selectable marker-free transgenic potato lines with increased resistance to pest or pathogens were created [177, 178], alleviating possibility of horizontal gene transfer.

On the other hand, vertical gene transfer, especially mediated by pollen, raises more concern. Transgene escapes have been documented for cotton, maize, soybean, oilseed rape, rice and wheat, indicating global dimensions of this problem [179]. In the case of Bt-plants, crop-to-crop gene flow can cause seed contamination, decrease efficiency of refuge strategies, or interfere with conventional or organic crop production. For instance, in Mexico where GM maize was not allowed for commercial cultivation, transgene escapes (*Bt-cry9C*, *Bt-cry-1Ab/1Ac* and *CP4 EPSPS* herbicide resistance transgene) have been found in traditional maize varieties [180]. An additional concern is the risk that pharmaceutical proteins, industrial enzymes, and vaccines produced by transgenic crops considered unsuitable for human consumption, can enter the food supply by outcrossing [181]. Transgenes can also move from GM crops to their wild relatives and alter their fitness, so that wild or weedy populations become more competitive and/or invasive, especially with introgression of insect-resistance or herbicide-tolerance genes. Although this invasiveness is more hypothesized than proven, GM crops or their volunteers often grow in vicinity of their wild variants, and hybridization with these plants has frequently occurred. Examples include cotton and oilseed rape, where traits of insect and herbicide resistance, even stacked in combinations that do not exist in commercially available crops, were found in their wild relatives [179].

Cross-pollination between GM and non-GM potato should be less worrying, since vegetative propagation by tubers (rather than true seeds) is the dominant reproduction strategy of potato, and tubers are not affected by the plant fertilization with "foreign" pollen. Outcrossing has been observed to occur only between adjacent potato fields, with rapid decreasing rate with distance, and no cross-pollination detection when the pollen-receiving plants were separated by more than 20 meters from the GM plants [182]. Additionally, majority of modern cultivars that evolved from complex hybridizations among several diploid and polyploid potato species, suffer from different types of male sterility and produce little or no viable pollen. Also, *S. tuberosum* is not able to hybridize with any of the non-tuber bearing *Solanum* species outside of the section *Petota* [183], and in most parts of the world, crosses with wild or cultivated relatives are highly unlikely, due to geographical isolation from potential crossing partners with a suitable endosperm balance number [184]. In contrast, from Southwestern USA to Southern Chile, in areas of potato diversity, natural hybridization occurs between wild and cultivated *Solanum* species [185], bearing risk of the gene flow from transgenic potato to neighboring plants of related species. Nevertheless, with measures such are increased isolation distance and development of transgenic lines from male sterile potato varieties [186], undesirable introgenesis in these wild species can be prevented or minimized. Besides, other biological means of confinement, including chloroplast transformation, apomixis, cleistogamy and diverse genetic barriers [179], can further minimize risks of transgene escapes.

8. Unintended traits

Crop improvement by genetic engineering requires obtaining transgenic lines with adequate expression of the heterologous gene and simultaneous preservation of all elite parental genetic attributes. One of the main limitations in achieving these requirements is the emergence of atypical plants – most often as a result of insertional mutagenesis or somaclonal variations that may occur in the tissue culture itself and/or during transformation.

In many plant species, including potato [187], the frequency of heterologous DNA insertions within coding or regulatory gene sequences exceeds 50% upon genetic transformation. Additionally, insertion-site mutations can alter the expression patterns of neighboring genes, especially if the heterologous gene is under the control of a strong promoter [188]. Another type of mutation, related to the transformation process itself, can occur in any part of the plant genome (genome-wide mutations) and is reflected in DNA polymorphism between transgenic and non-transgenic plants [189]. These latter changes are of epigenetic nature: the transformation process can activate transposon elements (TEs – whose activity is normally prevented by DNA hypermethylation), which then increase mutation rates and genomic rearrangements [190]. It is assumed that the same mechanism – activation of TEs – underlies somaclonal variations, a phenomenon associated with *in vitro* tissue culture and particularly pronounced during the callus phase which is characterized by a general reduction in cytosine methylation levels [191].

Insertional mutagenesis is not expected to be manifested in potato, being autotetraploid and possessing three other alleles that can potentially compensate for the insertional effect of a gene functional deletion. Even when insertional mutagenesis produces visible phenotypic changes due to the high heterozygosity of commercial potato cultivars, such phenomena are considered an extremely rare event [192]. On the other hand potato is quite susceptible to somaclonal variations in tissue culture even in the absence of transformation [193]. The incidence of atypical plants attributed to somaclonal variations, ranges between 15% and 80% in the population of transgenic potato lines, depending on cultivar [192, 194]. These are often manifested as reduced growth, deformed leaf shape, lower yield and other changes in development, clearly visible in changing field conditions, rather than in uniform ones such as greenhouses or *in vitro* cultures [192, 195]. Elimination of these variations by sexual hybridization is impossible without the simultaneous loss of the genetic integrity of the initial line, while asexual reproduction permanently fixes the status of the transgene within potato genetic background. Thus, the emergence of atypical plants is most often overcome by creating a large population of transgenic lines and selection of several lines with the desired phenotype and high transgene expression.

Beside insertional mutagenesis or somaclonal variations, the unexpected changes in transgenic lines may be a consequence of the transgene expression itself. It is especially expected with PIs, that may interact with plant endogenous protease targets structurally and functionally related to insect digestive proteases, bringing both positive and negative pleiotropic effects *in planta* [196]. For example, metabolic interference of introduced resistance factors in potato can impact protein levels in leaves, positively or negatively [197, 198], reduce glycoalkaloid levels naturally involved in host-plant resistance [199] or, on the contrary, trigger constitutive expression of naturally abiotic or biotic stress-responsive proteins, unexpectedly providing wider protection than the transgene itself [200].

Unintended traits have been identified in commercial GM crops, including insect or herbicide resistant maize, cotton, soybean and oilseed rape – that can exhibit different agronomic and compositional changes relative to their non-GM parental lines [201]. For example, Mon810 maize, carrying *cry1Ab*, exhibits compositional differences such as increased lignin, altered sugar and protein content, and a slight but significant delay in seed and plant maturation, connected with differential expression of 140 genes compared to its near-isogenic variety [202, 203]. On the other hand, plants protected by introduced insect or pathogen resistance are expected to reduce upregulation of self-defense proteins and metabolites compared to less protected near-isoline plants [204] since they experience a different level of biotic or abiotic stress related to transgenic traits. Furthermore, occurrence of unintended effects is not unique to the introduction of recombinant DNA. Traditional breeding is also confronted with undesired changes that result from hybridization, natural genetic recombination and chromosomal rearrangements or activity of transposable elements in plant genomes. There are a number of examples where conventional methods resulted in undesired effects, including potato cultivars with high level of glycoalkaloids [205], that were withdrawn from the market. Thus, for a safety assessment, it is necessary to ensure that transformation does not introduce new compounds, or cause changes in the levels or characteristics of endogenous compounds that may negatively impact human health [206]. Whether the transgenic line is as safe as its conventional variety is the fundamental safety issue to be addressed, rather than how much different they are.

9. Beyond transgenesis

Owing to public concern and reserved acceptance of transgenic crops in many parts of the world, two approaches, cisgenesis and intragenesis, are designed as an alternative to "old" transgene technology. Both concepts include introduction of genetic material derived from the species itself (intragenesis) or closely related, cross compatible species (cisgenesis). Although they use a genetic transformation step, the modified crop genome is designed to not contain any foreign gene, including selectable markers. Therefore, crops developed using these techniques correspond to plants generated through conventional breeding, but without unintentional introduction of undesired genetic elements. Intragenesis has been successfully used for developing potato with high amylopectin content by silencing of the granule-bound starch synthase gene, *GBSS* [207] or for potato with improved processing qualities, by specific tuber-silencing of several genes, *StAst*, *PhL* and *R1* (for low acrylamide) and *ppo* (reduction in black spot bruise development) [208, 209]. On the other hand, cisgenesis has been used for late blight resistance, with introduction of *Rpi-vnt1* gene from *Solanum venturii* in potato [210]. These intragenesis-generated potato varieties have been approved under different commercial names, including traits of modified tuber quality stacked with cisgenic late blight resistance (for instance Innate® Hibernate or Innate® Acclimate) [55].

Genome editing is the latest and most potent molecular technology. Using programmable endonucleases (Zincfinger, TALENs or CRISPR-Cas), alterations can be made at precise locations in the genome, including targeted insertion, replacement or disruption of genes in plants. Because of their precision, these techniques can produce fewer unintended effects, and therefore "edited" crops are considered potentially safer than those generated by random mutagenesis or insertion. In case of potato, both TALENs and CRISPR/Cas9 technologies have been mainly used to improve tuber quality (glycoalkaloids reduction, low acrylamide content and altered starch metabolism) or for herbicide resistance [211], but despite unlimited potential in genetic engineering, no pest-resistance gene incorporation has been reported yet. Importantly, CRISPR-based gene drives could be implemented to spread desirable genetic elements through pest populations themselves. For CPB, there is only one such report to date, where CRISPR/Cas9 was used for *vest* gene knockout, which resulted in a wingless phenotype [212]. However, this potential of gene editing for pest control or even pest eradication is currently highly controversial.

10. Looking into the future

For the growing world population that is expected to reach 10 billion by 2050, food production should be increased by 25–70% and, at the same time, it is necessary to reduce nutritional losses, greenhouse gas emissions from agriculture, pesticide overuse and address other environmental concerns [213].

Potato is now the world's third most important crop for human food consumption, after wheat and rice, but its production in the last 10 years stopped between 360 and 370 million tons annually [214]. Additionally, yield potential of potato has remained relatively unchanged, despite intensive breeding efforts [215], and century-old varieties (i.e., Russet Burbank and Bintje) are still cultivated due to lack of significant genetic improvements in potato. Narrow genetic base as a result of clonal propagation, multiple constraints such as inbreeding depression, self-incompatibility and incorporation of undesirable traits, limit the progress in conventional development of inbred potato lines [216]. On the other hand, genetic engineering has shown potential for fast, feasible, economic and environment-friendly introduction of resistance (and other beneficial) traits in commercially grown crops. However, to make full use of that potential it is necessary to improve existing and bring about new, more sustainable and cleaner gene manipulation technologies. By optimization of transgene expression level, its temporal or spatial programming (i.e., by use of wound-inducible or tissue-specific promoters), generating marker-free modified plants and exploiting new approaches such as cisgenesis/intragenesis or genome editing – it is possible to both decrease unintended effects and increase efficiency and public acceptability of transgenic crops.

For potato, there are no GM varieties with insect resistance traits in the markets, and strategies that rely on insecticides cannot be avoided – as well as their failure in pest control. For instance, imidacloprid, a neonicotinoid successfully used for almost 10 years in CPB control, started being ineffective at the beginning of this century [217]. On the other hand, as global population continues to expand, food production, including potato, has to increase by many folds and with wild potato varieties as only source of resistance traits and their introduction by breeding, that seems unattainable. Also it is questionable whether all potato pests, CPB especially, could be stopped by resistance factors existing in *Solanum* species [218] while these resistance traits combined with heterologous sources such as Cry-toxins can offer more extensive and durable protection [219]. Moreover, there are other Bt-toxins, such as Vip, Cyt and Sip [220], or toxins from other bacteria, waiting to prove their usefulness in pest control. So far only Vip3A has been commercialized in Bt cotton and maize [55]. Additionally, RNAi and even PIs can efficiently supplement and strengthen such protection. However, insects are exceptionally adaptable and evolution of resistance to any of these control measures, including combinations of different traits, is inevitable – but the rate of resistance evolution can be slowed down by efficient management strategies.

As a crucial concept of insect resistance management, refuges are essential for durability of both stacked and single-toxin crops, and where resistance is rare, 20% (or at least 10% for stacked traits) of a pest host plant refuge may be sufficient to delay resistance by a decade or more [31]. Smaller refuges are insecure even under highly effective toxins (or other traits) and all cases of field-evolved resistance are associated with low refuge presence, as one of the main factors [221]. Additionally, within IPM context, refuges also provide better support to populations of natural enemies, that are not only important in target pest control, but to prevent non-target secondary pest outbreaks that can seriously reduce benefits from introduced traits and bring production back to running on the insecticide treadmill [222]. Adding pheromone disruption, mass trapping or intercropping arrangements – integrated into scientifically supported management and adapted to the pest biology – can efficiently reduce pest population size, keeping damage below the economical threshold. Experiences with combination of the simplest practices in potato fields in some parts of the USA, such as rational use of chemical insecticides, trap rows and crop rotation [223] proved a potential of well-structured IPM approach to balance one technology with other complementary strategies. Such avoidance of relying on only one means of control would require complex pest adaptations that are less likely to happen compared to the occurrence and fixation of random single gene mutations that can render resistance to insecticides, Bt-toxin, PIs, RNAi or any other measure that may be implemented in the future.

The benefits of pest-resistant GM crops, incorporated in well-balanced IPM strategies, are clear – but it is also necessary to define and understand their limitations and risks. Heavy dependence and overuse of insecticides undoubtedly had many consequences: food poisoning, reduction in biodiversity, negative effects on non-target species and other formidable impacts on environment – and genetic engineering provides a chance to not repeat all those mistakes. However, we cannot expect that Bt or other pest-resistant modified crops will not have long-term ecological or evolutionary consequences, as well as that small or substantial compositional changes, as intended (or unintended) quantitative or qualitative alterations of metabolites, nutrients or toxins, cannot impact ecological interactions and/or food or feed safety. Such risks are present and inevitable, can vary depending on traits introduced and strategy used for its introduction – and with a precautionary approach, at least some of them can be avoided or mitigated. Additionally, every generated crop line is created in a unique event and should be evaluated for risks, benefits and sustainability only on a case-by-case basis.

So, taking all together, is genetic modification of plants a thrift or a threat? It only depends on how carefully and advisedly we use that tool in our hands.

Acknowledgements

This work was supported by the Ministry of Education, Science and Technological Development of the Republic of Serbia, contract number 451-03-9/2021-14/200007.

Conflict of interest

The authors declare that they have no conflict of interest.

Author details

Martin Raspor* and Aleksandar Cingel
Department of Plant Physiology, Institute for Biological Research "Siniša Stanković", National Institute of Republic of Serbia, University of Belgrade, Belgrade, Serbia

*Address all correspondence to: martin@ibiss.bg.ac.rs

References

[1] Herrera-Estrella L, Depicker A, Montagu MV, Schell J: Expression of chimaeric genes transfered into plant cells using a Ti-plasmid-derived vector. Nature. 1983;303:209-213. DOI: 10.1038/303209a0

[2] ISAAA Brief 55-2019 Global Status of Commercialized Biotech/GM Crops in 2019. Available from: https://www.isaaa.org [Accessed: 2021-04-11]

[3] Brookes G, Barfoot P: GM crop technology use 1996-2018: farm income and production impacts. GM Crops Food. 2020;11:242-261. DOI: 10.1080/21645698.2020.1779574

[4] Brookes G, Barfoot, P: Environmental impacts of genetically modified (GM) crop use 1996-2018: impacts on pesticide use and carbon emissions. GM Crops Food. 2020;11:215-241. DOI: 10.1080/21645698.2020.1773198

[5] OECD 1997. Consensus Document on the Biology of *Solanum tuberosum subsp. tuberosum* (potato). Available from: https://www.oecd.org/env/ehs/biotrack/46815598.pdf

[6] Oerke E: Crop losses to pests. The Journal of Agricultural Science. 2006;144(1),31-43: DOI:10.1017/S0021859605005708

[7] Hare DJ: Impact of defoliation by the Colorado potato beetle on potato yields. Journal of Economic Entomology. 1980;73:369-373. DOI: 0022-0493/80/0303-6905$00.75/0

[8] Rondon SI, Gao Y. The Journey of the Potato Tuberworm around the World. In: FK Perveen, editor. Moths – Pests of Potato, Maize and Sugar Beet. Intech Open; 2018. p. 17-52. DOI: 10.5772/intechopen.81934

[9] Mota-Sanchez D, Hollingworth, RM: The arthropod pesticide resistance database. Available from: http://www.pesticideresistance.org [Accessed: 2021-04-21]

[10] Grafius E: Economic Impact of Insecticide Resistance in the Colorado Potato Beetle (Coleoptera: Chrysomelidae) on the Michigan Potato Industry. Journal of Economic Entomology. 1997;90(5):1144-1151. DOI: 10.1093/jee/90.5.1144

[11] Casagrande RA: The Colorado potato beetle: 125 years of mismanagement. ESA Bulletin. 1987;33:142-150. DOI: 10.1093/besa/33.3.142

[12] Ooms G, Karp A, Roberts J: From tumour to tuber; tumour cell characteristics and chromosome numbers of crown-gall-derived tetraploid potato plants *(Solanum tuberosum* cv. Maris Bard). Theoretical and Applied Genetics. 1983;66:169-172. DOI: 10.1007/BF00265193

[13] Feitelson JS, Payne J, Kim L: *Bacillus thuringiensis*: insects and beyond. Nature Biotechnology. 1992;10(3):271-275. DOI: 10.1038/nbt0392-271

[14] Vaeck M, Reynaerts A, Höfte H, Jansens S, De Beuckeleer M, Dean C, Zabeau M, Van Montagu M, Leemans J: Transgenic plants protected from insect attack. Nature. 1987;328:33-37. DOI: 10.1038/328033a0

[15] Economic Research Service, U.S. DEPARTMENT OF AGRICULTURE Available from: https://www.ers.usda.gov [Accessed: 2021-05-01]

[16] Bravo A, Likitvivatanavong S, Gill SS, Soberón M: *Bacillus thuringiensis*: A story of a successful bioinsecticide. Insect Biochemistry and Molecular Biology. 2011;41(7):423-431. DOI: 10.1016/j.ibmb.2011.02.006

[17] Perlak FJ, Stone TB, Muskopf YM, Petersen LJ, Parker GB, McPherson SA, Wyman J, Love S, Reed G, Biever D, Fischhoff DA: Genetically improved potatoes: Protection from damage by Colorado potato beetles. Plant Molecular Biology. 1993;22:313-321. DOI: 10.1007/BF00014938

[18] Grafius E, Douches D. The present and future role of insect-resistant genetically modified potato cultivars in IPM. In: Romeis J, Shelton A, Kennedy G, editors. Integration of insect-resistant genetically modified crops within IPM programs. Springer: Dordrecht; 2008. pp. 195-221. DOI: 10.1007/978-1-4020-8373-0_7

[19] Chan MT, Chen LJ, Chang HH: Expression of *Bacillus thuringiensis* (B.t.) insecticidal crystal protein gene in transgenic potato. Botanical Bulletin of Academia Sinica. 1996;37:17-23.

[20] Kumar M, Chimote V, Singh R, Mishra GP, Naik PS, Pandey SK, Chakrabarti SK: Development of Bt transgenic potatoes for effective control of potato tuber moth by using *cry1Ab* gene regulated by GBSS promoter. Crop Protection. 2010;29:121-127. DOI: 10.1016/j.cropro.2009.11.001

[21] Estrada MA, Zarka K, Cooper S, Coombs J, Douches DS, Grafius EJ: Potato tuberworm (Lepidoptera: Gelichiidae) resistance in potato lines with the *Bacillus thuringiensis cry1Ac* gene and natural resistance. HortScience. 2007;42:1306-1311. DOI: 10.21273/HORTSCI.42.5.1306

[22] Amiri AN, Bakhsh A: An effective pest management approach in potato to combat insect pests and herbicide. 3 Biotech. 2019;9(1):16. DOI: 10.1007/s13205-018-1536-0

[23] Beuning LL, Mitra DS, Markwick NP, Gleave AP: Minor modifications to the cry1Ac9 nucleotide sequence are sufficient to generate transgenic plants resistant to *Phthorimaea operculella*. Annals of Applied Biology. 2001;138:281-292. DOI: 10.1111/j.1744-7348.2001.tb00113.x

[24] Davidson MM, Jacobs JME, Reader JK, Butler RC, Frater CM, Markwick NP, Wratten SD, Conner AJ: Development and evaluation of potatoes transgenic for a *cry1*Ac9 gene conferring resistance to potato tuber moth. Journal of the American Society for Horticultural Science. 2002;127:590-596. DOI: 10.21273/JASHS.127.4.590

[25] Mohammed A, Douches DS, Pett W, Grafius E, Coombs J, Liswidowati J, Li W, Madkour MA: Evaluation of potato tuber moth (Lepidoptera, Gelechiidae) resistance in tubers of *Bt-cry5* transgenic potato lines. Journal of Economic Entomology. 2000;193: 472-476. DOI:10.1603/0022-0493-93.2.472

[26] Douches DS, Li W, Zarka K, Coombs J, Pett W, Grafius E, El-Nasr T: Development of *Bt-cry5* insect-resistant potato lines 'Spunta-G2' and 'Spunta-G3'. HortScience. 2002;37:1103-1107. DOI: 10.21273/HORTSCI.37.7.1103

[27] Jacobs JME, Takla MFG, Docherty LC, Frater CM, Markwick NP, Meiyalaghan S, Conner AJ: Potato transformation with modified nucleotide sequences of the *cry9*Aa2 gene improves resistance to potato tuber moth. Potato Research. 2009;52:367-378. DOI: 10.1007/s11540-009-9140-7

[28] Mi X, Ji X, Yang J, Liang L, Si H, Wu J, Wang D: Transgenic potato plants expressing *cry3A* gene confer resistance to Colorado potato beetle. Comptes Rendus Biologies. 2015;338:443-450. DOI: 10.1016/j.crvi.2015.04.005

[29] Meiyalaghan S, Jacobs JME, Butler RC, Wratten SD, Conner AJ: Expression of *cry1*Ac9 and *cry9*Aa2 genes under a potato light-inducible *Lhca3* promoter in transgenic potatoes

for tuber moth resistance. Euphytica. 2006;147: 297-309. DOI: 10.1007/s10681-005-9012-4

[30] Ghasimi Hagh Z, Rahnama H, Panahandeh J, Baghban Kohneh Rouz B, Arab Jafari KM, Mahna N: Green-tissue-specific, C(4)-PEPC-promoter-driven expression of *Cry1*Ab makes transgenic potato plants resistant to tuber moth (*Phthorimaea operculella*, Zeller). Plant Cell Reports. 2009;28(12):1869-79. DOI: 10.1007/s00299-009-0790-3

[31] Tabashnik BE, Carrière Y: Surge in insect resistance to transgenic crops and prospects for sustainability. Nature Biotechnology. 2017;35(10):926-935. DOI: 10.1038/nbt.3974

[32] Ferro DN: Potential for resistance to *Bacillus thuringiensis*: Colorado potato beetle (Coleoptera: Chrysomelidae) – a model system. American Entomologist. 1993;39:38-44. DOI: 10.1093/ae/39.1.38

[33] Whalon ME, Miller DL, Hollingworth RM, Grafius EJ, Miller JR: Selection of a Colorado potato beetle (Coleoptera: Chrysomelidae) strain resistant to *Bacillus thuringiensis*. Journal of Economic Entomology. 1993;86:226-233. DOI: 10.1093/jee/86.2.226

[34] Raharadja U, Whalon ME: Inheritance of resistance to *Bacillus thuringiensis subsp. tenebrionis* CryIIIA δ-Endotoxin in Colorado potato beetle (Coleoptera: Chrysomelidae). Journal of Economic Entomology. 1995;88:21-26. DOI: 10.1093/jee/88.1.21

[35] Gunning RV, Dang HT, Kemp FC, Nicholson IC, Moores GD: New resistance mechanism in *Helicoverpa armigera* threatens transgenic crops expressing *Bacillus thuringiensis* Cry1Ac toxin. Applied and Environmental Microbiology. 2005;71:2558-2563. DOI: 10.1128/AEM.71.5.2558-2563.2005

[36] Zhang H, Tang M, Yang F, Yang Y, Wu Y: DNA-based screening for an intracellular cadherin mutation conferring non-recessive Cry1Ac resistance in field populations of *Helicoverpa armigera*. Pesticide Biochemistry and Physiology. 2013;107:148-152. DOI: 10.1016/j.pestbp.2013.06.007

[37] Wu Y: Detection and Mechanisms of Resistance Evolved in Insects to Cry Toxins from *Bacillus thuringiensis*. Advances in Insect Physiology. 2014;47:297-342. DOI: 10.1016/B978-0-12-800197-4.00006-3

[38] Loseva O, Ibrahim M, Candas M, Koller CN, Bauer LS, Bulla LA Jr: Changes in protease activity and Cry3Aa toxin binding in the Colorado potato beetle: implications for insect resistance to *Bacillus thuringiensis* toxins. Insect Biochemistry and Molecular Biology. 2002;32:567-577. DOI: 10.1016/S0965-1748(01)00137-0

[39] Hoffmann JA, Reichhard J-M, Hetru C: Innate immunity in higher insects. Current Opinion in Immunology. 1996;8:8-13. DOI: 10.1016/S0952-7915(96)80098-7

[40] Ma G, Roberts H, Sarjan M, Featherstone N, Lahnstein J, Akhurst R, Schmidt O: Is the mature endotoxin Cry1Ac from *Bacillus thuringiensis* inactivated by a coagulation reaction in the gut lumen of resistant *Helicoverpa armigera* larvae? Insect Biochemistry and Molecular Biology. 2005;35:729-739. DOI: 10.1016/j.ibmb.2005.02.011

[41] Shelton AM, Zhao JZ, Roush RT: Economic, ecological, food safety and social consequences of the deployment of Bt transgenic plants. Annual Review of Entomology. 2002;47:845-881. DOI: 10.1146/annurev.ento.47.091201.145309

[42] Alyokhin AV, Ferro DN: Modifications in flight and oviposition of Bt-resistant and Bt-susceptible Colorado potato beetles as a result of exposure to *Bacillus thuringiensis subsp.*

tenebrionis Cry3A toxin. Entomologia Experimentalis et Applicata. 1999;90:93-101. DOI: 10.1023/A:1003597616214

[43] Jin L, Yin W, Zhang L, Yang Y, Tabashnik BE, Wu Y: Dominant resistance to Bt cotton and minor cross-resistance to Bt toxin Cry2Ab in cotton bollworm from China. Evolutionary Applications. 2013;6:1222-1235. DOI: 10.1111/eva.12099

[44] Campagne P, Kruger M, Pasquet R, Le Ru B, Van den Berg J: Dominant inheritance of field-evolved resistance to Bt corn in *Busseola fusca*. PLoS One. 2013;8:e69675. DOI: 10.1371/journal.ponc.0069675

[45] Rajamohan F, Alzate O, Cotrill JA, Curtiss A, Dean DH: Protein engineering of *Bacillus thuringiensis* delta-endotoxin: mutations at domain II of CryIAb enhance receptor affinity and toxicity toward gypsy moth larvae. PNAS. 1996;93:14338-14343. DOI: 10.1073/pnas.93.25.14338

[46] Wu SJ, Koller CN, Miller DL, Bauer LS, Dean DH: Enhanced toxicity of *Bacillus thuringiensis* Cry3A delta-endotoxin in coleopterans by mutagenesis in a receptor binding loop. FEBS Letters. 2000;473:227-232. DOI: 10.1016/s0014-5793(00)01505-2

[47] Mandal CC, Gayen S, Basu A, Ghosh KS, Dasgupta S, Maiti MK, Sen SK: Prediction based protein engineering of domain I of Cry2A entomocidal toxin of *Bacillus thuringiensis* for the enhancement of toxicity against lepidopteran insects. Protein Engineering, Design and Selection. 2007;20:599-606. DOI: 10.1093/protein/gzm058

[48] Soberón M, Pardo-López L, López I, Gómez I, Tabashnik B, Bravo A: Engineering modified Bt toxins to counter insect resistance. Science. 2007;318:1640-1642. DOI: 10.1126/science.1146453

[49] Park Y, Abdullah MA, Taylor MD, Rahman K,Adang MJ: Enhancement of *Bacillus thuringiensis* Cry3Aa and Cry3Bb toxicities to coleopteran larvae by a toxin-binding fragment of an insect cadherin. Applied and Environmental Microbiology. 2009;75(10):3086-3092. DOI: 10.1128/AEM.00268-09

[50] Ge AZ, Rivers D, Milne R, Dean DH: Functional domains of *Bacillus thuringiensis* insecticidal crystal proteins. Journal of Biological Chemistry. 1991;266:17954-17958. DOI: 10.1016/S0021-9258(18)552211-2

[51] deMaagd RA, Kwa MSG, van der Klei H, Yamamoto T, Schipper B, Vlak JM, Stiekema WJ, Bosch D: Domain III substitution in *Bacillus thuringiensis* delta-endotoxin Cry1Ab results in superior toxicity for *Spodoptera exigua* and altered membrane protein recognition. Applied and Environmental Microbiology. 1996;62:1537-1543. DOI: 10.1128/AEM.62.5.1537-1543.1996

[52] Naimov S, Dukiandjiev S, de Maagd RA: A hybrid *Bacillus thuringiensis* delta-endotoxin gives resistance against a coleopteran and a lepidopteran pest in transgenic potato. Plant Biotechnology Journal. 2003;1:51-57. DOI: 10.1046/j.1467-7652.2003.00005.x

[53] Ahmed HA, Onarıcı S, Bakhsh A, Ahmed HA, Akdogan G, Karakoç Ö, Özcan S, Aydin G, Aasim M, Ünlü L, Sancak C, Naimov S, Özcan S: Targeted expression of insecticidal hybrid SN19 gene in potato leads to enhanced resistance against Colorado potato beetle (*Leptinotarsa decemlineata* Say) and tomato leafminer (*Tuta absoluta* Meyrick). Plant Biotechnology Reports. 2017;11:315-329. DOI: 10.1007/s11816-017-0453-8

[54] Salim, M, Bakhsh A, Gökçe A: Stacked insecticidal genes in potatoes exhibit enhanced toxicity against Colorado potato beetle, *Leptinotarsa*

decemlineata (Coleoptera: Chrysomelidae). Plant Biotechnology Reports. 2021;15:197-215. DOI: 10.1007/s11816-021-00668-3

[55] ISAAA. GM Approval Database. 2021. Available from: https://www.isaaa.org/[Accessed: 2021-05-11]

[56] Brévault T, Heuberger S, Zhang M, Ellers-Kirk C, Ni X, Masson L, Li X, Tabashnik BE, Carrière Y: Potential shortfall of pyramided transgenic cotton for insect resistance management. PNAS. 2013;110:5806-5811. DOI: 10.1073/pnas.1216719110

[57] Losey J, Raynor L, Carter ME: Transgenic pollen harms Monarch larvae. Nature. 1999;399:214. DOI: 10.1038/20338

[58] Riddick EW, Barbosa P: Impact of Cry-3A-intoxicated *Leptinotarsa decemlineata* (Coleoptera: Chrysomelidae) and pollen on consumption, development, and fecundity of *Coleomegilla maculata* (Coleoptera: Coccinellidae). Annals of the Entomological Society of America. 1998;91(3):303-30. DOI: 10.1093/aesa/91.3.303

[59] Ferry N, Mulligan EA, Majerus ME, Gatehouse AM: Bitrophic and tritrophic effects of Bt Cry3A transgenic potato on beneficial, non-target, beetles. Transgenic Research. 2007;16:795-812. DOI: 10.1007/s11248-007-9088-9

[60] Mendelsohn M, Kough J, Vaituzis Z, Matthews K: Are *Bt* crops safe? Nature Biotechnology. 2003;21(9):1003-1009. DOI: 10.1038/nbt0903-1003

[61] Kalushkov P, Nedved O: Genetically modified potatoes expressing Cry3A protein do not affect aphidophagous coccinellids. Journal of Applied Entomology. 2005;129:401-406. DOI: 10.1111/j.1439-0418.2005.00987

[62] Duan JJ, Marvier M, Huesing J, Dively G, Huang ZY: A meta-analysis of effects of Bt toxins on honeybees (Hymenoptera: Apidae). PLoS One. 2008;3(1):e1415. DOI: 10.1371/journal.pone.0001415

[63] Riddick EW, Dively G, Barbosa P: Effect of a seed mix deployment of Cry3A-transgenic and non-transgenic potato on the abundance of *Lebia grandis* (Coleoptera: Carabidae) and *Coleomegilla maculata* (Coleoptera: Coccinellidae). Annals of the Entomological Society of America. 1998;91(5):647-653. DOI: 10.1093/aesa/91.5.647

[64] Ashouri A : Transgenic-Bt potato plant resistance to the Colorado potato beetle affect the aphid parasitoid *Aphidius nigripes*. Communications in Agricultural and Applied Biological Sciences. 2004;69(3):185-189.

[65] Svobodová Z, Shu Y, Skoková Habuštová O, Romeis J, Meissle M: Stacked Bt maize and arthropod predators: exposure to insecticidal Cry proteins and potential hazards. Proceedings of the Royal Society B: Biological Sciences. 2017;284(1859):20170440. DOI: 10.1098/rspb.2017.0440

[66] Meissle M, Kloos S, Romeis J: Fate of multiple Bt proteins from stacked Bt maize in the predatory lady beetle *Harmonia axyridis* (Pallas) (Coleoptera: Coccinellidae). Environmental Pollution. 2021;268:115421. DOI: 10.1016/j.envpol.2020.115421

[67] Saxena D, Flores S, Stotzky G: Bt toxin is released in root exudates from 12 transgenic corn hybrids representing three transformation events. Soil Biology and Biochemistry. 2002;34: 133-137. DOI: 10.1016/S0038-0717(01)00161-4

[68] Saxena D, Pushalkar S, Stotzky G: Fate and Effects in Soil of Cry Proteins

from *Bacillus thuringiensis*: Influence of Physicochemical and Biological Characteristics of Soil. The Open Toxinology Journal. 2010;3:133-153. DOI: 10.2174/1875414701003010151

[69] Betz FS, Hammond BG, Fuchs L: Safety and Advantages of *Bacillus thuringiensis*-Protected Plants to Control Insect Pests. Regulatory Toxicology and Pharmacology. 2000;32:156-173. DOI: 10.1006/rtph.2000.1426

[70] Koch MS, Ward JM, Levine SL, Baum JA, Vicini JL, Hammond BG: The food and environmental safety of Bt crops. Frontiers in Plant Science. 2015;6:283. DOI: 10.3389/fpls.2015.00283

[71] Bucchini L, Goldman LR: Starlink Corn: A Risk Analysis. Environmental Health Perspectives. 2002;110(1):5-13. DOI: 10.1289/ehp.021105.

[72] Seralini GE, Cellier D, de Vendomois JS: New analysis of a rat feeding study with a genetically modified maize reveals signs of hepatorenal toxicity. Archives of Environmental Contamination and Toxicology. 2007;52(4):596-602. DOI: 10.1007/s00244-006-0149-5

[73] deVendomois JS, Roullier F, Dominique C, Seralini G: A comparison of the effects of three GM corn varieties on mammalian health. International Journal of Biological Sciences. 2009;5(7):706-26. DOI: 10.7150/ijbs.5.706

[74] Domingo JL: Safety assessment of GM plants: an updated review of the scientific literature. Food and Chemical Toxicology. 2016;95:12-18. DOI: 10.1016/j.fct.2016.06.013

[75] Brzin J, Kidric M: Proteinases and their inhibitors in plants: role in normal growth and in response to various stress conditions. Biotechnology and Genetic Engineering Reviews. 1995;13:420-467. DOI: 10.1080/02648725.1996.10647936

[76] Solomon M, Belenghi B, Delledonne M, Menachem E, Levine A: The involvement of cysteine proteases and protease inhibitor genes in the regulation of programmed cell death in plants. The Plant Cell. 1999;11:431-443. DOI: 10.1105/tpc.11.3.431

[77] De Leo F, Volpicella M, Licciulli F, Liuni S, Gallerani R, Ceci LR: PLANT-PIs: a database for plant protease inhibitors and their genes. Nucleic Acids Research. 2002;30:347-348. DOI: 10.1093/nar/30.1.347

[78] Bode W, Huber R: Structural basis of the endoproteinase-protein inhibitor interaction. Biochimica et Biophysica Acta. 2000;1477:241-252. DOI: 10.1016/S0167-4838(99)00276-9

[79] Reese JC. In: Hedin PA, editor. Nutrient-allelochemical interactions in host plant resistance. Plant Resistance to Insects. Washington, DC: American Chemical Society; 1983. p. 231-244.

[80] Broadway RM, Duffey SS: Plant proteinase inhibitors: mechanism of action and effect on the growth and digestive physiology of larval *Heliothis zea* and *Spodoptera exiqua*. Journal of Insect Physiology. 1986;32:827-833. DOI: 10.1016/0022-1910(86)90097-1

[81] Shukle RH, Murdock LL, Gallun RL: Identification and partial characterization of a major gut proteinase from larvae of the Hessian fly, *Mayetiola destructor* (Say) (Diptera: Cecidomyiidae). Insect Biochemistry. 1985;15(1):93-101. DOI: 10.1016/0020-1790(85)90049-6

[82] Marchetti S, Delledonne M, Fogher C, Chiaba C, Chiesa F, Savazzini F, Giordano A: Soybean Kunitz, C-II and PI-IV inhibitor genes confer different levels of insect

resistance to tobacco and potato transgenic plants. Theoretical and Applied Genetics. 2000;101(4):519-526. DOI: 10.1007/s001220051511

[83] Haq SK, Atif SM, Khan RH: Protein proteinase inhibitor genes in combat against insects, pests, and pathogens: natural and engineered phytoprotection. Archives of Biochemistry and Biophysics. 2004;431:145-159. DOI: 10.1016/j.abb.2004.07.022

[84] Azzouz H, Cherqui A, Campan EDM, Rahbe Y, Duport G, Jouanin L, Kaiser L, Giordanengo P: Effects of plant protease inhibitors, oryzacystatin I and soybean Bowman–Birk inhibitor, on the aphid *Macrosiphum euphorbiae* (Homoptera, Aphididae) and its parasitoid *Aphelinus abdominalis* (Hymenoptera, Aphelinidae). Journal of Insect Physiology. 2005;51:75-86. DOI: 10.1016/j.jinsphys.2004.11.010

[85] Mickel, CE, Standish J: Susceptibility of processed soy flour and soy grits in storage to attack by *Tribolium castaneum*. University of Minnesota. Minnesota Agricultural Experiment Station Bulletin. 1947;178:1-20.

[86] Green TR, Ryan CA: Wound-induced proteinase inhibitor in plant leaves: a possible defense mechanism against insects. Science. 1972;175:776-777. DOI: 10.1126/science.175.4023.776

[87] Hilder VA, Gatehouse AMR, Sheerman SE, Baker RF, Boulter D: A novel mechanism of insect resistance engineered into tobacco. Nature. 1987;330:160-163. DOI: 10.1038/330160a0

[88] Cingel A, Savić J, Lazarević J, Ćosić T, Raspor M, Smigocki A, Ninković S: Extraordinary Adaptive Plasticity of Colorado Potato Beetle: "Ten-Striped Spearman" in the Era of Biotechnological Warfare. International Journal of Molecular Sciences. 2016;*17*:1538. DOI: 10.3390/ijms17091538

[89] Bolter CJ, Latoszek-Green M: Effect of chronic ingestion of the cysteine proteinase inhibitor, E-64, on Colorado potato beetle gut proteinases. Entomologia Experimentalis et Applicata. 1997;83:295-303. DOI: 10.1046/j.1570-7458.1997.00184.x

[90] Michaud D, Nguyen-Quoc B, Yelle S: Selective inhibition of Colorado potato beetle cathepsin H by oryzacystatins I and II. FEBS Letters. 1993;331:173-176. DOI: 10.1016/0014-5793(93)80320-T

[91] Cloutier C, Fournier M, Jean C, Yelle S, Michaud D: Growth compensation and faster development of Colorado potato beetle (Coleoptera: Chrysomelidae) feeding on potato foliage expressing oryzacystatin I. Archives of Insect Biochemistry and Physiology. 1999;40:69-79. DOI: 10.1002/(SICI)1520-6327(1999)40:2<69::AID-ARCH1>3.0.CO;2-M.

[92] Cingel A, Savić J, Vinterhalter B, Vinterhalter D, Kostić M, Šešlija-Jovanović D, Smigocki A, Ninković S: Growth and development of Colorado potato beetle larvae, *Leptinotarsa decemlineata*, on potato plants expressing the oryzacystatin II proteinase inhibitor. Transgenic Research. 2015;24:729-740. DOI: 10.1007/s11248-015-9873-9.

[93] Brunelle F, Cloutier C, Michaud D: Colorado potato beetles compensate for tomato cathepsin D inhibitor expressed in transgenic potato. Archives of Insect Biochemistry and Physiology. 2004;55:103-113. DOI: 10.1002/arch.10135

[94] Cingel A, Savić J, Ćosić T, Zdravković-Korać S, Momčilović I, Smigocki A, Ninković S. Pyramiding rice cystatin OCI and OCII genes in transgenic potato (*Solanum tuberosum* L.) for resistance to Colorado potato beetle (*Leptinotarsa decemlineata* Say). Euphytica. 2014;198:425-438. DOI: 10.1007/s10681-014-1119-z.

[95] Cingel A, Savić J, Lazarević J, Ćosić T, Raspor M, Smigocki A, Ninković S: Co-expression of the proteinase inhibitors oryzacystatin I and oryzacystatin II in transgenic potato alters Colorado potato beetle larval development. Insect Science. 2017;24(5):768-780. DOI: 10.1111/1744-7917.12364

[96] Kondrák M, Kutas J, Szenthe B, Patthy A, Bánfalvi Z, Nádasy M, Gráf L, Asbóth B: Inhibition of Colorado potato beetle larvae by a locust proteinase inhibitor peptide expressed in potato. Biotechnology Letters. 2005;27:829-834. DOI: 10.1007/s10529-005-6190-z

[97] Gruden K, Strukelj B, Popovic T, Lenarcic B, Bevec T, Brzin J, Kregar I, Herzog-Velikonja J, Stiekema WJ, Bosch D, Jonqsma MA. The cysteine protease activity of Colorado potato beetle (*Leptinotarsa decemlineata* Say) guts, which is insensitive to potato protease inhibitors, is inhibited by thyroglobulin type-1 domain inhibitors. Insect Biochemistry and Molecular Biology. 1998;28:549-560. DOI: 10.1016/S0965-1748(98)00051-4

[98] Outchkourov NS, Rogelj B, Strukelj B, Jongsma MA: Expression of sea anemone equistatin in potato. Effects of plant proteases on heterologous protein production. Plant Physiology. 2003;133:379-390. DOI: 10.1104/pp.102.017293

[99] Brunelle F, Girard C, Cloutier C, Michaud D: A hybrid, broad-spectrum inhibitor of Colorado potato beetle aspartate and cysteine digestive proteinases. Archives of Insect Biochemistry and Physiology. 2005;60:20-31. DOI: 10.1002/arch.20078.

[100] Šmid I, Gruden K, Gašparič MB, Koruza K, Petek M, Pohleven J, Brzin J, Kos J, Zel J, Sabotič J: Inhibition of the growth of Colorado potato beetle larvae by macrocypins, protease inhibitors from the parasol mushroom. Journal of Agricultural and Food Chemistry. 2013;61:12499-12509. DOI: 10.1021/jf403615f

[101] Šmid I, Rotter A, Gruden K, Brzin J, Gašparič MB, Kos J, Žel J, Sabotič J: Clitocypin, a fungal cysteine protease inhibitor, exerts its insecticidal effect on Colorado potato beetle larvae by inhibiting their digestive cysteine proteases. Pesticide Biochemistry and Physiology. 2015;122:59-66. DOI: 10.1016/j.pestbp.2014.12.022

[102] Gatehouse AM, Davison GM, Newell CA, Merryweather A, Hamilton WD, Burgess EP, Gatehouse JA: Transgenic potato plants with enhanced resistance to the tomato moth, *Lacanobia oleracea*: growth room trials. Molecular Breeding. 1997;3(1):49-63. DOI: 10.1023/A:1009600321838

[103] Goulet MC, Dallaire C, Vaillancourt LP, Khalf M, Badri AM, Preradov A, Duceppe MO, Goulet C, Cloutier C, Michaud D: Tailoring the specificity of a plant cystatin toward herbivorous insect digestive cysteine proteases by single mutations at positively selected amino acid sites. Plant Physiology. 2008;146(3):1010-1019. DOI: 10.1104/pp.108.11574

[104] Rasoolizadeh A, Munger A, Goulet MC, Sainsbury F, Cloutier C, Michaud D: Functional proteomics-aided selection of protease inhibitors for herbivore insect control. Scientific Reports. 2016;6:38827. DOI: 10.1038/srep38827

[105] Álvarez-Alfageme F, Martínez M, Pascual-Ruiz S, Castañera P, Diaz I, Ortego F: Effects of potato plants expressing a barley cystatin on the predatory bug *Podisus maculiventris* via herbivorous prey feeding on the plant. Transgenic Research. 2007;16:1-13. DOI: 10.1007/s11248-006-9022-6

[106] MacIntosh SC, Kishore GM, Perlak FJ, Marrone PG, Stone TB, Sims SR, Fuchs RL: Potentiation of *Bacillus thuringiensis* insecticidal activity by serine protease inhibitors. Journal of Agricultural and Food Chemistry. 1990;38:1145-1152. DOI: 10.1021/jf00094a051

[107] Zhao J-Z, Fan YL, Fan XL, Shi XP, Lu MG: Evaluation of transgenic tobacco expressing two insecticidal genes to delay resistance development of *Helicoverpa armigera*. Chinese Science Bulletin. 1999;44:1871-1874. DOI: 10.1007/BF02886343

[108] Pardo-López L, Muñoz-Garay C, Porta H, Rodríguez-Almazán C, Soberón M, Bravo A: Strategies to improve the insecticidal activity of Cry toxins from *Bacillus thuringiensis*. Peptides. 2009;30(3):589-595. DOI: 10.1016/j.peptides.2008.07.027

[109] Bell HA, Down RE, Fitches EC, Edwards JP, Gatehouse AMR: Impact of genetically modified potato expressing plant-derived insect resistance genes on the predatory bug *Podisus maculiventris* (Heteroptera: Pentatomidae). Biocontrol Science and Technology. 2003;13:729-741. DOI: 10.1080/09583150310001606543

[110] Ferry N, Raemaekers RJM, Majerus MEN, Jouanin L, Port G, Gatehouse JA, Gatehouse AMR: Impact of oilseed rape expressing the insecticidal cysteine protease inhibitor oryzacystatin on the beneficial predator *Harmonia axyridis* (Multicoloured Asian ladybeetle). Molecular Ecology. 2003;12:493-504. DOI: 10.1046/j.1365-294x.2003.01736.x

[111] Schuler TH, Denholm I, Jouanin L, Clark SJ, Clark AJ, Poppy GM: Population-scale laboratory studies of the effect of transgenic plants on nontarget insects. Molecular Ecology. 2001;10:1845-53. DOI: 10.1046/j.0962-1083.2001.01309.x

[112] Bouchard E, Michaud D, Cloutier C: Molecular interactions between an insect predator and its herbivore prey on transgenic potato expressing a cysteine proteinase inhibitor from rice. Molecular Ecology. 2003;12(9):2429-2437. DOI: 10.1046/j.1365-294x.2003.01918.x.

[113] Ashouri A, Michaud D, Cloutier C: Recombinant and classically selected factors of potato plant resistance to the Colorado potato beetle, *Leptinotarsa decemlineata*, variously affect the potato aphid parasitoid *Aphidius nigripes*. BioControl. 2001;46:401-418. DOI: 10.1023/A:1014123712776

[114] Atkinson HJ, Johnston KA, Robbins M: Prima facie evidence that a phytocystatin for transgenic plant resistance to nematodes is not a toxic risk in the human diet. The Journal of Nutrition. 2004;134:431-434. DOI: 10.1093/jn/134.2.431

[115] Santos AFS, Silva MDCda, Napoleão TH, Paiva PMG, Correia MTS, Coelho LCBB: Lectins: Function, structure, biological properties and potential applications. Current Topics in Peptide and Protein Research. 2014;15:41-62.

[116] Reyes-Montaño EA, Vega-Castro NA. Plant Lectins with Insecticidal and Insectistatic Activities. In: Ghousia Begum, editor. Insecticides-Agriculture and Toxicology; IntechOpen: London, UK; 2018. p. 17-42. DOI: 10.5772/intechopen.74962

[117] Gatehouse AMR, Down RE, Powell KS, Sauvion N, Rahb Y, Newell CA, Merryweather A, Boulter D, Gatehouse JA: Effects of GNA-expressing transgenic potato plants on peach-potato aphid, *Myzus persicae*. Entomologia Experimentalis et Applicata. 1996;79:295-307.

[118] Down RE, Gatehouse AMR, Hamilton WDO, Gatehouse JA: Snowdrop lectin inhibits development and decreases fecundity of the glasshouse potato aphid (*Aulacorthum solani*) when administered in vitro and via transgenic plants both in laboratory and glasshouse trials. Journal of Insect Physiology. 1996;42:1035-1045. DOI: 10.1016/S0022-1910(96)00065-0

[119] Aasen SS, Hågvar EB: Effect of potato plants expressing snowdrop lectin (GNA) on the performance and colonization behaviour of the peach-potato aphid *Myzus persicae*, Acta Agriculturae Scandinavica, Section B — Soil & Plant Science. 2012;62:352-361. DOI: 10.1080/09064710.2011.619996

[120] Mi X, Liu X, Yan H, Liang L, Zhou X, Yang J, Si H, Zhang N: Expression of the *Galanthus nivalis* agglutinin (GNA) gene in transgenic potato plants confers resistance to aphids. Comptes Rendus Biologies. 2017;340(1):7-12. DOI: 10.1016/j.crvi.2016.10.003

[121] Gatehouse JA, Gatehouse AMR, Fitches E: Effects of snowdrop lectin (GNA) delivered via artificial diet and transgenic plants on the development of tomato moth (*Lacanobia oleracea*) larvae in laboratory and glasshouse trials. Journal of Insect Physiology. 1997;43:727-739. DOI: 10.1016/S0022-1910(97)00042-5.

[122] Gatehouse AMR, Davison GM, Stewart JN, Gatehouse LN, Kumar A, Geoghegan IE, Birch ANE, Gatehouse JA: Concanavalin A inhibits development of tomato moth (*Laconobia oleracea*) and peach-potato aphid (*Myzus persicae*) when expressed in transgenic potato plants. Molecular Breeding. 1999;5:153-165. DOI: 10.1023/A:1009681705481

[123] Birch ANE, Geoghegan IE, Majerus MEN, McNicol JW, Hackett C, Gatehouse AMR, Gatehouse JA: Tri-trophic interactions involving pest aphids, predatory 2-spot ladybirds and transgenic potatoes expressing snowdrop lectin for aphid resistance. Molecular Breeding. 1999;5:75-83. DOI: 10.1023/A:1009659316170

[124] Bell HA, Fitches EC, Marris GC, Bell J, Edwards JP, Gatehouse JA, Gatehouse AMR, et al. Transgenic GNA expressing potato plants augment the beneficial biocontrol of *Lacanobia oleracea* (Lepidoptera; Noctuidae) by the parasitoid *Eulophus pennicornis* (Hymenoptera; Eulophidae). Transgenic Research. 2001;10:35-42. DOI: 10.1023/a:1008923103515

[125] Ewen SW, Pusztai A: Effect of diets containing genetically modified potatoes expressing *Galanthus nivalis* lectin on rat small intestine. Lancet. 1999;354(9187):1353-4. DOI: 10.1016/S0140-6736(98)05860-7

[126] Fire A, Xu S, Montgomery MK, Kostas SA, Driver SE, Mello CC: Potent and specific genetic interference by double-stranded RNA in *Caenorhabditis elegans*. Nature. 1998;391:806-811. DOI: 10.1038/35888

[127] Christiaens O, Niu J, Nji Tizi Taning C: RNAi in Insects: A Revolution in Fundamental Research and Pest Control Applications. *Insects*. 2020;11(7):415. DOI: 10.3390/insects11070415

[128] Baum JA, Bogaert T, Clinton W, Heck GR, Feldmann P, Ilagan O, Johnson S, Plaetinck G, Munyikwa T, Pleau M, Vaughn T, Roberts J: Control of coleopteran insect pests through RNA

interference. Nature Biotechnology. 2007;25:1322-1326. DOI: 10.1038/nbt1359

[129] Spit J, Philips A, Wynant N, Santos D, Plaetinck G, Vanden Broeck J: Knockdown of nuclease activity in the gut enhances RNAi efficiency in the Colorado potato beetle, *Leptinotarsa decemlineata*, but not in the desert locust, *Schistocerca gregaria*. Insect Biochemistry and Molecular Biology. 2017;81:103-116. DOI: 10.1016/j.ibmb.2017.01.004

[130] Shukla JN, Kalsi M, Sethi A, Narva KE, Fishilevich E, Singh S, Mogilicherla K, Palli SR: Reduced stability and intracellular transport of dsRNA contribute to poor RNAi response in lepidopteran insects. RNA Biology. 2016;13:656-669. DOI: 10.1080/15476286.2016.1191728

[131] Miyata K, Ramaseshadri P, Zhang Y, Segers G, Bolognesi R, Tomoyasu Y: Establishing an in vivo assay system to identify components involved in environmental RNA interference in the western corn rootworm. PLoS ONE. 2014;9:e101661. DOI: 10.1371/journal.pone.0101661

[132] Cappelle K, de Oliveira CFR, Van Eynde B, Christiaens O, Smagghe G: The involvement of clathrin-mediated endocytosis and two Sid-1-like transmembrane proteins in double-stranded RNA uptake in the Colorado potato beetle midgut. Insect Molecular Biology. 2016;25:315-323. DOI: 10.1111/imb.12222

[133] Guo WC, Fu K, Yang S, Li X, Li GQ: Instar-dependent systemic RNA interference response in *Leptinotarsa decemlineata* larvae. Pesticide Biochemistry & Physiology. 2015;123:64-73. DOI: 10.1016/j.pestbp.2015.03.006

[134] Yoon JS, Shukla J, Zhong JG, Mogilicherla K, Palli SR: RNA interference in the Colorado potato beetle, *Leptinotarsa decemlineata*: Identification of key contributors. Insect Biochemistry and Molecular Biology. 2016;78:78-88. DOI: 10.1016/j.ibmb.2016.09.002

[135] Gu L, Knipple DC: Recent advances in RNA interference research in insects: implications for future insect pest management strategies. Crop Protection. 2013;45:36-40. DOI: 10.1016/j.cropro.2012.10.004

[136] Kumar A, Congiu L, Lindström L, Piiroinen S, Vidotto M, Grapputo A: Sequencing, *de novo* assembly and annotation of the Colorado potato beetle, *Leptinotarsa decemlineata*, transcriptome. PLoS ONE. 2014;9:e86012:1- e86012:17. DOI: 10.1371/journal.pone.0086012

[137] Zhu F, Xu J, Palli R, Ferguson J, Palli SR. Ingested RNA interference for managing the populations of the Colorado potato beetle, *Leptinotarsa decemlineata*. Pest Management Science. 2011;67:175-182. DOI: 10.1002/ps.2048

[138] Zhou LT, Jia S, Wan PJ, Kong Y, Guo WC, Ahmat T, Li GQ: RNA interference of a putative S-adenosyl-L-homocysteine hydrolase gene affects larval performance in *Leptinotarsa decemlineata* (Say). Journal of Insect Physiology. 2013;59:1049-1056. DOI: 10.1016/j.jinsphys.2013.08.002

[139] Liu XP, Fu KY, Lü FG, Meng QW, Guo WC, Li GQ: Involvement of FTZ-F1 in the regulation of pupation in *Leptinotarsa decemlineata* (Say). Insect Biochemistry and Molecular Biology. 2014;55:51-60. DOI: 10.1016/j.ibmb.2014.10.008

[140] Lü FG, Fu KY, Guo WC, Li GQ: Characterization of two juvenile hormone epoxide hydrolases by RNA interference in the Colorado potato beetle. Gene. 2015;570:264-271. DOI: 10.1016/j.gene.2015.06.032

[141] Guo WC, Liu XP, Fu KY, Shi JF, Lü FG, Li GQ: Nuclear receptor ecdysone-induced protein 75 is required for larval-pupal metamorphosis in the Colorado potato beetle *Leptinotarsa decemlineata* (Say). Insect Molecular Biology. 2016;25:44-57. DOI: 10.1111/imb.12197

[142] Wan PJ, Fu KY, Lü FG, Guo WC, Li GQ: Knockdown of a putative alanine aminotransferase gene affects amino acid content and flight capacity in the Colorado potato beetle *Leptinotarsa decemlineata*. Amino Acids. 2015;47:1445-1454. DOI: 10.1007/s00726-015-1978-1

[143] Wan PJ, Fu KY, Lü FG, Wang XX, Guo WC, Li GQ: Knocking down a putative *Δ1-pyrroline-5-carboxylate dehydrogenase* gene by RNA interference inhibits flight and causes adult lethality in the Colorado potato beetle *Leptinotarsa decemlineata* (Say). Pest Management Science. 2015;71:1387-1396. DOI: 10.1002/ps.3941

[144] Gaddelapati SC, Kalsi M, Roy A, Palli SR: Cap 'n' collar C regulates genes responsible for imidacloprid resistance in the Colorado potato beetle, *Leptinotarsa decemlineata*. Insect Biochemistry and Molecular Biology. 2018;99:54-62. DOI: 10.1016/j.ibmb.2018.05.006

[145] Ochoa-Campuzano C, Martínez-Ramírez C, Contreras E, Rausell C, Real MD: Prohibitin, an essential protein for Colorado potato beetle larval viability, is relevant to *Bacillus thuringiensis* Cry3Aa toxicity. Pesticide Biochemistry and Physiology. 2013;107:299-308. DOI: 10.1016/j.pestbp.2013.09.001

[146] Guo W, Bai C, Wang Z, Wang P, Fan Q, Mi X, Wang L, He J, Pang J, Luo X, Fu W, Tian Y, Si H, Zhang G, Wu J: Double-Stranded RNAs High-Efficiently Protect Transgenic Potato from *Leptinotarsa decemlineata* by Disrupting Juvenile Hormone Biosynthesis. Journal of Agricultural and Food Chemistry. 2018;66(45): 11990-11999. DOI: 10.1021/acs.jafc.8b03914

[147] Hussain T, Aksoy E, Çalışkan ME, Bakhsh A: Transgenic potato lines expressing hairpin RNAi construct of molting-associated EcR gene exhibit enhanced resistance against Colorado potato beetle (*Leptinotarsa decemlineata, Say*). Transgenic Research. 2019;28:151-164. DOI: 10.1007/s11248-018-0109-7

[148] Bolognesi R, Ramaseshadri P, Anderson J, Bachman P, Clinton W, Flannagan R, et al: Characterizing the Mechanism of Action of Double-Stranded RNA Activity against Western Corn Rootworm (*Diabrotica virgifera virgifera* LeConte). PLoS ONE. 2012;7(10):e47534. DOI: 10.1371/journal.pone.0047534

[149] Kota M, Daniel H, Varma S, Garczynski SF, Gould F, Moar WJ: Overexpression of the *Bacillus thuringiensis* (Bt) Cry2Aa2 protein in chloroplasts confers resistance to plants against susceptible and Bt-resistant insects. PNAS. 1999;96:1840-1845. DOI: 10.1073/pnas.96.5.1840

[150] Bally J, Fishilevich E, Bowling AJ, Pence HE, Narva KE, Waterhouse PM: Improved insect-proofing: Expressing double-stranded RNA in chloroplasts. Pest Management Science. 2018;74:1751-1758. DOI: 10.1002/ps.4870

[151] Jin S, Singh ND, Li L, Zhang X and Daniell H: Engineered chloroplast dsRNA silences *cytochrome p450 monooxygenase*, *V-ATPase* and *chitin synthase* genes in the insect gut and disrupts *Helicoverpa armigera* larval development and pupation. Plant Biotechnology Journal. 2015;13:435-446. DOI: 10.1111/pbi.12355

[152] Bally J, McIntyre GJ, Doran RL, Lee K, Perez A, Jung H, Naim F,

Larrinua IM, Narva KE, Waterhouse PM: In-Plant Protection against *Helicoverpa armigera* by Production of Long hpRNA in Chloroplasts. Frontiers in Plant Science. 2016;7:1453. DOI: 10.3389/fpls.2016.01453

[153] Zhang J, Khan SA, Hasse C, Ruf S, Heckel DG, Bock R: Full crop protection from an insect pest by expression of long double-stranded RNAs in plastids. Science. 2015;347:991-994. DOI: 10.1126/science.1261680

[154] Yu Y, Yu P-C, Chang W-J, Yu K, Lin C-S: Plastid Transformation: How Does it Work? Can it Be Applied to Crops? What Can it Offer? International Journal of Molecular Sciences. 2020;21(14):4854. DOI: 10.3390/ijms21144854

[155] He W, Xu W, Fu K, Guo W, Zhang J: Low Mismatch Rate between Double-Stranded RNA and Target mRNA Does Not Affect RNA Interference Efficiency in Colorado Potato Beetle. Insects. 2020;11(7):449. DOI: 10.3390/insects11070449

[156] Khajuria C, Ivashuta S, Wiggins E, Flagel L, Moar W, Pleau M, Miller K, Zhang Y, Ramaseshadri P, Jiang C, et al: Development and Characterization of the First Insect Population From Western Corn Rootworm, *Diabrotica virgifera virgifera* LeConte. PLoS ONE. 2018;13:e0197059. DOI: 10.1371/journal.pone.0197059

[157] Mishra S, Dee J, Moar W. et al: Selection for high levels of resistance to double-stranded RNA (dsRNA) in Colorado potato beetle (*Leptinotarsa decemlineata* Say) using non-transgenic foliar delivery. Scientific Reports. 2021;11:6523. DOI: 10.1038/s41598-021-85876-1

[158] Chen J, Peng Y, Zhang H, Wang K, Zhao C, Zhu G, Palli RS, Han Z: Off-target effects of RNAi correlate with the mismatch rate between dsRNA and non-target mRNA. RNA Biology. 2021;1-13. DOI: 10.1080/15476286.2020.1868680

[159] Haller S, Widmer F, Siegfried BD, Zhuo X, Romeis J: Responses of two ladybird beetle species (Coleoptera: Coccinellidae) to dietary RNAi. Pest Management Science. 2019;75:2652-2662. DOI: 10.1002/ps.5370

[160] Powell M, Pyati P, Cao M, Bell H, Gatehouse J, Fitches E: Insecticidal effects of dsRNA targeting the *Diap1* gene in dipteran pests. Scientific Reports. 2017;7:1-13. DOI: 10.1038/s41598-017-15534-y

[161] Sigoillot FD, King RW: Vigilance and validation: Keys to success in RNAi screening. ACS Chemical Biology. 2012;6:47-60. DOI: 10.1021/cb100358f

[162] Alemán LM, Doench J, Sharp PA: Comparison of siRNA-induced off-target RNA and protein effects. RNA. 2007;13:385-395. DOI: 10.1261/rna.352507

[163] Whyard S, Singh AD, Wong S: Ingested double-stranded RNAs can act as species-specific insecticides. Insect Biochemistry and Molecular Biology. 2009;39:824-832. DOI: 10.1016/j.ibmb.2009.09.007

[164] Bachman PM, Bolognesi R, Moar WJ et al: Characterization of the spectrum of insecticidal activity of a double-stranded RNA with targeted activity against Western Corn Rootworm (*Diabrotica virgifera virgifera* LeConte). Transgenic Research. 2013;22:1207-1222. DOI: 10.1007/s11248-013-9716-5

[165] Arora AK, Chung SH, Douglas AE: Non-Target Effects of dsRNA Molecules in Hemipteran Insects. Genes. 2021;12(3):407. DOI: 10.3390/genes12030407

[166] Dubelman S, Fischer J, Zapata F, Huizinga K, Jiang C, Uffman J, Levine S,

Carson D: Environmental fate of double-stranded RNA in agricultural soils. PloS One. 2014;9:e93155. DOI: 10.1371/journal.pone.0093155

[167] Fischer JR, Zapata F, Dubelman S, Mueller GM, Uffman JP, Jiang C, Jensen PD, Levine SL: Aquatic fate of a double-stranded RNA in a sediment–water system following an overwater application. Environmental Toxicology and Chemistry. 2017;36:727-734. DOI: 10.1002/etc.3585

[168] Jensen PD, Zhang Y, Wiggins BE, Petrick JS, Zhu J, Kerstetter RA, Heck GR, Ivashuta SI: Computational sequence analysis of predicted long dsRNA transcriptomes of major crops reveals sequence complementarity with human genes. GM Crops Food. 2013;4(2):90-7. DOI: 10.4161/gmcr.25285.

[169] Petrick JS, Frierdich GE, Carleton SM, Kessenich CR, Silvanovich A, Zhang Y, Koch M: Corn rootworm-active RNA DvSnf7: Repeat dose oral toxicology assessment in support of human and mammalian safety. Regulatory Toxicology and Pharmacology. 2016;81:57-68. DOI: 10.1016/j.yrtph.2016.07.009

[170] Dávalos A, Henriques R, Latasa MJ, Laparra M, Coca M: Literature Review of Baseline Information on Non-Coding RNA (ncRNA) to Support the Risk Assessment of ncRNA-Based Genetically Modified Plants for Food and Feed. EFSA Supporting Publication. 2019;16:EN-1688. DOI: 10.2903/sp.efsa.2019.EN-1688

[171] Andersson JO: Lateral gene transfer in eukaryotes. Cellular and Molecular Life Sciences. 2005;62:1182-1197. DOI: 10.1007/s00018-005-4539-z

[172] Pontiroli A, Simonet P, Frostegard A, Vogel TM, Monier JM: Fate of transgenic plant DNA in the environment. Environmental Biosafety Research. 2007;6:15-35. DOI: 10.1051/ebr:2007037

[173] Miki B, McHugh S: Selectable marker genes in transgenic plants: applications, alternatives and biosafety. Journal of Biotechnology. 2004;107:193-232. DOI: 10.1016/j.jbiotec.2003.10.011

[174] De Vetten N, Wolters AM, Raemakers K, Van Der Meer I, Ter Stege R, Heeres E, Heeres P, Visser R: A transformation method for obtaining marker-free plants of a cross-pollinating and vegetatively propagated crop. Nature Biotechnology. 2003;21:439-442. DOI: 10.1038/nbt801

[175] Cingel A. Vinterhalter B. Vinterhalter D. Ćalić-Dragosavac D. Smigocki A. Ninković S: *Agrobacterium*-mediated transformation of two Serbian potato cultivars (*Solanum tuberosum* L. cv. Dragačevka and cv. Jelica). African Journal of Biotechnology. 2010;9(30):4644-4650.

[176] Chong-Pérez B, Angenon G. Strategies for generating marker-free transgenic plants. In: Sithole-Niang I, editor. Genetic Engineering. IntechOpen: London, UK; 2013. p. 17-48. DOI: 10.5772/55573

[177] Guo WC, Wang ZA, Luo XL, Jin X, Chang J, He J, Tu EX, Tian YC, Si HJ, Wu JH: Development of selectable marker-free transgenic potato plants expressing *cry3A* against the Colorado potato beetle (*Leptinotarsa decemlineata* Say). Pest Management Science. 2016;72(3):497-504. DOI: 10.1002/ps.4013.

[178] Orbegozo J, Solorzano D, Cuellar WJ, Bartolini I, Roman ML, Ghislain M, Kreuze J: Marker-free PLRV resistant potato mediated by Cre-*loxP* excision and RNAi. Transgenic Research. 2016;25:813-828. DOI: 10.1007/s11248-016-9976-y

[179] Ryffel GU: Transgene flow: Facts, speculations and possible

countermeasures. GM Crops and Food. 2014;5(4):249-258. DOI: 10.4161/21645698.2014.945883

[180] Mercer KL, Wainwright JD: Gene flow from transgenic maize to landraces in Mexico: an analysis. Agriculture, Ecosystems & Environment. 2008;123:109-115. DOI: 10.1016/j.agee.2007.05.007

[181] Stewart PA, Knight AJ: Trends affecting the next generation of U.S. agricultural biotechnology: Politics, policy, and plant-made pharmaceuticals. Technological Forecasting & Social Change. 2005;72:521-534. DOI 10.1016/j.techfore.2004.03.001

[182] McPartlan HC, Dale PJ: An assessment of gene transfer by pollen from field grown transgenic potatoes to nontransgenic potatoes and related species. Transgenic Research. 1994;3:216-225. DOI: 10.1007/BF02336774

[183] Love S: Ecological risk of growing transgenic potatoes in the United States and Canada. American Potato Journal. 1994;71(10):647-658. DOI: 10.1007/BF02851433

[184] Eijlander R, Stiekema WJ: Biological containment of potato *(Solanum tuberosum)*: outcrossing to the related wild species black nightshade *(Solanum nigrum)* and bittersweet *(Solanum dulcamara)*. Sexual Plant Reproduction. 1994;7:29-40. DOI: 10.1007/BF00241885

[185] Scurrah M, Celis-Gamboa C, Chumbiauca S, Salas A, Visser RGF: Hybridization between wild and cultivated potato species in the Peruvian Andes and biosafety implications for deployment of GM potatoes. Euphytica. 2008;164:881-892. DOI: 10.1007/s10681-007-9641-x

[186] Celis C, Scurrah M, Cowgill S, Chumbiauca S, Green J, Franco J, Main G, Kiezebrink D, Visser RG, Atkinson HJ: Environmental biosafety and transgenic potato in a centre of diversity for this crop. Nature. 2004;432(7014):222-225. DOI: 10.1038/nature03048

[187] Linsdey K, Wei W, Clarke MC, Mcardele HF, Rooke LM, Topping JF: Tagging genomic sequences that direct transgene expression by activation of a promoter trap in plants. Transgenic Research. 1993;2(1):33-47. DOI: 10.1007/BF01977679

[188] Wilson A, Jonathan R, Latham R, Steinbrecher AR: Transformation-induced mutations in transgenic plants: Analysis and biosafety implications. Biotechnology and Genetic Engineering Reviews. 2006;23:209-234. DOI: 10.1080/02648725.2006.10648085

[189] Sala F, Arencibia AD, Castiglione S, Yifan H, Labra M, Savini C, Bracale M, Pelucci P: Somaclonal variation in transgenic plants. Acta Horticulturae. 2000;530:411-419. DOI: 10.17660/ActaHortic.2000.530.48

[190] Tanurdzic M, Vaughn MW, Jiang H, Lee TJ, Slotkin RK, Sosinski B, Thompson WF, Doerge RW, Martienssen RA: Epigenomic consequences of immortalized plant cell suspension culture. PLoS Biology. 2008;6:2880-2895. DOI: 10.1371/journal.pbio.0060302

[191] Ngezahayo F, Xu C, Wang H, Jiang L, Pang J, Liu B: Tissue culture-induced transpositional activity of mPing is correlated with cytosine methylation in rice. BMC Plant Biology. 2009;9:91. DOI: 10.1186/1471-2229-9-91

[192] Conner AJ, Williams MK, Abernethy DJ, Fletcher PJ, Genet RA: Field performance of transgenic potatoes. New Zealand Journal of Crop and Horticultural Science. 1994;22:361-371. DOI: 10.1080/01140671.1994.9513847

[193] Belknap WR, Corsini D, Pavek JJ, Snyder GW, Rockhold DR, Vayda ME: Field performance of transgenic Russet Burbank and Lemhi Russet potatoes. American Potato Journal. 1994;71:285-296. DOI: 10.1007/BF02849055

[194] Meiyalaghan S, Barrell PJ, Jacobs ME, Conner AJ: Regeneration of multiple shoots from transgenic potato events facilitates the recovery of phenotypically normal lines: assessing a *cry9*Aa2 gene conferring insect resistance. BMC Biotechnology. 2011;11:93. DOI: 10.1186/1472-6750-11-93

[195] Cingel A, Savić J, Ćosić T, Raspor M, Ghalawenji N, Smigocki A, Ninković, S: Phenotypic performance of transgenic potato (*Solanum tuberosum* L.) plants with pyramided rice cystatin genes *OCI* and *OCII*. Archives of Biological Sciences. 2015;67(3):957-964. DOI: 10.2298/ABS141201058C

[196] Schlüter U, Benchabane M, Munger A, Kiggundu A, Vorster J, Goulet MC, Cloutier C, Michaud D. Recombinant protease inhibitors for herbivore pest control: a multitrophic perspective. Journal of Experimental Botany. 2010;61(15):4169-4183. DOI: 10.1093/jxb/erq166

[197] Badri MA, Rivard D, Coenen K, Michaud D: Unintended molecular interactions in transgenic plants expressing clinically-useful proteins– The case of bovine aprotinin travelling the potato leaf cell secretory pathway. Proteomics. 2009;9:746-756. DOI: 10.1002/pmic.200700234

[198] Goulet C, Benchabane M, Anguenot R, Brunelle F, Khalf M, Michaud D: A companion protease inhibitor for the protection of cytosol-targeted recombinant proteins in plants. Plant Biotechnology Journal. 2010;8:142-154. DOI: 10.1111/j.1467-7652.2009.00470.x.

[199] Birch ANE, Geoghegan IE, Griffiths DW, McNicol JW: The effect of genetic transformations for pest resistance on foliar solanidine-based glycoalkaloids of potato (*Solanum tuberosum*). Annals of Applied Biology. 2002;140:143-149. DOI: 10.1111/j.1744-7348.2002.tb00166.x

[200] Munger A, Coenen K, Cantin L, Goulet C, Vaillancourt LP, Goulet MC, Tweddell R, Sainsbury F, Michaud D: Beneficial 'unintended effects' of a cereal cystatin in transgenic lines of potato, *Solanum tuberosum*. BMC Plant Biology. 2012;12:198. DOI: 10.1186/1471-2229-12-198

[201] Wilson KA. Will gene-edited and other GM crops fail sustainable food systems? In: Kassam A, Kassam L, editors. Rethinking Food and Agriculture. Woodhead Publishing; 2021. p. 247-284. DOI: 10.1016/B978-0-12-816410-5.00013-X

[202] Barros E, Lezar S, Anttonen MJ, Van Dijk JP, Röhlig RM, Kok EJ, Engel KH. Comparison of two GM maize varieties with a near-isogenic non-GM variety using transcriptomics, proteomics and metabolomics. Plant Biotechnology Journal. 2010;8:436-451. DOI: 10.1111/j.1467-7652.2009.00487.x

[203] La Paz JL, Pla M, Centeno E, Vicient CM, Puigdomènech P: The use of massive sequencing to detect differences between immature embryos of MON810 and a comparable non-GM maize variety. PLoS One. 2014;9(6):e100895. DOI: 10.1371/journal.pone.0100895

[204] Mattsson JL: Mixtures in the real world: the importance of plant self-defense toxicants, mycotoxins, and the human diet. Toxicology and Applied Pharmacology. 2007;223(2):125-132. DOI: 10.1016/j.taap.2006.12.024

[205] Hellenas KE, Branzell C, Johnsson H, Slanina P: High levels of

glycoalkaloids in the established Swedish potato variety Magnum Bonum. Journal of the Science of Food and Agriculture. 1995;68:249-255. DOI: 10.1002/jsfa.2740680217

[206] Konig A, Cockburn A, Crevel R, Debruyne E, Grafstroem R, Hammerling U, Kimber I, Knudsen I, Kuiper H, Peijnenburg A, Penninks A, Poulsen M, Schauzu M, Wal, J: Assessment of the safety of foods derived from genetically modified (GM) crops. Food and Chemical Toxicology. 2004;42:1047-1088. DOI: 10.1016/j.fct.2004.02.019

[207] de Vetten, N, Wolters A, Raemakers K, van der Meer I, ter Stege R, Heeres E, Heeres P, Visser R: A transformation method for obtaining marker-free plants of a cross-pollinating and vegetatively propagated crop. Nature Biotechnology. 2003;21:439-442. DOI: 10.1038/nbt801

[208] Rommens CM, Ye J, Richael C, Swords K: Improving potato storage and processing characteristics through all-native DNA transformation. Journal of Agricultural and Food Chemistry. 2006;54:9882-9887. DOI: 10.1021/jf062477l

[209] Rommens CM, Ya H, Swords K, Richael C, Ye J: Low-acrylamide French fries and potato chips. Plant Biotechnology Journal. 2008;6:843-853. DOI: 10.1111/j.1467-7652.2008.00363.x

[210] Haverkort AJ, Struik PC, Visser RGF, Jacobsen E: Applied biotechnology to combat late blight in potato caused by *Phytophthora infestans*. Potato Research. 2009;52:249-264. DOI: 10.1007/s11540-009-9136-3

[211] Hameed A, Zaidi SS, Shakir S, Mansoor S: Applications of New Breeding Technologies for Potato Improvement. Frontiers in Plant Science. 2018;9:925. DOI: 10.3389/fpls.2018.00925

[212] Gui S, Taning CNT, Wei D, Smagghe G: First report on CRISPR/Cas9-targeted mutagenesis in the Colorado potato beetle, *Leptinotarsa decemlineata*. Journal of Insect Physiology. 2020;121:104013. DOI: 10.1016/j.jinsphys.2020.104013

[213] Hunter MC, Smith RG, Schipanski ME, Atwood LW, Mortensen DA: Agriculture in 2050: recalibrating targets for sustainable intensification. BioScience. 2017;67:385-390. DOI: 10.1093/biosci/bix010

[214] FAOSTAT database. 2021. Available from: http://www.fao.org/ [Accessed: 2021-04-11]

[215] Douches DS, Maas D, Jastrzebski K, Chase RW: Assessment of potato breeding progress in the USA over the last century. Crop Science. 1996;36:1544-1552. DOI: 10.2135/cropsci1996.0011183X003600060024x

[216] Lindhout P, Meijer D, Schotte T, Hutten RC, Visser RG, Van Eck J: Towards F1 hybrid seed potato breeding. Potato Research. 2011;54:301-312. DOI: 10.1007/s11540-011-9196-z

[217] Weber, D: Colorado beetle: pest on the move. Pest Outlook. 2003;14:256-259. DOI: 10.1039/b314847p

[218] França FH, Plaisted RL, Roush RT, Via S, Tingey WM: Selection response of the Colorado potato beetle for adaptation to the resistant potato, *Solanum berthaultii*. Entomologia Experimentalis et Applicata. 1994;73:101-109. DOI: 10.1111/j.1570-7458.1994.tb01844.x

[219] Coombs JJ, Douches DS, Li W, Grafius EJ, Pett WL: Combining engineered (*Bt-cry3A*) and natural resistance mechanisms in potato for

control of Colorado potato beetle. Journal of the American Society for Horticultural Science. 2002;127:62-68. DOI: https://doi.org/10.21273/JASHS.127.1.62

[220] Domínguez-Arrizabalaga M, Villanueva M, Escriche B, Ancín-Azpilicueta C, Caballero P: Insecticidal activity of *Bacillus thuringiensis* proteins against coleopteran pests. Toxins. 2020;12:430. DOI: 10.3390/toxins12070430.

[221] Tabashnik BE, Brévault T, Carrière Y: Insect resistance to Bt crops: lessons from the first billion acres. Nature Biotechnology. 2013;31:510. DOI: 10.1038/nbt.2597

[222] Zhao JH, Ho P, Azadi H: Benefits of Bt cotton counterbalanced by secondary pests? Perceptions of ecological change in China. Environmental Monitoring & Assessment. 2011;173:985-994. DOI: 10.1007/s10661-010-1439-y

[223] Alyokhin A, Mota-Sanchez D, Baker M, Snyder WE, Menasha S, Whalon M, Dively G, Moarsic WF: The Red Queen in a potato field: Integrated pest management versus chemical dependency in Colorado potato beetle control. Pest Management Science. 2015;71:343-356. DOI: 10.1002/ps.3826

Chapter 10

Visiting Potato from a Breeding Perspective: Accomplishments and Prospects

Navjot Singh Brar, Sat Pal Sharma and Prashant Kaushik

Abstract

Several enhancements to the conventional potato breeding are possible though they have encouragement as well as limitations. I n this direction, the marker-assisted selection may be utilized to stack major genes as well as QTLs. Whereas the genetic transformation and genome editing methods accelerate the process of ricking of genes/transgenes. Moreover, these methodologies supplemented with the next-generation sequencing (NGS) platforms and pipelines further aid in reaching the potato ideotype. Here, we overviewed the critical topics that are related to potatoes, from general background, breeding behavior, breeding approaches employed to the potato improvement. Overall, this information complied might serve as background information that is important for potato breeders.

Keywords: Potato, varieties, heterosis, heterosis, polyploidy, wild relatives

1. Introduction

Potato is among the most important food staples that rank overall fourth after cereals (maize, wheat and rice), belonging to one of the largest genus Solanum (over 1500 species) of family Solanaceae [1, 2]. Solanaceae family comprises of about 90 genera consisting of 3000–4000 species. Potato offers a considerable component of the world's food source. From unknown until the sixteenth century in the six following centuries, potato cultivation had spread from its centre of origin, in South America into the rest of the world [3, 4]. The genetic diversity is harboured in wild relatives and landraces considered to be valuable sources of deviation for genetic enhancement and crop improvement because the genetic foundation of the modern cultivated potato is quite narrow [5]. At present, the collected developed to guarantee the long-term upkeep of potato hereditary resources and reaffirms the benefits of potato genetic resources [3]. Collaboration between potato researchers and gene bank curators promotes the utilization of the genetic resources [6].

Moreover, there are over 5000 cultivar varieties of potato-based on its size, color, shape, texture, flavor, taste, storage quality and cooking quality [7]. These varieties are differing in physiochemical properties (carotenoids and ascorbic acid content) because of the location, agronomy practices, climatic and degree of stress conditions of that area [8, 9]. The potato crop is affordable (the poorest and most undernourished households can afford), high in quality nutrients (potential high food security crop), matures rapidly (4–6 months), need moderate care (irrigation at an interval of 6–7 days), easy to cook, protect itself against microbes (impermeable to gases, water and chemicals), feed entire populations from hunger (high food security crop), easy to digest (quick breakdown with high glycemic index), used for some byproducts production (starch and alcohol) and also consumed by the animal as fodder [10]. Potato has high dietary fibers, magnesium, manganese, potassium, phosphorus, amino acids, proteins, carbohydrates, minerals, moisture, starch content, vitamins (Vit-C, B_6) as well as other antioxidants like polyphenols and carotenoids and low in fats [11, 12].

Worldwide, economic losses occur in potato because of diseases like late blight although these diseases are controlled by regular application of fungicides [13, 14]. Recent improvement in next-generation sequencing (NGS) technologies has resulted in a major reduction in the sequencing costs that makes genotyping with NGS systems cheaper and achievable [15]. Massive genotyping of the gene bank collections as well as posting the info will be a strategy to show the prospective utilization of germplasm collections in gene banks. Some gene banks have started distribution of germplasm collections together with all the genotyping information by NGS datasets [16]. NGS technologies are particularly helpful in the taxonomy that depends considerably on the herbarium specimen conceived from wild plants from the wild [17]. In this review, we have gathered the information from the general background, breeding behaviour, conventional breeding, genetic engineering to NGS methodologies employed to the potato improvement. This information is going to be a useful resource for potato breeders, offering information about the development made and prospects of reading a potato ideotype.

2. Taxonomy

S. tuberosum further diverges into two subspecies:

a. *andigena*: It is a diploid grown mainly in the Central and South American regions and is adapted to short day conditions.

b. *tuberosum*: It is a tetraploid potato with worldwide cultivation [18–20]. A general belief is that the subspecies *tuberosum* has descended from subspecies *andigena* introduced to Europe that later adapted to longer day lengths [21].

The section Petota is splits in to 8 cultivated and 228 wild species of potato, which are further grouped into 21 taxonomic series (19 tuber bearing+2 non-tuberous) [22]. Out of the cultivated species, only *S. tuberosum* ssp. *tuberosum* is extensively cultivated all around the globe, while others are cultivated especially in the Andean nations.

Sr No	Solanum species	Chromosome number	Ploidy level
1	*Solanum ajanhuiri* *Solanum goniocalyx* *Solanum phureja* *Solanum stenotomum*	2n = 2x = 24	Diploid
2	*Solanum chaucha* *Solanum juzepczukii*	2n = 3x = 36	Triploid
3	*Solanum tuberosum ssp. tuberosum* *Solanum tuberosum ssp. andigena*	2n = 4x = 48	Tetraploid
4	*Solanum curtilobum*	2n = 5x = 60	Pentaploid

3. Origin and evolution

Series *tuberosa* (containing *S. tuberosum*) and other series of subsection *potatoe* have two centres of diversity. One is a long-stretching Andean terrain in Argentina, Bolivia, Colombia, Ecuador, Peru, and Venezuela, while the other is in central Mexico [18, 23–31]. This theory is based on the fact that the plants originally introduced into Europe were late flowering and tuberising, and the morphological description [32]. Such transition can take place in a fairly short period of approximately ten years of selection [33]. An alternative school of thought is that, after the potato blight epidemic in Europe, new germplasm of *S. tuberosum subsp. tuberosum, which* originated from Chile was introduced into Europe [18].

Hawkes [18] and Grun [34] opined that the cultigenic species *S. stenotomum* is the most primitive and progenitor of all other cultivated material. *S. leptophyes* Bitt. has been theorized as the probable progenitor of *S. stenotomum* based on morphological similarity [18]. The first cultivated material of *S. stenotomum*, has also been considered to be domesticated from *S. brevicaule* complex genepool [25, 34–37]. With advent of molecular techniques seven different chloroplast haplotypes were distinguished in a selection of wild and cultivated species [38]. Kardolus et al. [39] revealed that *S. tuberosum* subsp. *tuberosum* forms a cluster with *S. multidissectum* and *S. canasense* in the Brevicaule complex. *S. tuberosum* is believed to be a straight tetraploid of *S. stenotomum* by some workers but some evidence strongly support the allotetraploid origin of *S. tuberosum* [40]. As per another report the cultivated species are in the same clade as the northern Brevicaule clade that consists of *S. bukasovii*, *S. ambosinum*, *S. canasense*, *S. leptophyes*, *S. achacachense* Card. and *S. multidissectum* [41]. Multiple origin from *S. stenotomum* is believed to be cause of rising of initial populations of *S. tuberosum* subsp. *andigena* [42].

To summarize, first diploid cultivated material (*S. stenotomum*) has probably descended from one of the species in the Brevicaule complex. Sexual polyploidization, accompanied by hybridization and human selection led to the development of tetraploid landraces (*S. tuberosum* subsp. *andigenum*). However, there is an absence of sufficient molecular data to point out a particular wild ancestral species.

4. Domestication of potato

Spanish conquerors introduced potato into the European countries by the 16th century [18, 43–46]. There are two competing theories about the nature of the first material to be introduced into Europe. Grun [34] and Hawkes [18] suggested the very first potato material brought to Europe consisted of *S. tuberosum* subsp. *andigena* from the Andes, quite probably from Colombia. The late blight epidemic in Europe during the 1840s led to the destruction of most of the original stock of potato. In the post epidemic period, new introductions consisted mainly of *S. tuberosum* subsp. *tuberosum.* Whereas Juzepczuk and Bukasov [47] were of the opinion that the subsp. *tuberosum* germplasm from Chile was already a part of early introductions in Europe, as morphology and growing conditions of early European plants and Chilean material bore similarities. Chilean potatoes were suitable for growing in Europe as they were adapted for tuberization under long-day conditions. DNA analysis of the historical herbarium specimens suggested that although Andean potato arrived first but Chilean potato was present long before late blight epidemics in Europe [44].

Introduction of potato to the Bengal floodplains, Nile delta, Morocco and Nigeria was made by European colonizers, colonial governors, missionaries [48, 49].

Emigrant farmers carried the potato to Australia and South America that led to the establishment of the potato in Argentina and Brazil. The tuber spread was along the old Asian routes through the Caucasus to Turkey, and from Russian federation to western China [31].

During the 20th century, potato emerged as a truly global food. After the Second World War, the potato was grown on a huge span of arable land in Germany and Britain, and potato has surpassed cereal production in Belarus and Poland. Since 1960s, cultivation of potato has been expanding in the ever-developing world [50, 51], it is grown as a cash crop in Bangladesh.

5. Floral biology

Potato inflorescence is terminal comprising 1–30 (but usually 7–15) flowers, depending on the type of cultivar [52–55]. The inflorescence is cymose, and flowers are actinomorphic and hypogynous. Arrangement of floral parts is regular. Five petal arrangement of the flower gives it a star shape [56]. Depending upon the cultivar, shape and size of lobes of sepals vary. The androecium comprises of five stamens alternating with the petals. The anthers collectively form a cone shaped structure to conceal the ovary [55]. Anthers are bright yellow or orange coloured except in case of male sterile plants in which the colour of anthers is light yellow or yellow green [57]. The ovary is superior and bilocular with ovules arranged at the periphery of the placenta.

Details of the *S. tuberosum* inflorescence are given below:

Inflorescence	**Solitoryor cymose**
Flower	Bisexual, actinomorphic
Calyx	Sepals five, united, persistent valvate aestivation
Corolla	Petals five, united, velvate aestivation
Androecium	Stamens five, epipetalous
Gynoecium	Bicarpellary syncarpoes, ovary superior bilocular, placenta with many vacuoles
Fruits	Berry or capsule
Seeds	Many, endospermous

Colour of the corolla varies from white to complex range of blue, red, and purple [53]. Opening of flowers start near the base of the inflorescence and proceed upward at the rate of about 2–3 flowers each day [54]. Long day length accompanied by high humidity and low temperature are conducive for potato flowering [57, 58]. Flower production and berry setting is favoured by 12–14-hour photoperiod and night temperature of 12-15°C [59, 60]. Short day duration at the time of flowering may result in abscission of floral bud [58]. Flower and fruit production in potato is influenced by several factors such as genotype, temperature, photoperiod, inflorescence position, plant/stem density, competition between flower and tuber, precipitation, date of planting and nutrient level [61–65]. Flowers remain open for 2–4 days, and out of this duration, pollen production and stigma remains receptive for about 2 days [57]. The fruit type is a berry, and are spherical to ovoid in shape, about 14 cm in diameter. Berries are green in colour or green-tinged, and upon ripening bear white or purple spots or bands [53, 66].

Floral bud abscission occurs in case of short days at the time of flowering, hence giving the impression of poor flowering of a cultivar [58]. Thus, conditions favourable for flowering and fruiting in tropics and subtropics can be found at higher altitudes (1500 m above sea level) [67]. Characteristics like days to flowering, flowering duration, the intensity of flowering and fruit set have wide genetic diversity [60]. A survey on flowering behaviour, male sterility and berry set was conducted across 25 countries by Gopal [67]. Flowering initiated after 6–15 weeks of planting and duration of flowering ranged from 1 to 10 weeks. The setting of berries ranged from 0 to more than 5 berries/plant, while there no setting in 31.8% of accessions in blooming. Production of flowers and fruits is influenced by several factors like temperature, photoperiod, genotype, inflorescence position, plant/stem density, flower and tuber competition, precipitation, date of planting and nutrient level [61–65]. The number of primary flowers increased with increase in plant density while the proportion of flowers on lateral stems reduced [62].

6. Pollination

Potato is predominantly a self-pollinated plant and is occasionally cross-pollinated [54, 56]. Generally, diploid wild species are insect-pollinated and cross-breeding in nature. Presence of insects is imperative in facilitating cross-breeding and selfing in potato. Bumblebee species like *Bombus terricola* and *B. impatiens* are particularly good pollinators for potatoes [68, 69]. European honey bee (*Apis mellifera*) and *B. fervidus* do not contribute to the pollination of potato, as the flowers are devoid of nectar [70]. Despite the lack of pollinator resources provided by the crop, a great diversity of bees was recorded in a potato-dominated agroecosystem [71]. Wind does not play any role in the pollination of potato, and no seed set was observed [68]. There are no detailed studies of pollination behavior of potato in India. Controlled pollination can be achieved under field or greenhouse. However, crosses made under the field conditions are prone to losses from the environmental factors like wind, rain, heat and drought. Therefore, breeders prefer crossing in the greenhouse. The crossing should preferably be done during the early morning hours when the temperature is moderate [54].

7. Wild relatives of potato

Comprehensive taxonomic treatment by Hawkes [18] found there are 235 potato species in total, 228 outdoors and 7 cultivated potato species. Various studies, implementing advanced molecular resources with a considerable amount of samples covering a broad range of species have advised that a reconsideration of the taxonomic classification is necessary [72]. As previously, potato species are hugely sophisticated in taxonomic classification. A broad area of distribution, together with an extensive selection of altitudinal division, from sea level up to 4500 MSL, indicates a broad range of adaptation this has resulted in the huge diversity and adaptations in the potatoes [73].

Genetic diversity of the germplasm and usefulness has been the drive to incorporate wild genes into cultivated types. The achievements of the application of wild relatives for genetic improvement relies a great deal on crossability with developed species. The gene pool is essentially the most often used concept determining the level of relatedness between species [74]. Though the genepool concept has been generally accepted, efforts to utilize the genepool concept to the potato was also presented [75]. Manipulation methods to alter the ploidy level in potato have been discovered. Even important genes from the tertiary genepool could be unveiled using bridge species in the crossing, embryo rescue, and somatic hybridization [76]. Currently, potato genetic materials are preserved in gene banks around the planet and therefore, are offered for potato breeders as well as researchers [77]. Cultivated potatoes are conserved primarily as clonal collections, like a tuber, *in vitro* and cryopreservation; on the flip side, wild potato species are primarily gathered up and also retained in the type of botanical seeds [78, 79].

8. Fertility issues in potato breeding

Potato is propagated sexually by seeds and asexually by tubers [80]. Most of *Solanum* species are diploid in nature with obligate allogamy (cross-pollination) which is result of multi-allelic gametophytic self-incompatibility (S) locus, thus preventing self-fertilization among Solanum species. In contrast to this, tetraploid cultivated potato (*Solanum tuberosum* ssp. *tuberosum* L.) [81]. However, their highly heterozygous nature (interlocus and intralocus) with tetrasomic inheritance pose difficulty in genetic complexity and challenge in potato breeding, and this is further aggravated by, high genetic load due to accumulation of deleterious alleles as a result of its vegetative propagation. Severe inbreeding depression is anticipated upon selfing, which results in the reduction of seedling germination and many reproductive complexities like reduction in flowering [57].

Conventionally potato varieties are developed through hybridization and selection, with a huge investment of time and resources because of its complex multi-locus inheritance and tetraploid genome. Successful hybridization programme between different potato populations have to deal with many barriers like pre-zygotic barriers including pollen and pistil incompatibility and post-zygotic barriers like embryo and endosperm abortion, sterility and hybrid breakdown in segregating generations [82], that leads to the hindering of the breeding programmes [60]. In male-sterile plants, flowers do not produce functional anthers or viable pollen, but the ovaries usually function [57]. The failure to produce pollen may be an inherent characteristic with sterility being dominant over fertility [83]. Even after successful fertilization by overcoming these issues, development of seeds requires proper endosperm development.

Male sterility is the result of nuclear cytoplasm interactions; the predominant Ms. gene interacts with the cytoplasm, for instance, the diploid hybrids between *S. tuberosum* Group Tuberosum haploids × Group Phureja yield all or perhaps nearly all-male sterile progeny [84]. The occurrence of male sterility typically leads to issues for potato breeders, as the option of parental lines can limit the introgression of characteristics [85]. The frequency of male fertile offspring in a hybrid between the group Tuberosum and Phureja are different because of their different ploidy levels [86–88].

In the last couple of years, a pattern emerged in a group of potato breeders to reconsider the pick as a diploid species constructed from a compilation of inbred lines that capture the favourable genetic diversity accessible in cultivated and wild potatoes [89]. Inbreeding due to selfing might be useful for organizing the entire gene pool into different favourably interacting and healthy epistatic systems. Whatever the nature of its, self-compatible 2x cultivars will offer an even more appealing self-compatible source than *S. chacoense* since they will avoid the undesirable linkage drag regarding the usage of an untamed species within the development of 2x inbred lines. Loss of S-RNase functionality is a standard route to self-compatibility [90].

9. Unilateral compatibility

The endosperm balance number (EBN) seems to be very likely that a mechanism related to a loss of protein functionality results in the formation of 2n gametes. Although it is not complete, the consistency of the self-incompatible self-compatible rule indicates a link between inter- and intraspecific pollen rejection [91, 92]. The EBN concept was helpful to elucidate the nature of the pollinator result in haploid removal. The triploid block is a reproductive screen resulting from endosperm malfunction due to the epigenetic event of genomic imprinting. Evidence implies that the endosperm dosage devices are imprinted within the gametes; therefore, the similar gene being functionally different in paternal and maternal chromosomes [93].

Spooner et al. [22] proposed a concept particularly for the Potato, implementing 5 crossability groups based on self-compatible/self-incompatible systems and endosperm balance number (EBN). The main genepool of potato contains *S. tuberosum* ssp. *tuberosum* with all cultivars and landraces. All the cultivated potatoes are tetraploid (2n = 4x = forty eight) with 4EBN. Potato has a vast secondary gene pool comprising of related wild species that gives a rich, distinctive, and different supply of hereditary variation. The EBN is a unit identifying the realizations of interspecific crosses [94]. Hybridization within every group is anticipated to achieve success rather than hybridization across groups, and therefore the executions of hybridization may be predicted. Whereas the genepool principle, as well as the EBN model, provides assistance in the utilization of wild genetic resources, additionally, they provide insight into phylogenetic connection and also taxonomy. Nevertheless, species crossability are always crucial to offer concrete evidence. Potato researchers have developed strategies to conquer the hybridization screen to transfer genes from wild species of the secondary and even tertiary genepool [95].

Haploids exhibited disomic inheritance, that implies that every chromosome combined with its homolog, thus giving means for simplifying genetic research in potato. They can furthermore be well utilized for research on natural mutation and chromosome pairing accumulated at the tetraploid fitness level. In this direction, the reason behind the generation of haploids was acquiring a genetic bridge between the different genomes of Solanum species [96]. Haploids from tetraploids usually don't flower and can also be male sterile because of inbreeding throughout the tasks of haploidization [97]. Selection of haploids can result in diploid breeding lines; additionally, a particular kind of haploids are accustomed for understanding the segregation of characteristics at the tetraploid level if numerous haploids are made of a single tetraploid genotype [98]. Whereas, tuberization in potato is controlled by day length [99, 100], and plant hormones, such as gibberellin and jasmonic acid also play a crucial role in defining tuberization. Although, specific potato genotype tuberizes under a particular day length condition along with specific physiological requirements that vary from genotype to genotype. In *in vitro* studies, no particular method of tuberization is found, and it's regarded as a complex trait. Utilizing the genome sequence [101] as well as info on Ft, it was determined that the potato genomic locus StSP6A, induces movable tuberization signal. The StSP6A signal led to the induction of tuber growth at the stolon termini. They've postulated that diverse allelic deviation of this gene is connected with the domestication of potatoes.

10. Breeding behavior of potato: from conventional to new breeding technologies

Potato breeding and improvement is an uphill task owing to its complex genetic structure and multi-allelic gene action arising due to its tetraploid genome [102].

Any breeding programme relies on the objective of the programme, germplasm availability and breeding method/technique. Genetic resources of potato are quite rich as compared to any other cultivated plant consisting of about 190 wild and primitive species [103], resulting in great amount of genetic diversity readily available for exploitation. Its rich variations are also attributed due to its reproductive biology which shows there can be 40% (range 21–74%) natural cross-pollination [104]. Besides this, its tetraploid nature (2n = 4x = 48) having four sets of chromosomes entirely homologous shows random pairing at meiosis [57] further adding to its diversity and genetic variations. This sexual reproduction generates ample amount of diversity by recombining the variants of genes that arose by mutation. As a consequence, potatoes are highly heterozygous individuals that display inbreeding depression on selfing and thus become the major impediment for the exploitation of its heterosis [105, 106].

Despite the broad genetic base, progress in efforts for potato breeding is quite slow, and its genetic gains are not fixable due to the obligatory out-breeding nature. Several conventional, as well as modern breeding techniques, have been utilized for improvement in yield, processing, storage-quality [107] and against biotic stresses [108, 109]. Although conventional breeding approaches like hybridization, clonal selection, irradiation/mutagens and introgression has been successfully employed [57]. But the progress is limited and slower due to demanding tasks of introgression and phenotypic characterization of better performing individuals in successive generations. Apart from this, intraspecific incompatibilities and inbreeding depression lead to failure of trait incorporations in the polyploid crop.

Although conventional breeding has played an important role in potato improvement by developing coloured potatoes and potatoes with improved nutrients [110], but the progress is very slow. In order to overcome these challenges, biotechnological, molecular breeding and genome editing tools, considered as new breeding techniques, have played an important role to facilitate interspecies crosses, and towards augmenting and broadening of the genetic base of gene pool of cultivated material. Biotechnological techniques like *in vitro* meristem shoot tips culture have been successfully eliminated potato virus Y [111]. This method was crucial and reliable for supplying pathogen-free seed potatoes to farm [112]. Embryo culture technique has been used successfully for improving resistance to potato leafroll virus so as to circumvent interspecific incompatibility [113]. Utilization of somaclonal variation resulting heritable phenotypic changes arise during the cell culture and regeneration of potato tissue culture was reported from leaf protoplasts of 'Russet Burbank' cultivar [114] along with improved resistance to pathogens like *Phytophthora infestans*; *Alternaria solani* [114, 115] and tuber morphology [116]. The somatic fusion of potato protoplasts with protoplasts of wild relatives has also been extensively exploited for introgression of novel sources of disease and pest resistance [105, 109, 117–119].

Potato is a model crop in which transgenic or genetic engineering technology has been exploited to the maximum extent, and it is one of the first crops for which transgenic plants were regenerated [120]. Genetic engineering is an important and highly effective tool for incorporating single gene or pyramiding gene into elite potato cultivars with minimal or no disturbances to their genetic background [121]. Numerous transgenic genotypes have been developed for a wide range of traits, including pest and disease resistances; abiotic stress resistance; quality attributes for improved processing, nutrition and appearance etc. Gene silencing is another novel technique which uses RNAi for traits like increased carotenoid content and reducing cold-induced sweetening [122–124].

Apart from transgenics/genetic engineering techniques, marker-assisted breeding (MAB) has been successfully demonstrated in tetraploid potato [125] for potato cyst nematode resistance trait. Several other examples like resistance to the nematode *Globodera rostochiensis*, resistance to potato virus X and resistance to potato wart [126] are the success stories of the application of MAB in potato. But the progress in MAB is negligible as compared to other crops due to its complex tetrasomic inheritance and high allelic variation [127]. However, in the current era of genomic breeding, prediction of genomic information is the best method to use for making breeding decisions [128]. Rather than using only significant marker-trait associations to build a prediction version, genomic prediction makes simultaneous usage of all markers [129]. In potato, genomic selection (GS) models are being utilized for predicting the accuracies of prediction models for various traits like for maturity [130], tuber starch content and chipping quality [131], *Phytophthora infestans* infection, plant maturity, tuber starch yield and tuber yield have been successfully predicted using GS models [132].

For the successful application of genome editing technologies, the first and foremost requirement is the availability of efficient transformation systems. Since potato has excellent availability of genomic resources as well as genome sequence and efficient transformation systems, several workers used various genome editing approaches *viz.* zinc-finger nucleases (ZFNs) [133–137] for improving traits like herbicide resistance, modification of starch, bio-fortification and reducing anti-nutritional factors to enhance overall increased quality of produce. Earlier for targeting traits like insect resistance, proteins, vitamins and carotenoids, transgenic technology was extensively used. Still, due to their off-target, copy number variations and other drawbacks, the trend has been shifted towards these new breeding technologies whereby TALENs and more recently CRISPR/CAS9 genome editing technologies were used for targeting traits like alteration of starch composition or hormonal expressions, reduction of anti-nutritional elements, imparting herbicide resistance, improving starch quality and overcoming self-incompatibility issues.

11. Conclusions and future prospects

The genetic improvement of potato depends on germplasm sources. In the genomics era, germplasm development can be easily performed by incorporating noval alleles from wild species, landraces, cultivated varieties, and even from distantly related species. In corporating the genomics equipment will substantially enhance the effectiveness of introgressing multi genic characteristics. Introgression may be possible through sexual hybridization, or molecular manipulations. In the context of molecular manipulations, different breeding technologies as TALEN and CRISPR/Cas9 are already used to improve the potato ideotype as per the market requirements. Moreover, the potato genome sequence, as well as useful potato hereditary transformation strategies, have hugely facilitated potato genetic engineering. The commercialization of these engineered goods is challenging because of regulatory/ethical restrictions and consumer preferences.

Breeding objectives like bio-fortification, as well as the removal of anti-nutritional factors like steroidal glycoalkaloids as already achieved. Additionally, incorporation of abiotic (environmental, salinity, drought, temperature) anxiety resistance that comes with improved nutrition can facilitate potato to acclimatize in varied agro-ecological zones, therefore impeding food shortage in less fertile/ water deficit farming lands. Further expansion of food studies can establish several preliminary values to rationalize the health advantages of potato derived foods. Indeed, the potato genome sequence has facilitated the relative genomic analyses to determine the genes helpful for improving several agronomically significant characteristics as tuberization, damage of bitterness, along with ailments opposition.

Whereas, the studies concentrating on food safety and protection can offer considerable means to meet up the soaring food demands, particularly in the food-deficit countries. The rapid advancement of growing genetic engineering has supplied brand new exciting resources to produce crops with nutritional traits and enhanced yield. In this particular context, potato harvest has potential that is enormous to help with food security as it can offer inexpensive, energy food that is high at a sustainable basis.

Conflicts of interest

The authors declare no conflict of interest.

Funding

This research received no external funding.

Author details

Navjot Singh Brar[1], Sat Pal Sharma[1] and Prashant Kaushik[2*]

1 Department of Vegetable Science, Punjab Agricultural University, Ludhiana, India

2 Instituto de Conservación y Mejora de la Agrodiversidad Valenciana, Universitat Politècnica de València, Valencia, Spain

*Address all correspondence to: prakau@doctor.upv.es

References

[1] FAO . http://faostat3.fao.org/home/E.org/ [Internet]. 2019.

[2] Myers, N. A Wealth Of Wild Species. New York: Routledge, 2019, https://doi.org/10.4324/9780429048197.

[3] Machida-Hirano, R.; Niino, T. Potato genetic resources. In *The Potato Genome*; Springer, 2017; pp. 11-30.

[4] Christenhusz, M.J.; Fay, M.F.; Chase, M.W. *Plants of the World: an illustrated encyclopedia of vascular plants*; University of Chicago Press, 2017; 816 pp.

[5] Bhandari, H.R.; Bhanu, A.N.; Srivastava, K.; Singh, M.N.; Shreya, H.A. Assessment of genetic diversity in crop plants-an overview. *Adv. Pl. Agric. Res.* 2017, *7,* 00255.

[6] Halewood, M.; Chiurugwi, T.; Sackville Hamilton, R.; Kurtz, B.; Marden, E.; Welch, E.; Michiels, F.; Mozafari, J.; Sabran, M.; Patron, N. Plant genetic resources for food and agriculture: opportunities and challenges emerging from the science and information technology revolution. *New Phytol.* 2018, *217*, 1407-1419.

[7] Singh, J.; Kaur, L. *Advances in potato chemistry and technology*; Academic press, 2016; 732 pp.

[8] Saini, R.K.; Keum, Y.-S. Significance of genetic, environmental, and pre-and postharvest factors affecting carotenoid contents in crops: a review. *J. Agric. Food Chem.* 2018, *66*, 5310-5324.

[9] Escuredo, O.; Seijo-Rodríguez, A.; Rodríguez-Flores, M.S.; Meno, L.; Seijo, M.C. Changes in the Morphological Characteristics of Potato Plants Attributed to Seasonal Variability. *Agriculture* 2020, *10*, 95.

[10] Joardder, M.U.; Masud, M.H. Food Preservation in Developing Countries: Challenges and Solutions; Springer, 2019; pp 27-55.

[11] Yadav, V.; Jha, R.; Kaushik, P.; others Biochemical Composition of Potato Biowaste of Popular North Indian Cultivars. *Indian Journal of Biochemistry and Biophysics (IJBB)* 2021, *58*, 100-104.

[12] Brown, C.R.; Haynes, K.G.; Moore, M.; Pavek, M.J.; Hane, D.C.; Love, S.L.; Novy, R.G.; Miller, J.C. Stability and broad-sense heritability of mineral content in potato: iron. *Am. J. Pot. Res.* 2010, *87*, 390-396.

[13] Kirk, W.W.; Felcher, K.J.; Douches, D.S.; Coombs, J.; Stein, J.M.; Baker, K.M.; Hammerschmidt, R. Effect of host plant resistance and reduced rates and frequencies of fungicide application to control potato late blight. *Pl. Dis.* 2001, *85*, 1113-1118.

[14] Tsedaley, B. Late blight of potato (*Phytophthora infestans*) biology, economic importance and its management approaches. *J. Biol. Agric. Healthcare* 2014, *4*, 215-225.

[15] Zhang, J.; Chiodini, R.; Badr, A.; Zhang, G. The impact of next-generation sequencing on genomics. *J. Genet. Genomics* 2011, *38*, 95-109.

[16] Aksoy, E.; Demirel, U.; ÖZTÜRK, Z.N.; ÇALIŞKAN, S.; ÇALIŞKAN, M.E. Recent advances in potato genomics, transcriptomics, and transgenicsunder drought and heat stresses: A review. *Turkish J. Bot.* 2015, *39*, 920-940.

[17] Castañeda-Álvarez, N.P.; de Haan, S.; Juarez, H.; Khoury, C.K.; Achicanoy, H.A.; Sosa, C.C.; Bernau, V.; Salas, A.; Heider, B.; Simon, R. Ex situ conservation priorities for the wild relatives of potato (Solanum L. section Petota). *PloS One* 2015, *10(6),* e0129873.

[18] Malhi, G.S.; Kaur, M.; Kaushik, P. Impact of Climate Change on

Agriculture and Its Mitigation Strategies: A Review. *Sustainability* 2021, *13*, 1318.

[19] Hanneman Jr., R.E. The testing and release of transgenic potatoes in the North American Center of diversity. In *Biosafety of Sustainable Agriculture: Sharing Biotechnology Regulatory Experiences of the Western Hemisphere*, Krattiger, A.F.; Rosemarin, A. Eds.; ISAAA, Ithaca and SEI, Stockholm, 1994; pp. 47-67.

[20] OECD. Consensus document on the biology of *Solanum tuberosum subsp. tuberosum* (potato). Organization for Economic Co operation and Development, Paris, 1997; pp. 38.

[21] Anonymous. The biology of *Solanum tuberosum* (potato). Regulation Directive of the plant products Division, Agriculture and Agri–Food Canada. Organization for economic co–operation and development (OECD), 1996, *11*, 1-38.

[22] Spooner, D.M.; Ghislain, M.; Simon, R.; Jansky, S.H.; Gavrilenko, T. Systematics, Diversity, Genetics, and Evolution of Wild and Cultivated Potatoes. *Bot. Rev.* 2014, *80*, 283-383.

[23] Graves, C. The potato treasure of the Andes: From agriculture to culture. International Potato Center, 2001; pp 208.

[24] F.A.O. 2008 http://www.fao.org/potato–2008/en/potato/origins.html

[25] Ugent, D. The potato in Mexico: geography and primitive culture. *Econ. Bot.* 1968, *22*, 109-123.

[26] Child, A. A synopsis of Solanum subgenus Potatoe (G. DON) (D'ARCY) (tuberarium) (DUN.) BITTER (s.l.). *Feddes Repertorium* 1990, *101*, 209-235.

[27] Swaminathan, M.; Magoon, M. Origin and cytogenetics of the commercial potato. *Adv. Genet.* 1961, *10*, 217-256.

[28] Hawkes, J.G. Origins of cultivated potatoes and species relationships. In *Potato Genetics*, Bradshaw J.E.; Mackay G.R. Eds.; CAB International, Wallingford, 1994; pp. 3-42.

[29] Spooner, D.M.; Núñez, J.; Trujillo, G.; Del Rosario Herrera, M.; Guzmán, F.; Ghislain, M. Extensive simple sequence repeat genotyping of potato landraces supports a major reevaluation of their gene pool structure and classification. *Proc. Natl. Acad. Sci. USA* 2007, *104*, 19398-19403.

[30] OECD Safety assessment of transgenic organisms. OECD Publishing, 2008; pp 744.

[31] Smith, A.F. *Potato: A global history*. Reaktion Books, 2012; pp 142.

[32] Howard, H.W. Genetics of the potato. Logos Press, London, 1970; https://doi.org/10.1002/star.19700220812

[33] Simmonds, N.W. Studies of the tetraploid potatoes. III. Progress in the experimental re–creation of the Tuberosum group. *J. Linnean Soc. Bot.* 1966, *59*, 279-288.

[34] Grun, P. The evolution of cultivated potatoes. *Econ. Bot.* 1990, *44*, 39-55.

[35] Correll, D.S. The potato and its wild relatives. Texas Research Foundation, Renner, Texas, 1962; pp. 1-606.

[36] Miller, J.T.; Spooner, D.M. Collapse of species boundaries in the wild potato *Solanum brevicaule* complex (*Solanaceae, S.* sect. *Petota*): molecular data. *Pl. Syst. Evol.* 1999, *214*, 103-130.

[37] van den Berg, R.G.; Miller, J.T.; Ugarte, M.L.; Kardolus, J.P.; Villand, J.; Nienhuis, J.; Spooner, D.M. Collapse of morphological species in the wild potato

Solanum brevicaule complex (Solanaceae: sect. *Petota*). *Am. J. Bot.* 1998, *85*, 92-109.

[38] Hosaka, K. Who it the mother of the potato? Restriction endonuclease analysis of chloroplast DNA of cultivated potatoes. *Theor. Appl. Genet.* 1986, *72*, 606-618.

[39] Kardolus, J.P.; van Eck, H.J.; van den Berg, R.G. The potential of AFLPs in biosystematics: A first application in *Solanum* taxonomy (*Solanaceae*). *Pl. Syst. Evol.* 1998, *210*, 87-103.

[40] Kaushik, P. *Precision Vegetable Farming Technologies: An Update*; IntechOpen, 2021; ISBN 978-1-83881-922-4.

[41] Spooner, D.M.; McLean, K.; Ramsay, G.; Waugh, R.; Bryan, G. J. 2005a. A single domestication for potato based on mutilocus amplified fragment length polymorphism genotyping. *Proc. Nat. Acad. Sci. USA* 2005a, *102*, 14694-14699.

[42] Sukhotu, T.; Hosaka, K. Origin and evolution of Andigena potatoes revealed by chloroplast and nuclear DNA markers. *Genome* 2006, *49*, 636-647.

[43] Ríos, D.; Ghislain, M.; Rodriguez, F.; Spooner, D.M. What is the origin of the European potato? Evidence from Canary Island landraces. *Crop Sci.* 2007, *47*, 1271-1280.

[44] Ames, M.; Spooner, D.M. DNA from herbarium specimens settles a controversy about origins of the European potato. *Am. J. Bot.* 2008, *95*, 252-257.

[45] Peralta, I.; Spooner, D.; Knapp, S. Taxonomy of wild tomatoes and their relatives (Solanum sect. Lycopersicoides, sect. Juglandifolia, sect. Lycopersicon; Solanaceae. *Syst. Bot. Monogr.* 2008, *84*, 1-186.

[46] Knapp, S.; Peralta, I. The tomato (*Solanum lycopersicum* L.) and its botanical relatives. In *The Tomato Genome*; Causse, M., Giovannoni, J., Bouzayen, M., Zouine, M., Eds.; Springer: Berlin, Germany, 2016, pp. 7-21.

[47] Juzepczuk, S.W.; Bukasov, S.M. A contribution to the question of the origin of the potato. *Trudy. Vsesoyuzn. Szeda. Genet. Selek.* 1929, *3*, 593-611.

[48] Laufer, B. The American plant migration. Part I: The potato. Field Museum of Natural History Anthropological Series, 1938, pp. 1-132.

[49] Sauer J.D. Historical Geography of Crop Plants: A Select Roster. CRC Press, Boca Raton, FL, USA, 2017, 320 p.

[50] I.Y.P. International Year of Potato, 2008 http://www.fao.org/potato–2008/en/potato/diffusion.html

[51] Hijamns, R.J. Global distribution of the potato crop. *Am. J. Pot. Res.* 2001, *78*, 403-412.

[52] Plaisted, R.L. Potato. In: *Hybridization of crop plants*. Fehr, W.R.; Hadley, H.H., Eds.; American Society of Agronomy and Crop Science Society of America, publishers, Madison, Wisconsin, 1980, pp. 483-494.

[53] Spooner, D.M., Knapp, S. K. *Solanum stipuloideum* Rusby, the correct name for *Solanum circaeifolium* Bitter. *Am. J. Pot. Res.* 2013, *90*, 301-305.

[54] Acquaah, G. Principles of plant genetics and breeding. Blackwell publishing, Maden, MA, USA, 2007, 758 p

[55] Struik, P.C. The canon of potato science: 40. Physiological age of seed tubers. *Pot. Res.* 2007, *50*, 375-377

[56] Caligari, P.D.S. Breeding new varieties. In: *The potato crop: scientific*

basis for improvement, Harris, P.M. Ed.; Chapman and Hall, London, 1992, 909 pp.

[57] Sleper, D.A.; Poehlman, J.M. *Breeding field crops*. Blackwell publishing, 2006; pp 424.

[58] Almekinders, C.J.M.; Struik P.C. Shoot development and flowering in potato (*Solanum tuberosum* L). *Pot. Res.* 1996, *39*, 581-607.

[59] Almekinders, C.J.M. The effect of photoperiod on flowering and TPS production in the warm tropics. *Pot. Res.* 1992, *35*, 433-442.

[60] Gopal, J. Considerations for successful breeding. In: Gopal J, Khurana SMP (eds) *Handbook of potato production, improvement and postharvest management*, Haworth's Press, New York, 2006, pp 77-108

[61] Almekinders, C.J.M.; Wiersema, S.G. Flowering and true seed production in potato (*Solanum tuberosum* L.) 1. Effects of inflorescence position, nitrogen treatment and harvest date of berries. *Pot. Res.* 1991, *34*, 365-377.

[62] Almekinders, C.J.M. Flowering and true seed production in potato (*Solanum tuberosum* L.) 2. Effects of stem density and pruning of lateral stems. *Pot. Res.* 1991, *34*, 379-388.

[63] Thijn, G.A. Observations on flower induction with potatoes *Euphytica* 1954, *3*, 28-34

[64] Jauhnl, G. Zur Kenntnis des Bluuhverhaltens der Kartoffel. *Verouff. Bundesanst. alp. Landw. Admont*. 1954, *9*, 40-74.

[65] Saini, I.; Kaushik, P.; Al-Huqail, A.A.; Khan, F.; Siddiqui, M.H. Effect of the Diverse Combinations of Useful Microbes and Chemical Fertilizers on Important Traits of Potato. *Saudi Journal of Biological Sciences* 2021, doi:10.1016/j.sjbs.2021.02.070.

[66] Spooner, D.M.; Salas, A. Structure, biosystematics, and genetic resources. In: *Handbook of potato production, improvement, and post–harvest management*, Gopal, J.; Khurana, S.M.P. Eds.; Haworth's Press, Binghampton, 2006, pp. 1-39.

[67] Gopal, J. Flowering behaviour, male sterility, and berry setting in tetraploid *Solanum tuberosum* germplasm. *Euphytica* 1994, *72*, 133-142.

[68] White, J.W. Pollination of potatoes under natural conditions. *CIP Circular* 1983 *11*, 1-2.

[69] Batra, S.W.T. Male–fertile potato flowers are selectively buzz–pollinated only by *Bombus terricola* Kirby in upstate New York. *J. Kansas Entomol. Soc.* 1993, *66*, 252-254.

[70] Sanford, J.C.; Hanneman, R.E. The use of bees for the purpose of inter–mating in potato. *Am. Pot. J.* 1981, *58*, 481-485.

[71] Buchanan, A.L.; Gibbs, J.; Komondy, L.; Szendrei, Z. Bee community of commercial potato fields in Michigan and *Bombus impatiens* visitation to neonicotinoid–treated potato plants. *Insects* 2017, *8*, 30.

[72] De Haan, S.; Rodriguez, F. Potato origin and production. In *Advances in Potato Chemistry and Technology*; Elsevier, 2016, pp. 1-32.

[73] Hijmans, R.J. Diversity and ecology of the potato: the use of spatial analysis in crop science; Wageningen University. 2002.

[74] Hunter, D. Crop wild relatives: a manual of in situ conservation; Routledge, 2012; 440 p

[75] Jansky, S.H.; Charkowski, A.O.; Douches, D.S.; Gusmini, G.; Richael, C.; Bethke, P.C.; Spooner, D.M.; Novy, R.G.; De Jong, H.; De Jong, W.S. Reinventing

potato as a diploid inbred line–based crop. *Crop Science* 2016, *56*, 1412-1422.

[76] Ortiz, R. Potato breeding via ploidy manipulations. *Pl. Breed. Rev.* 1998, *16*, 15-86.

[77] Jansky, S.H.; Dempewolf, H.; Camadro, E.L.; Simon, R.; Zimnoch-Guzowska, E.; Bisognin, D.A.; Bonierbale, M. A case for crop wild relative preservation and use in potato. *Crop Science* 2013, *53*, 746-754.

[78] Maxted, N.; Ford-Lloyd, B.V.; Hawkes, J.G. *Plant genetic conservation: the* in situ *approach*; Springer Netherlands, 2013, 250 pp.

[79] Plucknett, D.L.; Smith, N.J. *Gene banks and the world's food*; Princeton University Press, 2014.

[80] Harris, P.M. The Potato Crop. The scientific basis for improvement. Chapman & Hall, London, 1978, 730 pp.

[81] Watanabe, K.N.; Orrillo, M.; Iwanaga, M.; Ortiz, R.; Freyre, R.; Perez, S. Diploid potato germplasm derived from wild and landrace genetic resources. *Am. Pot. J.* 1994, *71*, 599-604.

[82] Camadro, E.L.; Carputo, D.; Peloquin, S.J. Substitutes for genome differentiation in tuber–bearing *Solanum*: interspecific pollen– pistil incompatibility, nuclear–cytoplasmic male sterility, and endosperm. *Theor. Appl. Genet*. 2004,*109*, 1369-1376.

[83] Pushkarnath; Dwivedi, R.S. Potato floral biology series. II. Pollen and pollination. *Indian Pot. J.* 1961, *3*, 76-85.

[84] Muthoni, J.; Shimelis, H.; Melis, R. Genetics and Reproductive Biology of Cultivated Potato (*Solanum tuberosum* L.): Implications in Breeding. *Rep. Biol. Pl.* 2014, *164*-194

[85] Kempe, K.; Gils, M. Pollination control technologies for hybrid breeding. *Molec. Breed.* 2011, *27*, 417-437.

[86] Jansky, S. Overcoming hybridization barriers in potato. *Pl. Breed.* 2006, *125*, 1-12.

[87] Newbigin, E.; Uyenoyama, M.K. The evolutionary dynamics of self-incompatibility systems. *TRENDS Genet.* 2005, *21*, 500-505.

[88] Zuriaga, E.; Muñoz-Sanz, J.V.; Molina, L.; Gisbert, A.D.; Badenes, M.L.; Romero, C. An S-Locus Independent Pollen Factor Confers Self-Compatibility in 'Katy' Apricot. *PLOS One* 2013, *8*, e53947.

[89] Borrelli, G.M.; De Vita, P.; Mastrangelo, A.M.; Cattivelli, L. *Integrated views in plant breeding: modern approaches for an old topic*. Crop Physiology: Applications for Genetic improvement and Agronomy. Elsevier: New York, NY, USA, 2009, 327-354.

[90] Bradshaw, J.E. *Root and tuber crops*; Springer Science & Business Media, 2010, *7*, 10.1007/978-0-387-92765-7

[91] Ramawat, K.; Merillon, J.M.; Shivanna, K. Reproductive Biology of Plants. Boca Raton: CRC Press, 2014, https://doi.org/10.1201/b16535

[92] Brownfield L, Köhler C. Unreduced gamete formation in plants: mechanisms and prospects. *J. Exp. Bot*. 2011, *62(5)*, 1659-1668. doi:10.1093/jxb/erq37

[93] Rodrigues, J.A.; Zilberman, D. Evolution and function of genomic imprinting in plants. *Genes Dev.* 2015, *29*, 2517-2531.

[94] Machida-Hirano, R. Diversity of potato genetic resources. *Breed. Sci.* 2015, *65*, 26-40.

[95] Vincent, H.; Wiersema, J.; Kell, S.; Fielder, H.; Dobbie, S.; Castaneda Alvarez, N.; Guarino, L.; Eastwood, R.;

León, B.; Maxted, N. A prioritized crop wild relative inventory to help underpin global food security. *Biol. Conser.* 2013, *167*, 265-275.

[96] Watanabe, K. Potato genetics, genomics, and applications. *Breed. Sci.* 2015, *65*, 53-68.

[97] Gross, K.; Schiestl, F.P. Are tetraploids more successful? Floral signals, reproductive success and floral isolation in mixed-ploidy populations of a terrestrial orchid. *Ann. Bot.* 2015, *115*, 263-273.

[98] Bartkiewicz, A.M.; Chilla, F.; Terefe-Ayana, D.; Lübeck, J.; Strahwald, J.; Tacke, E.; Hofferbert, H.-R.; Linde, M.; Debener, T. Maximization of Markers Linked in Coupling for Tetraploid Potatoes via Monoparental Haploids. *Front. Pl. Sci.* 2018, *9*, 620.

[99] Rodríguez-Falcón, M.; J. Bou and S. Prat. Seasonal control of tuberization in potato: Conserved elements with the flowering re-sponse. *Ann. Rev. Pl. Biol.* 2006, *57,* 151-18

[100] Seabrook, J.E.A.; S. Coleman and D. Levy. Effect of photoperi-od on in vitro tuberization of potato (Solanum tubersoum L.). *Pl. Cell Tissue Organ Cult.* 1999, *34,* 43-51.

[101] P.G.S.C. Genome sequence and analysis of the tuber crop potato. *Nature* 2011, *475*, 189.

[102] Carputo, D.; Barone, A. Ploidy level manipulations in potato through sexual hybridization. *Ann. Appl Biol.* 2005,*146*, 71-79.

[103] Jacobs, M.M.; Smulders, M.J.; van den Berg, R.G.; Vosman, B. What's in a name; Genetic structure in Solanum section Petota studied using population–genetic tools. *BMC Evol. Biol.* 2011, *11*, 42.

[104] Brown, C.R. Outcrossing rate in cultivated autotetraploid potato. *Am. Pot. J.* 1993, 70, 725-734.

[105] Bradshaw, J.E. Potato–breeding strategy. In *Potato biology and biotechnology*. Elsevier Science BV, 2007, pp 157-177.

[106] Lindhout, P.; Meijer, D.; Schotte, T.; Hutten, R.C.; Visser, R.G.; van Eck, H.J. Towards F1 hybrid seed potato breeding. *Pot. Res.* 2011, *54*, 301-312.

[107] Halterman, D.; Guenthner, J.; Collinge, S.; Butler, N.; Douches, D. Biotech potatoes in the 21st century: 20 years since the first biotech potato. *Am. J. Pot. Res.* 2016, *93*, 1-20.

[108] Naess, S.K.; Bradeen, J.M.; Wielgus, S.M.; Haberlach, G.T.; Mcgrath, J.M.; Helgeson, J.P. Resistance to late blight in *Solanum bulbocastanum* is mapped to chromosome 8. *Theor. Appl. Genet.* 2000, *101*, 697-704.

[109] Tek, A.L.; Stevenson, W.R.; Helgeson, J.P., Jiang, J. Transfer of tuber soft rot and early blight resistances from Solanum brevidens into cultivated potato. *Theor. Appl. Genet.* 2004, *109*, 249-254.

[110] Patil, V.U.; Sundaresha, S.; Kawar, P.G.; Bhardwaj, V. Biology of *Solanum tuberosum* (Potato). Series of Crop Specific Biology Documents, 2016. https://krishi.icar.gov.in/jspui/handle/123456789/16821

[111] Kassanis, B. The use of tissue cultures to produce virus-free clones from infected potato varieties. *Ann. Appl. Biol.* 1957, *45*, 422-427.

[112] Brar, N.S.; Saini, D.K.; Kaushik, P.; Chauhan, J.; Kamboj, N.K. Directing for Higher Seed Production in Vegetables. *Agronomy-Climate Change & Food Security* 2020, 55.

[113] Eijlander, R.; Stiekema, W.J. Biological containment of potato (Solanum tuberosum): outcrossing to the related wild species black nightshade (*Solanum nigrum*) and bittersweet

(*Solanum dulcamara*). *Sexual Pl. Rep.* 1994, *7*, 29-40.

[114] Shepard, J.F.; Bidney, D.; Shahin, E. Potato protoplasts in crop improvement. *Science*, 1980, *208*, 17-24.

[115] Matern, U.; Strobel, G.; Shepard, J. Reaction to phytotoxins in a potato population derived from mesophyll protoplasts. *Proc. Nat. Acad. Sci.* 1978, *75*, 4935-4939.

[116] Thieme, R.; Griess, H. Somaclonal variation in tuber traits of potato. *Pot. Res.* 2005, *48*, 153-165.

[117] Helgeson, J.P.; Pohlman, J.D.; Austin, S.; Haberlach, G.T.; Wielgus, S.M.; Ronis, D.; Zambolim, L.; Tooley, P.; McGrath, J. M.; James, R.V.; Stevenson, W.R. Somatic hybrids between *Solanum bulbocastanum* and potato: a new source of resistance to late blight. *Theoret. Appl. Genet.* 1998, *96*, 738-742.

[118] Brown, C.R., Mojtahedi, H., James, S., Novy, R.G. and Love, S. Development and evaluation of potato breeding lines with introgressed resistance to Columbia root-knot nematode (*Meloidogyne chitwoodi*). *Am. J. Pot. Res.* 2006, 83, 1-8.

[119] Thieme, R.; Rakosy–Tican, E.; Nachtigall, M.; Schubert, J.; Hammann, T.; Antonova, O.; Gavrilenko, T.; Heimbach, U.; Thieme, T. Characterization of the multiple resistance traits of somatic hybrids between *Solanum cardiophyllum* Lindl. and two commercial potato cultivars. *Pl. Cell Rep.* 2010, *29*, 1187-1201.

[120] An, G.; Watson, B.D.; Chiang, C.C. Transformation of tobacco, tomato, potato, and *Arabidopsis thaliana* using a binary Ti vector system. *Pl. Physiol*, 1986, *81*, 301-305.

[121] Conner, A.J.; Jacobs, J.M.E.; Genet, R.A. Transgenic potatoes versus "traditional" potatoes: what's the difference. In *Commercialization of Transgenic Crops: Risk, benefit and trade considerations*, McLean, G.D.; Waterhouse, P.M.; Evans, G.; Gibbs, M.J.; Australian Government Publishing Services, Canberra, 1997, pp. 23-36.

[122] Chen, S.; Hajirezaei, M.R.; Zanor, M.I.; Hornyik, C.; Debast, S.; Lacomme, C.; Fernie, A.R.; Sonnewald, U.; Boernke, F. RNA interference mediated repression of sucrose phosphatase in transgenic potato tubers (*Solanum tuberosum*) strongly affects the hexose to sucrose ratio upon cold storage with only minor effects on total soluble carbohydrate accumulation. *Pl. Cell Environ.* 2008, *31*, 165-176.

[123] Bhaskar, P.B.; Wu, L.; Busse, J.S.; Whitty, B.R.; Hamernik, A.J.; Jansky, S.H.; Buell, C.R.; Bethke, P.C.; Jiang, J. Suppression of the vacuolar invertase gene prevents cold–induced sweetening in potato. *Pl. Physiol.* 2010, *154*, 939-948.

[124] Chawla, R.; Shakya, R.; Rommens, C.M. Tuber specific silencing of asparagine synthetase reduces the acrylamideforming potential of potatoes grown in the field without affecting tuber shape and yield. *Pl. Biotechnol. J.* 2012, *10*, 913-924.

[125] Moloney, C.; Griffin, D.; Jones, P.W.; Bryan, G.J.; McLean, K.; Bradshaw, J.E.; Milbourne, D. Development of diagnostic markers for use in breeding potatoes resistant to *Globodera pallida* pathotype Pa2/3 using germplasm derived from *Solanum tuberosum* ssp. *andigena* CPC 2802. *Theor. Appl. Genet.* 2010, *120*, 679-689.

[126] Gebhardt, C.; Bellin, D.; Henselewski, H.; Lehmann, W.; Schwarzfischer, J.; Valkonen, J.P.T. Marker–assisted combination of major genes for pathogen resistance in potato. *Theor. Appl. Genet.* 2006, *112*, 1458-1464.

[127] Luo, Z.W.; Hackett, C.A.; Bradshaw, J.E.; McNicol, J.W.; Milbourne, D. Construction of a genetic linkage map in tetraploid species using molecular markers. *Genetics* 2011, *157*, 1369-1385.

[128] Meuwissen, T.H.E., Hayes, B.J., and Goddard, M.E. Prediction of total genetic value using genome–wide dense marker maps. *Genetics* 2001, *157*, 1819-1829.

[129] Windhausen, V.S.; Atlin, G.N.; Hickey, J.M.; Crossa, J.; Jannink, J.L.; Sorrells, M.E.; et al. Effectiveness of genomic prediction of maize hybrid performance in different breeding populations and environments. *G3: Genes, Genomes, Genetics* 2012, *2*, 1427-1436.

[130] Slater, A.T.; Cogan, N.O.; Forster, J.W.; Hayes, B.J.; Daetwyler, H.D. Improving genetic gain with genomic selection in autotetraploid potato. *Pl. Genome* 2016, *9*, 3.

[131] Sverrisdóttir, E.; Byrne, S.; Sundmark, E.H.R.; Johnsen, H.Ø.; Kirk, H.G.; Asp, T.; Janss, L.; Nielsen, K.L. Genomic prediction of starch content and chipping quality in tetraploid potato using genotyping–by–sequencing. *Theor. Appl. Genet.* 2017, *130*, 2091-2108.

[132] Stich, B.; van Inghelandt, D. Prospects and potential uses of genomic prediction of key performance traits in tetraploid potato. *Front. Pl. Sci.* 2018; *9*, 159.

[133] Jinek, M.; Chylinski, K.; Fonfara, I.; Hauer, M.; Doudna, J.A.; Charpentier, E. A programmable dual–RNA–guided DNA endonuclease in adaptive bacterial immunity. *Science* 2012, *337*, 816-821.

[134] Schaart, J.G.; van de Wiel, C.C.; Lotz, L.A.; Smulders, M.J. Opportunities for products of new plant breeding techniques. *Trends Pl. Sci.* 2016, *21*, 438-449.

[135] Weeks, D.P. Gene editing in polyploid crops: wheat, camelina, canola, potato, cotton, peanut, sugar cane, and citrus. In *Progress in molecular biology and translational science*, Volume 149, Academic Press, 2017, pp. 65-80.

[136] Nadakuduti, S.S.; Buell, C.R.; Voytas, D.F.; Starker, C.G.; Douches, D.S. Genome editing for crop improvement–applications in clonally propagated polyploids with a focus on potato (*Solanum tuberosum* L.). *Front. Pl. Sci.* 2018, *9*, 1607.

[137] Korobko, I.V.; Georgiev, P.G.; Skryabin, K.G.; Kirpichnikov, M.P. GMOs in Russia: research, society and legislation. *Acta Nat.* 2016, *8*, 6-13.

Chapter 11

Impact and Management of Diseases of *Solanum tuberosum*

Olusola L. Oyesola, Oluwadurotimi S. Aworunse, Margaret I. Oniha, Onyemaechi H. Obiazikwor, Oluwakemi Bello, Olubunmi M. Atolagbe, Ayodele A. Sobowale, Jacob O. Popoola and Olawole O. Obembe

Abstract

Solanum tuberosum (Potato) is one of the essential economic crops with the potential to reduce hunger due to its high yield per unit area of land compared with many economic crops. However, its yield losses due to pest and disease attacks could be as high as 100%, depending on its tolerance level and pest and disease. Over the years, several disease management strategies have been researched, ranging from synthetic pesticides to the formulation of biopesticides as disease control measures. Moreso, recent breakthroughs in genetic engineering have simplified plant disease management strategies by developing techniques for conferring resistance on plants. Potato is a vital food crop worldwide, and with the struggle to suppress world food insecurity, effective disease management strategies must be employed for high production of quality and quantity potato, enough to feed the ever-increasing world population. Therefore, attention must be given to how disease-free potatoes can be produced to meet the unending demand for food by the continually increasing world population.

Keywords: Potato, Disease-free, Crops, Pathogens, Biocontrol, Resistance

1. Introduction

Potato (*Solanum tuberosum* L.) is the most popular vegetable crop of great importance worldwide and follows only wheat and rice as a food crop [1]. It is a source of carbohydrate being a starchy vegetable; it is, however, as a vegetable a very significant source of vitamin C, potassium, and dietary fibre as well as magnesium, vitamin B6, iron, carotenoids, and phenolic acids [2, 3]. It grows in a wide range of climates and is adopted by a broad range of cultures [4]. Potato is a critical alternative to the major cereal crops for feeding the world's population [5]. However, its production has two main challenges: disease and nutrient management [6]. Pathogens such as bacteria, fungi, viruses, nematodes, and phytoplasmas attack potato plants, causing diseases, which result in a significant loss of yield [7]. Naming a pathogen that negatively affects the host's health is the primary means to define any disease [8]. For instance, potato plays host to heterothallic species, *Phytophthora infestans* [9], which causes late blight disease. This single pathogen caused severe devastation in the late 1840s in Europe [10] and cost Ireland 25% of its population in just four years [11]. Potatoes still have many diseases, but many other alternative crops make most countries not depend on potatoes like Ireland in the 1800s [12]. In recent time, potato crop loss due to late blight disease alone is estimated at $6.7 billion annually worldwide [5]. A significant challenge to the management of *P. infestans* is the rate at which it adapts to control strategies [5]. More research on epidemiology and the host-pathogen interaction is needed to devise the most appropriate management strategy [7]. Also, insight into pathogen population dynamics offers an essential input for effective disease management [13].

Meanwhile, effective management of the disease requires implementing an integrated disease management approach [14]. Guchi [7] proposed investigating several control options and implementing an integrated management strategy based on local needs [7]. Therefore, this chapter aims to discuss the general/overall impact of the diseases of *Solanum tuberosum* as well as their management. This would increase awareness and awaken researchers' intervention to develop globally effective control or management strategies.

2. Some host pathogens and diseases of potato

Diverse host-pathogens are associated with the different diseases of potatoes, among which are bacteria and fungi. Plant pathogens responsible for diseases in potatoes include viruses, fungi, oomycetes, and bacteria [15]. A pathogenic bacterium known as *Ralstonia solanacearum* is responsible for the devastating bacterial wilt of potato and other solanaceous plants [16, 17]. The bacterium, *Ralstonia solanacearum,* is a gram-negative, non-spore-forming, aerobic, soil-borne motile pathogen that hinders tuber production resulting in economic losses [16–18]. It is distributed worldwide, affecting more than 200 economically essential crops, including potato [19]. The pathogen, usually disseminated by infected seed tuber, soil, water, and farm machinery [20], penetrates to infect the roots through wounds or natural openings and rapidly propagates within the host to attack the plant's vascular system. Consequently, it forestalls the translocation of nutrients and water, culminating in wilt, collapse and complete deadening of the plant and its decay [21, 22]. The ubiquitous plant pathogenic fungus of *Colletotrichum coccodes* is responsible for the blemish disease of potatoes called black dot [23].

The typical characteristic of black dot disease is the microsclerotia on infected tissue present in all potato parts. These microsclerotia, which usually survive in the soil for lengthy periods, lead to high disease incidence when soil inoculum levels increase [24]. Sequel to fungal colonisation of roots are colonisations in the stems, stolons, and tubers [25], and fungal contamination of tubers with *C. coccodes* leads to the development of lesions on the epicarp and loss of water during storage [26, 27]. The potato late blight disease is caused by *Phytophthora infestans* [28]. It affects the potato foliage and tubers. The foliage symptoms begin with brown to black, water-soaked lesions on leaves and stem that produce visible white spores at the lesion margins under humid conditions. This may result in the rapid collapse of the entire plants and orchards. Sporangia in the soil from the foliage initiate the tuber infection that starts from the wounds, eyes, or lenticels. The lesions appear as copper brown, red, or purplish, and white spores appear on tuber surfaces in storage.

Streptomyces spp. is the bacterial pathogen responsible for common scab in potato, and characteristic tan to dark brown, circular or irregular lesions rough in texture are produced. The scab may be superficial (russet scab), slightly raised (erumpent scab), or sunken (pitted scab). Its lesion type is determined by potato cultivar, maturity of tuber at infection, soil organic matter content, pathogen strain, and the environment [29]. Another disease caused by the bacterium is soft rot, which is the most destructive of all storage diseases caused by *Erwinia carotovora*. The disease symptoms include tan- to brown-coloured water-soaked areas of granular, mushy tissue often outlined by brown to black margins. During storage periods, soft rot bacteria penetrate tubers already infected with other potato diseases. The rottening from bacterial penetration is accelerated by the heat generated from the intense respiration in the storage environment.

Early blight of potatoes, caused by *Alternaria solani*, usually affects its leaves, but tuber infections can also occur. The lesions found in the tubers are dark, sunken, and circular, usually surrounded by purple to grey raised tissue. Its underlying tissues are void of moisture, leathery and these brown lesions may have increased during storage with shrivelled tubers [29]. *Fusarium sambucinum* or *F. coeruleum* is responsible for dry rot that causes inner light to be dark brown or black dry crumbly rot of potato with collapsed tissue often lined with secondary white other-coloured fungal growth. This rot may commence at an injury site (bruise or cut), and the fungus penetrates the tuber to rot out its centre. In furtherance, the extensive rotting results in the shrinking and complete collapse of tissue and usually leaves a dark sunken area outside the tuber and internal cavities [29]. The silver scurf, caused by *Helminthosporium solani*, infects only the uber periderm (skin). The lesions appear first at the stolon end as small pale brown spots that may be difficult to detect at harvest but continues development during storage. While in storage, these lesions darken, sloughing off the skin occurs with many small circular lesions coalescing to form large lesions. The potato tubers tend to dry out and become wrinkled from excessive moisture loss during storage [29]. The fungus *Rhizoctonia solani* causes the black scurf disease, which does not reduce yield, even in storage. Fungal sclerotia develop in irregular, black hard masses on the tuber surface that harvesting tubers may reduce immediately after vine-kill and skin set. Sclerotia allow the pathogens to survive in the soil. Inside wet soils, *R. solani* may induce dark, sunken lesions on underground sprouts and stolons with consequent deprivation of nutrients, the complete killing of the potato tubers, reduction in transfer of starches (results to reduced sizes) [29].

Pink rot infections caused by *Phytophthora erythroseptica* commence at the stolon end and culminates in rotten, internal rubbery skin that turns pink after about 15 to 20 minutes of exposure to warm air (with a clear delineation between healthy and diseased tissue). On exposure to air, the tuber flesh turns pink and then brown-black. The fungal pathogen *Pythium* spp. is responsible for leak infections, penetrates tubers through harvest wounds, and continues to grow in transit and storage. Its infections develop into internal watery, grey, or brown rot, but the outer cortex remains intact, with well-defined red-brown lines demarcating healthy and infected tissue [29].

Viruses are among the predominant phytopathogens that cause approximately 50% of all emerging plant diseases [15]. Potato virus Y (PVY) is one of the most harmful viruses infecting potatoes across the globe since the 1980s [30].

3. The impacts of diseases on the yield (quality and quantity) of potato

In 2013, more than 368 million tonnes were produced from 19.4 million hectares [31]. Though hundreds of varieties of potato are grown in temperate and sub-tropical areas, its diversification in various agroclimatic conditions leads to a decrease in its production and productivity due to its low genetic base and various biotic factors, which makes it susceptible to many devastating diseases. The crop infection due to fungi, viruses, bacteria, and viroids alters its metabolism. These pathogens affect the crop's morphological, physiological, and biochemical characteristics leading to altered distribution of photoassimilates, with resultant effects on its quality ad quantity.

Viral diseases of potatoes are devastating because they are tough to manage and transmitted via the tubers to subsequent generations. Viruses have the potential to alter the physiology of potato plants drastically, causing disorders. These disorders of growth processes cause stunting, leaf deformation, dwarfing, and reduction in the yield of potato tubers and product quality up to 88% [32–34]. Tens of potato viruses have been discovered and characterised, and the most cataclysmic are: Potato virus M (PVM): Potato virus S (PVS); Potato virus X (PVX); Potato virus Y (PVY); and Potato leaf roll virus (PLRV, virus L). PVX can debilitate 10–40% of potato in a single infections cycle and possess enormous devastating effects when combined with other potato viruses; due to its synergistic interaction with potyviruses, tuber losses yield close to 80% [33]. For example, the yield of potato simultaneously infected with PVM and PVX will decline to 60%, and when it is a complex infection of PVM + PVX + PVY, it will decline by 83.7%, i.e., total loss of yield [35]. In potato tubers infected with viral diseases, the content nutrients become reduced compared to healthy ones. Other biochemical and physiological changes also occur, resulting in a decrease in the quantity and quality of starch grains in the debilitated tissues, the acidity of starch, and amylase content [36]. There are varying losses in potato production from viruses; they are determined by the variety's resistance, the viral pathocomplexity, the level of spread of a specific virus, and their combinations with other viruses [37].

Bacterial diseases are one significant biotic constraint of potato production in the subtropical and tropical regions. Several bacterial diseases devastate potato, resulting in severe damages, especially on tubers, leading to economic losses. The most acute diseases are bacterial wilt caused by *Ralstonia solanacearum* [38] and the backleg caused by *Pectobacterium atrosepticum, P. carotovorum* subsp. *brasiliensis, P. wasabiae, Dickeya solani* and *D. dianthicola* [39, 40]. Loss of yield in potato crop is due to bacterial diseases that could be direct and indirect. There are specific facets: short-term impacts like yield loss and unvendability, and others with long-term impacts with environmental, economic, and social effects [39].

To date, potato late blight is still one of the most devastating diseases in potato-producing regions worldwide and causes substantial economic losses of about 25–57%. Pathogenic fungus, *Phytophthora infestans,* are responsible for late blight disease in potato. Late blight disease is highly destructive and one of the diseases threatening global food security [41]. Its outbreak in Ireland resulted in famine, which led to millions of people's starvation and eventual death and subsequent continuous significant losses of potatoes worldwide. Therefore, it remains the most debilitating disease of the food crop, which causes annual potato losses sufficient enough to feed several millions of people [42]. Despite the apparent debilitating potential of late blight, it is tough to estimate losses because of other environmental factors that simultaneously affect potato yield.

Meanwhile, the economic impact of potato late blight in the USA was appraised to be around 210 million US dollars, while a worldwide assessment of potato loss by late blight in the second world countries based on an average production was about 15%. This represents approximately 2.75 billion US dollars loss in developing countries. However, a critical method of estimating the economic impact of potato late blight is by determining the usage of fungicide. With this method, the estimated fungicide currently used in developing countries stands at 750 million US Dollar. Therefore, about 1 billion US Dollar is spent on fungicides yearly to manage fungal disease worldwide [43].

4. Management strategies of the diseases of potato

Potato is among the high-income-yielding crops globally and can contribute to poverty reduction in developing regions [44]. However, Potato cultivation is beset with several diseases caused by diverse pathogens in the field and during storage, accounting for 50 to 60% of annual losses [45–47]. Control strategies that have been deployed to manage diseases in potato include the application of chemical fungicides, biological control agents, and cultural practices involving crop rotation.

4.1 Chemical control

Diseases caused by fungi are critical in potato production and require several synthetic fungicide options to reduce them to tolerable economic levels. Fungicides are preparations of different organic and inorganic compounds which can inhibit or destroy phytopathogenic fungi. These chemicals exert their effects by disrupting cell membranes of their targets or instigating catalytic enzymes in plant host tissue to suppress fungal growth and proliferation [48]. Practically, conventional management of potato diseases relies on the timely application of preventive fungicides [48, 49]. To control black rot disease, seed tubers are immersed in the fungicides thiabendazole, captofal, chloramizol sulphate, prochloraz, or a combination of each before field planting. Pencycuron and thiabendazole have also been documented to control black scurf and silver scurf effectively, respectively [26, 27]. Rahman et al. [50] demonstrated the effectiveness of Filthane M-45, Melody Duo, Secure, Metaril, and Ridomil gold to minimise Phytophthora infestans-induced late blight improve the yield of potato. More so, the application of dimethomorph, mancozeb, and fenamidone + mancozeb can significantly reduce the severity of late blight and increase potato yield [51]. The application of the antagonist *Trichoderma harzianum* combined with flutolanil seed dressing offers protection against *Rhizoctonia solani* damage throughout the growing season (Wilson et al., [52]. Although fungicides have been shown to manage potato diseases effectively, they are not without their attendant problems. It is now known that continuous application of fungicides results in resistance in many pathogenic fungi of potato. Whereas metalaxyl containing fungicides show good action against *Phytophthora infestans*, prolonged applications have resulted in resistant *P. infestans* [53]. Several metalaxyl-insensitive genotypes of *P. infestans* have been reported in different regions of the world. For example, in 1980, phenylamide resistant isolates of *P. infestans* were detected on field-grown potatoes in Netherlands, Switzerland, and Ireland [48, 49, 54]. In addition to fungicide resistance, the harmful consequences on non-target organisms, risk to soil environment, and carcinogenic potentials have discouraged the use of synthetic fungicides, thereby prompting the search for efficient, safe, and eco-friendly disease management options [55, 56].

4.2 Biological control

Disease management using biological control agents is touted as efficient alternatives to chemical fungicides as they are more eco-friendly and reduce the risk of the emergence of fungicide-resistant strains of plant pathogens [57, 58]. A biological control refers to the application of microbial antagonists or their by-products to inhibit plant diseases. Organisms that antagonise plant pathogens are known as biological control agents (BCAs). Such organisms are highly specific in their action against target pathogens, their products are biodegradable, and their mass production requires low cost [59, 60].

Here, we discuss the biological control strategies – microbial inoculants (beneficial, non-pathogenic single-strains of microorganisms that antagonise plant pathogens), microbial consortium (combination of different genera or species of symbiotically living microorganisms) isolated from the natural environment, and the application of phytoextracts [61–63].

4.3 Microbial inoculants

These are single strains of active beneficial microorganisms that offer protection against diverse pathogens or promote crop productivity and health when applied to crops or incorporated into the soil [63]. Microbial inoculants are an effective and cheap alternative strategy to reduce the severity of plant diseases [64–66]. *Agrobacterium*, *Pseudomonas*, *Bacillus*, *Alcaligenes*, *Streptomyces,* and others have been reported as effective bacterial control agents [16, 17, 60]. These organisms suppress bacterial and fungal pathogens by releasing active compounds, including siderophores, antibiotics, enzymes, and the plant hormone, indole-1,3-acetic acid. *Pseudomonas* strain has been widely investigated for their potential as BCAs because of their active nature and abundance in the rhizosphere [60]. Tariq et al. [67] demonstrated the antagonistic potential of *Pseudomonas* sp. StS3 against *Rhizoctonia solani,* which causes potato black scurf. *Streptomyces violaceusniger* AC12AB promoted growth by 26.8% and significantly reduced potato typical scab disease severity by up to 90% in field trials [66]. In addition to enhancing potato tuber biomass by 33% and 22% in two location field trials, *Bacillus amyloliquefaciens* strain BAC03 considerably reduced the severity of potato scab disease by 17–57% compared to control. BAC03 also enhanced potato tuber weight by 33% and 26% in the two locations [68].

4.4 Microbial consortium

This combination of BCAs consists of various microbial strains that synergistically confer enhanced plant growth activities and superior pathogen inhibition capabilities [69–71]. Compared to single-species microbial inoculants, the microbial consortium is more useful in field applications as it offers a wide range of biocontrol activities that promote inoculant efficiency and, in turn, improve plant growth and disease suppressability [56]. The application of a microbial product comprising a consortium of *Bacillus subtilis* and *Trichoderma harzianum* inhibited common scab disease in potato caused *Streptomyces* spp. by 30.6%–46.1%, and improved yield by 23.0%–32.2% [72]. Inoculation of *Fusaria* infested soil with a bacterial consortium of *Pseudomonas aeruginosa* (B4, B23, B25, and B35), *Alcaligenes feacalis* (B16), and *S. marcescens* (B8) was reported to not only suppress fusarium wilt of potato by 94% but also considerably improved plant biomass by 186.9% (Fresh weight) and 214. 75% (dry weight) [56]. Treatment with a consortium formulation comprising *Enterobacter amnigenus* strain A167, *Serratia plymuthica* strain A294, *Serratia rubidaea* strain H440, *S. rubidaea* strain H469 and *Rahnella aquatilis* strain H145 significantly reduced potato soft rot severity and incidence by 62–75% and 48–61%, respectively, when compared to a positive control with pathogens alone [73]. Also, a combination of rhizobacteria in combination with commercial arbuscular mycorrhiza fungi (AMF) have been reported to effective in abating bacterial wilt of potato [16, 17].

4.5 Phytoextracts

Green plants harbour a plethora of secondary metabolites that could serve as eco-friendly, natural alternatives to chemical fungicides [50, 74, 75]. Phytoextracts are botanicals, natural oils, and plant volatiles that show pest/pathogen control activities. They are usually extracted from fresh or dried plant parts using alcohol, water, or other solvents. Phytoextracts can be fungicidal or fungistatic in action and exert their effects by inducing conditions unfavourable for pathogen growth and proliferation [44]. The application of botanicals can significantly reduce the cost of crop protection and the occurrence of pathogen resistance [44]. Several phytoextracts have been widely tested and reported as effective suppressors of plant pathogens [50, 75]. Dried cheerota plant (*Swertia chirata* Ham.) and jute leaf (*Corchorus capsularis* L.) have been reported to exhibit *in vitro* antibacterial activity against *Erwinia carotovora* subsp. *carotovora* (Ecc) P-138 s, the causative pathogen of soft rot in potato. Under storage conditions, the plant extracts also considerably attenuated bacterial soft rot disease of different potato varieties [50]. Regardless of the mode of application (seed coating or soil inclusion), Canada milkvetch extract (MVE) effectively abated *Verticillium dahlia*-induced wilt by 55–84% in two potato cultivars – Kennebec and Russet Burbank compared to the control under growth room conditions. MVE also significantly reduced vascular discolouration and infection by 55% and 45%, respectively, in two potato cultivars in the first year of the field trial. In the second year, MVE reduced all wilt parameters by 19–31% while increasing yield by 18% on the cultivar Kennebec [76]. Soil drenching with aqueous leaf extracts of *Hibiscus sabdariffa*, *Eucalyptus globulus,* and *Punica granatum* substantially reduced the severity of bacterial wilt disease of potato relative to inoculated control under greenhouse and field conditions. While the reduction in disease severity under field conditions was similar (up to 63.23 to 68.39%) for all the three plant extracts, *E. globulus* leaf extract showed maximum abatement (94% reduction) of disease symptom development under greenhouse condition compared to extracts of *H. sabdariffa* and *P. granatum* [77]. Fumigation of seed tubers of potato with *Allium sativum* – derived essential oils has been shown to manage stem cancer, silver scurf, dry rot, black scurf, and gangrene in small–scale farming systems [78, 79].

4.6 Cultural control

A well-known cultural method to manage the diseases of potato is crop rotation. This refers to cultivating economic plants in recurrent succession and a sequential fashion on the same piece of land [80]. Rotation using different cover crops and suitable fallow periods can contribute to the attenuation of multiple soil-borne pathogens and diseases and enhance the diversity of beneficial soil microflora [81]. Evidence is mounting to show the use of *Brassica* spp. like cabbage, broccoli, cabbage, kale, cauliflower, turnip, rapeseed, canola, radish, different mustards, and other related plants as rotation or green manure crops [82, 83]. These crops produce sulphur-containing glycosinolates degraded as part of a biofumigation process to generate isothiocyanates deleterious to several soil pathogens. *Brassica* spp. have been effectively used to abate populations of soil-borne fungal pathogens, nematodes, and weeds and promote crop yield and soil properties [82]. Other non-brassica crops like ryegrass have good suppression ability over soil-borne pathogens. In several rotation studies, rapeseed and canola crops prior to potato cultivation significantly attenuated (in the range of 25–75%) soil-borne disease due to common scab and *Rhizoctonia* over many seasons to less successful rotations or no rotation [84, 85]. A field trial at a highly infested site with a powdery scab, ryegrass, rapeseed, canola, and Indian mustard grown as rotation crops and green manure suppressed powdery scab in the subsequent potato crop 15–40%. Additionally, rapeseed and canola abated black scurf by 70–80% compared to a standard oats rotation (**Figure 1**) [82].

S/N	Potato Diseases	Associated Pathogens	Pathogen Type
1	Bacterial wilt	*Ralstonia solanacearum*	Bacterium
2	Potato late blight	*Phytophthora infestans*	Fungus
3	Potato virus Y (PVY) disease	Potyvirus Y	Virus
4	Potato scab	*Streptomyces* spp	Bacterium
5	Early blight of potato	*Alternaria solani*	Bacterium
6	Internal blight	*Fusarium sambucinum or F. coeruleum*	Fungus
7	Silver scurf	*Helminthosprium solani*	Fungus
8	Black scurf disease	*Rhizoctonia solani*	Fungus
9	Pink rot	*Phytophthora erythroseptica*	Fungus
10	Soft rot	*Erwinia carotovora*	Bacterium
11	Yellow potato cyst nematode	*Globodera rostochiensis*	Nematode
12	White potato cyst nematode	*G. pallida*	
13	Root-knot nematodes	*Meloidogyne incognita, M.* spp, *Nacobbus aberrans*	Nematode
14	Potato rot nematode	*Ditylenchus destructor*	Nematode
15	Root lesion nematode	*Pratylenchus* spp.	Nematode
16	Stubby-root nematodes	*Trichodorus* spp. and *Paratrichodorus* spp.	Nematode
17	Lance nematode	*Hoplolaimus galeatus*	Nematode
18	Dagger nematode	*Xiphinema* spp.	Nematode

Figure 1.
Management strategies for potato diseases.

5. Methods for raising disease-free potato

Potato is affected by a wide range of fungal, viral, bacterial, and nematodal diseases [86]. These result in colossal yield loss annually. Therefore, it is imperative to exploit strategies for raising disease-free potato to reduce losses caused by pathogens, thus ensuring food security.

Some of the strategies for raising disease-free potato are:

5.1 Conventional plant breeding

The breeding of potato is a huge task due to inherent genetic and biological factors. Breeding for increased resistance to *Phytophthora infestans* (causal agent of late blight) is one of the most critical targets in potato breeding [87]. Plant breeders incorporated resistance against early and late blight disease by crossing hybrid lines with wild species (*S. brevidens* and *S. bulbocastanum*), which exhibited resistance against fungal pathogens [88, 89]. Potato plants resistant to diseases have been produced using conventional plant breeding. However, this process is tedious, and it takes time to achieve success.

5.2 Induced resistance

Resistance in plants can be induced by applying exogenous substances, or agents including living and non-living agents. Resistance to both fungal and viral diseases has been reported in potato. Quintanilla and Brishammar [90] reported systemic induced resistance to late blight in potato by treating with salicylic acid and *Phytophthora crptogea*. In their study, the non-pathogenic fungus *Phytophthora crptogea* and salicylic acid were used as inducer agents. Nadia *et al.*, [91] showed that chemicals under greenhouse and field conditions induced resistance against early and late blight diseases. The inducers used in this study were ascorbic acid, dichloro-isonicotinic acid, ethylene diamine tetraacetic acid, and calcium chloride. Chemicals and fungicides (at low concentration) can induce resistance [92]; similar reports include

Andreu *et al.*, [93]. Several studies have reported using biological agents as inducers of resistance in potato [94–98] reported mycorrhiza-induced resistance in potato. Induced resistance against potato virus Y (PVY^{NTN}) has also been achieved [99].

5.3 Genetic engineering approach

Genetic engineering has been used to raise-disease free transgenic potato plants. However, this technique requires specialised skill, sophisticated equipment, and technical know-how. However, the problem of acceptance and ethical issues may also arise.

Extreme resistance to late blight disease by transferring 3 *R* genes from wild relatives into African farmer-preferred potato varieties was reported by [100]. Three late blight resistance genes from wild potato species were transferred as a stack into the farmer-preferred varieties, Tigoni and Shangi. *R* gene expression analysis in 18 transgenic events showed different transgenic events exhibiting different expression levels in the three genes. Engineering virus resistance using a modified potato gene has been reported by [101]. They reported that the transgenic expression of the $pvrl^2$ gene from pepper confers resistance to potato virus Y (PVY) in potato. The development of late blight-resistant potato by cisgene stacking was studied by Jo *et al.*, [102].

RNA interference (RNAi) is an emerging post-transcriptional technique that has been used to produce crops resistant to diseases. Production of potato lines resistant to *P. infestans* through the RNAi technique has been reported [103]. RNAi technology can be directed to degrade the pathogen's mRNA that enters the host cell or silence endogenous genes of the host cell that aid pathogenicity. RNAi's mechanism of pathogen control is not dependent on producing a foreign protein that could be allergenic or toxic in the host plants. This makes this technology more acceptable than the typical transgenic approaches for disease control [104].

5.4 Plant tissue culture techniques

This technique can be used to produce disease-free pre-basic seeds. Disease-free pre-basic seed potato was produced through tissue culture in Nepal [105]. The use of disease-free seeds can help reduce the transmission of pathogens from propagating materials such as tuber to the field. It has been reported that quality seeds alone can increase yield by 15–20% in Bangladesh [106]. Therefore, micropropagation of potato can help reduce disease transmission through propagating materials; however, little has been achieved on the use of somatic embryos [107], and more researches are required for more remarkable breakthroughs in this regard.

5.5 New/advanced breeding techniques

Genome editing of potato using new technologies such as zinc-finger nucleases (ZFNs), transcription activator-like effector nucleases (TALENs), and clustered regularly interspaced palindromic repeats (CRISPR) associated nuclease 9 is currently been exploited. CRISPR/Cas9 has emerged as a breakthrough in gene editing; however, limited studies have been done on potatoes using this technique [108]. Genome editing using CRISPR/Cas9 has been used to engineer virus resistance in plants by targeting host genes directly involved in host-viral interactions [109–113]. This technique has been used to knock out potato genes/factors like eukaryotic translation initiation factors (*elf4E* and isoform *elf*(*iso*) *4E* that interact with viruses to assist viral infection [114]. Potato varieties resistant to viruses can be produced using this technique. Late blight resistance in potato has also been achieved using CRISPR/Cas9 genome editing. Functional knockouts of *stDND1*, *StCHL1*, and *DMG400000582* (*STDMR6*–1) genes generated increased resistance against late blight in potato [115].

Therefore, holistic and integrated approaches are required for raising disease-free potato in order to overcome the ever-evolving phytopathogens and mitigate losses; including post-harvest losses caused by these pathogens, therefore ensuring food security.

6. Conclusions

This chapter discusses the host-pathogens association of different diseases in potato and their impact on yield. The findings highlight management strategies of these diseases: chemical control, biological control, microbial inoculants, microbial consortium, phytoextracts, and cultural control. In addition, current methods for raising disease-free potatoes to reduce annual yield loss were reported in detail. Based on the presented findings, annual yield loss (pre-and post-harvest) is still high. Thus, the management strategies alone are promising but combining the different methods and exploiting disease-free potato can translate into an integrated management approach of potato diseases.

Acknowledgements

We wish to appreciate Covenant University for sponsoring this publication.

Conflict of interest

The authors declare no conflict of interests.

Author details

Olusola L. Oyesola[1], Oluwadurotimi S. Aworunse[1], Margaret I. Oniha[1], Onyemaechi H. Obiazikwor[2], Oluwakemi Bello[1], Olubunmi M. Atolagbe[1], Ayodele A. Sobowale[3], Jacob O. Popoola[1] and Olawole O. Obembe[1,4*]

1 Department of Biological Sciences, Covenant University, Otta, Ogun State, Nigeria

2 Faculty of Life Sciences, Department of Plant Biology and Biotechnology, University of Benin, Benin City, Nigeria

3 Botany Department, University Of Ibadan, Nigeria

4 UNESCO Chair on Plant Biotechnology, Covenant University, Nigeria

*Address all correspondence to: olawole.obembe@covenantuniversity.edu.ng

References

[1] Jansky S, Navarre R, Bamberg J. Introduction to the special issue on the nutritional value of potato. American Journal of Potato Research. 2019;**96**:95-97

[2] Beals K. Potatoes, nutrition, and health. American Journal of Potato Research. 2019;**96**:102-110

[3] Navarre, D. A., Brown, C. R. and Sathuvalli, V. R. 2019. Potato vitamins, minerals and phytonutrients from a plant biology perspective. American Journal of Potato Research 96:111-126.

[4] King, J.C. and Slavin, J.L., 2013. White vegetables: a forgotten source of nutrients. White potatoes, human health, and dietary guidance. Adv. Nutr. 4, 393S-401S

[5] Haas BJ, Kamoun S, Zody MC, Jiang RH, Handsaker RE, Cano LM, et al. Genome sequence and analysis of the Irish potato famine pathogen Phytophthora infestans. Nature. 2009;**461**(7262):393-398

[6] Möller K, Habermeyer J, Zinkernagel V, Reents H-J. Impact and interaction of nitrogen and Phytophthora infestans as yield-limiting and yield-reducing factors in organic potato (Solanum tuberosum L.) crops. Potato Res. 2006;**49**:281-301

[7] Guchi E. Disease management practice on potato (Solanum tuberosum L.) in Ethiopia. World. Journal of Agricultural Research. 2015;**3**(1):34-42

[8] Scholthof KBG. The disease triangle: Pathogens, the environment and society. Nature Reviews Microbiology. 2007;**5**(2):152-156

[9] Secor GA, Rivera-Varas VV. Emerging diseases of cultivated potato and their impact on Latin America. Revista Latinoamericana de la Papa (Suplemento). 2004;**1**:1-8

[10] Mizubuti ES, Fry WE. Potato late blight. In: The Epidemiology of Plant Diseases. Dordrecht: Springer; 2006. pp. 445-471

[11] King JF, editor. The History of the Irish Famine: Emigration and the Great Irish Famine. Routledge; 2019

[12] Regan, S., Gustafson, V., Rothwell, C., Sardana, R., Flinn, B., Mallubhotla, S., Bagchi, M., Siahbazi, M., Chakravarty, B., Wang-Pruski, G. and De Koeyer, D. (2006). Finding the perfect potato: Using functional genomics to improve disease resistance and tuber quality traits. Canadian Journal of Plant Pathology, 28(S1), S247-S255.

[13] Stellingwerf JS, Phelan S, Doohan FM, Ortiz V, Griffin D, Bourke A, et al. Evidence for selection pressure from resistant potato genotypes but not from fungicide application within a clonal Phytophthora infestans population. Plant Pathology. 2018;**67**(7):1528-1538

[14] Demissie YT. Integrated potato (*Solanum tuberosum* L.) late blight (Phytophthora infestans) disease management in Ethiopia. American Journal of BioScience. 2019;7(6):123-130

[15] Bernardo P, Charles-Dominique T, Barakat M, Ortet P, Fernandez E, Filloux D, et al. Geometagenomics illuminates the impact of agriculture on the distribution and prevalence of plant viruses at the ecosystem scale. ISME Journal. 2018;**12**:173-184. DOI: 10.1038/ismej.2017.155

[16] Aguk JA, Karanja N, Schulte-Geldermann E, Bruns C, Kinyua Z, Parker M. Control of bacterial wilt (Ralstonia solanacearum) in potato

(Solanum tuberosum) using rhizobacteria and arbuscular mycorrhiza fungi. African Journal of Food, Agriculture, Nutrition and Development. 2018a;*18*(2):13371-13387

[17] Aguk JA, Karanja N, Schulte-Geldermann E, Bruns C, Kinyua Z, Parker M. Control of bacterial wilt (*Ralstonia solanacearum*) in potato (*Solanum tuberosum*) using rhizobacteria and arbuscular mycorrhiza fungi. African journal food, agriculture and nutritional. Development. 2018b;**18**(2):13371-13387

[18] Elphinstone JG. The current bacterial wilt situation: A global overview. In: Allen C, Prior P, Hayward AC, editors. Bacterial Wilt Disease and the *Ralstonia Solanacearum* Species Complex. St. Paul, MN: APS Press; 2005. pp. 9-28

[19] Remenant B, Coupat-Goutaland B, Guidot A, Cellier G, Wicker E, Allen C, et al. Genomes of three tomato pathogens within the *Ralstonia solanacearum* species complex reveal significant evolutionary divergence. BMC Genomics. 2010;**11**(1):379. DOI: 10.1186/1471-2164-11-379

[20] Denny TP, Hayward AC. Ralstonia. In: Schaad NW et al., editors. Laboratory Guide for the Identification of Plant Pathogenic Bacteria. 3rd ed. St. Paul: APS Press; 2001. pp. 151-174

[21] McGarvey JA, Denny TP, Schell MA. Spatial-temporal and quantitative analysis of growth and EPS I production by *Ralstonia solanacearum* in resistant and susceptible tomato cultivars. Bacteriology. 1999;**89**(12):1233-1239

[22] Makarova S, Makhotenko A, Spechenkova N, Love AJ, Natalia O, Kalinina NO, et al. Interactive responses of potato (*Solanum tuberosum* L.) plants to heat stress and infection with potato virus Y. Frontiers in Microbiology. 2018;**9**:1-14

[23] Massana-Codina J, Schnee S, Allard P-M, Rutz A, Boccard J, Michellod E. Cle'roux M, Schürch S, Gindro K, Wolfender J-L. insights on the structural and metabolic resistance of potato (*Solanum tuberosum*) cultivars to tuber black dot (*Colletotrichum coccodes*). Frontiers in Plant Science. 2020;**11**:1-19

[24] Lees AK, Brierley JL, Stewart JA, Hilton AJ, Wale SJ, Gladders P, et al. Relative importance of seed-tuber and soil-borne inoculum in causing black dot disease of potato. Plant Pathology. 2010;**59**:693-702. DOI: 10.1111/j.1365-3059.2010.02284.x

[25] Andrivon D, Lucas J-M, Guérin C, Jouan B. Colonization of roots, stolons, tubers and stems of various potato (*Solanum tuberosum*) cultivars by the black-dot fungus *Colletotrichum coccodes*. Plant Pathology. 1998;**47**:440-445. DOI: 10.1046/j.1365-3059.1998.00267.x

[26] Lees AK, Hilton AJ. Black dot (Colletotrichum coccodes): An increasingly important disease of potato. Plant Pathology. 2003a;*52*(1):3-12

[27] Lees AK, Hilton AJ. Black dot (*Colletotrichum coccodes*): An increasingly important disease of potato. Plant Pathology. 2003b; **52**:3-12. DOI: 10.1046/ j.1365-3059.2003.00793.

[28] Aguilera-Galvez C, Champouret N, Rietman H, Lin X, Wouters D, Chu Z, et al. (2018). Two different R gene loci co-evolved with Avr2 of *Phytophthora infestans* and confer distinct resistance specificities in potato. Studies in mycology. 2018; 89, 105-115. doi: 10.1016/j.simyco.2018.01.002

[29] Scheufele SB. UMass Extension Vegetable Programme- Potato, identifying diseases. Retrieved from: https://ag.umass.edu/vegetable/fact-sheets/potato-identifying-diseases. 2021; Accessed: May 16, 2021.

[30] Scholthof KB, Adkins S, Czosnek H, Palukaitis P, Jacquot E, Hohn T, et al. Top 10 plant viruses in molecular plant pathology. Molecular Plant Patholology. 2011;**12**:938-954. DOI: 10.1111/j.1364-3703.2011.00752.x

[31] Dooh1 JPN, Boydoul1 FU, Madjerembe A, Tsouala DBT, Adagoro DBH, Ambang PKAZ. Inventory of the potato diseases and impact on growth and yield traits in far North Cameroon. Int. J. Biol. Chem. Sci. 2020;**14**(8):2826-2836

[32] Nasrollakhnedzad S, Romanenko N, Beloshapkina O. The harmfulness of mono- and mixed viral infections and environmentally friendly methods of their control *Theory and practice of combating parasitic diseases (zoonoses)*. 2002;**3**(Moscow):216-217

[33] Kreuze J, Souza-Dias J, Jeevalatha A, Figueira A, Valkonen J and Jones R 2020 Viral diseases in potato In: *The potato crop. Its agricultural, nutritional and social contribution to humankind* eds H Campos and O Ortiz (springer, Cham)

[34] Romanenko N, Nasrollakhnedzad S, Beloshapkina O. To the Question of the Prevalence and Harmfulness of Potato Viruses *Scientific Notes Crops for Sustainable Agriculture in the 21st Century*. Moscow: RAAS; 2002. pp. 294-304

[35] Ganshina E 1978 The impact of viral infection on yield and starchiness of potato tubers *Viral diseases of crops and their control* (Leningrad: All-Union Academy of Agricultural Sciences) pp 120-121

[36] Kirjukhin V, Gameva F, Parfenova A. Biosynthesis and starch quality during infection with viruses potato *J Potato and Vegetables*. 1971;**9**:39

[37] Kolychikhina M, Beloshapkina O. Efficiency of application of iodine-containing preparation and plant growth regulators in protection of potato against viruses in the field *J Potato and Vegetables*. 2017;**4**:27-30

[38] Khairy AM, Tohamy MRA, Zayed MA, Ali MAS. Detecting pathogenic bacterial wilt disease of potato using biochemical markers and evaluate resistant in some cultivars. Saudi Journal of Biological Sciences. 2021 ISSN 1319-562X, https://doi.org/10.1016/j.sjbs.2021.05.045

[39] Charkowski, A. Sharma, K. Parker, M. L. Secor, G. A. Elphinstone, J. 2020. Bacterial diseases of potato. In: Campos H., Ortiz O. (eds) The Potato Crop. Its agricultural, nutritional and social contribution to humankind. Cham (Switzerland). Springer, Cham. ISBN: 978-3-030-28683-5. pp. 351-388.

[40] De Boer, S. H. and Rubio, I. 2004. Blackleg of potato. *The Plant Health Instructor*. DOI:10.1094/PHI-I-2004-0712-01. Updated 2016

[41] Fry W E, Birch P R, Judelson H S, Grunwald N J, Danies G, Everts K L, Gevens A J, Gugino B K, Johnson D A, Johnson S B, McGrath M T, Myers K L, Ristaino J B, Roberts P D, Secor G, Smart C D. 2015. Five reasons to consider Phytophthora infestans a reemerging pathogen. Phytopathology, 105, 966-981.

[42] Liang KONG, Hui-bin WANG, Shuai-shuai WANG, Ping-ping XU, Ruo-fang ZHANG, Suomeng DONG, Xiao-bo ZHENG, Rapid detection of potato late blight using a loop-mediated isothermal amplification assay, Journal of Integrative Agriculture, Volume 19, Issue 5, 2020, Pages 1274-1282, https://doi.org/10.1016/S2095-3119(19)62816-9.

[43] Tsedaley B. Late blight of potato (Phytophthora infestans) biology, economic importance and its management approaches. *Journal of biology*. Agriculture and Healthcare. 2014;**4**(25):2014

[44] Mulugeta T, Muhinyuza JB, Gouws-Meyer R, Matsaunyane L, Andreasson E, Alexandersson E. Botanicals and plant strengtheners for potato and tomato cultivation in Africa. Journal of Integrative Agriculture. 2020;*19*(2):406-427

[45] Lastochkina O, Baymiev A, Shayahmetova A, Garshina D, Koryakov I. Shpirnaya. Effect: I., ... & Palamutoglu, R; 2020

[46] Secor GA, Gudmestad NC. Managing fungal diseases of potato. Canadian Journal of Plant Pathology. 1999;*21*(3):213-221

[47] Olsen N, Miller J, Nolte P. Diagnosis and management of potato storage diseases. In: Idaho Agricultural Experiment Station. University of Idaho Extension. 2006; CIS1131. https://www.extension.uidaho.edu/publishing/pdf/CIS/CIS1131.pdf.

[48] Majeed A, Muhammad Z, Ullah Z, Ullah R, Ahmad H. Late blight of potato (Phytophthora infestans) I: Fungicides application and associated challenges. Turkish Journal of Agriculture-Food Science and Technology. 2017;*5*(3):261-266

[49] Mehi Lal, Sanjeev Sharma, Saurabh Yadav and Santosh Kumar (June 6th 2018). Management of Late Blight of potato, Potato - From Incas to All Over the World, Mustafa Yildiz, IntechOpen, DOI: 10.5772/intechopen.72472. Available from: https://www.intechopen.com/books/potato-from-incas-to-all-over-the-world/management-of-late-blight-of-pota

[50] Rahman MM, Dey TK, Ali MA, Khalequzzaman KM, Hussain MA. Control of late blight disease of potato by using new fungicides. Int. J. Sustain. Crop Prod. 2008;*3*(2):10-15

[51] Khadka RB, Chaulagain B, Subedi S, Marasini M, Rawal R, Pathak N, et al. Evaluation of fungicides to control potato late blight (Phytophthora infestans) in the plains of Nepal. Journal of Phytopathology. 2020;*168*(5):245-253

[52] Wilson PS, Ahvenniemi PM, Lehtonen MJ, Kukkonen M, Rita H, Valkonen JPT. Biological and chemical control and their combined use to control different stages of the Rhizoctonia disease complex on potato through the growing season. Annals of Applied Biology. 2008;*153*(3):307-320

[53] Delen N. Fungicides. Turkiye, Nobel press. In: ISBN no:978-605-320-347-6. 2016

[54] Jmour W, Hamada W. First report of A2 mating type of Phytophthora infestans in Tunisia using molecular markers and some observations on its metalaxyl resistance. Tun. J. Plant Prot. 2006;**1**:85-92

[55] Villa F, Cappitelli F, Cortesi P, Kunova A. Fungal biofilms: Targets for the development of novel strategies in plant disease management. Frontiers in microbiology. 2017;*8*:654

[56] Devi AR, Sharma GD, Majumdar PB, Pandey P. A multispecies consortium of bacteria having plant growth promotion and antifungal activities, for the management of Fusarium wilt complex disease in potato (Solanum tuberosum L.). Biocatalysis and agricultural biotechnology. 2018;*16*:614-624

[57] Deepa J, Mathew KS. Evaluation of endophytic microbial consortium for the management of bacterial wilt of tomato caused by Ralstonia solanacearum. Journal of Biological Control. 2015;*29*(3):148-156

[58] Oyesola, O.L., Sobowale A.A., Obembe O.O. Effectiveness of Trichoderma koningii Extract on Aspergillus Species Isolated from Rotting Tomato (*Solanum lycopersicum*

Mill). *Trop J Nat Prod Res*. 2020; 4(11):961-965. doi.org/10.26538/tjnpr/v4i11.19

[59] Pal KK, McSpadden Gardener B. Biological control of plant pathogens. The Plant Health Instructor. 2006. DOI: 10.1094/PHI-A-2006-1117-02

[60] Shoda M. Bacterial control of plant diseases. Journal of bioscience and bioengineering. 2000;*89*(6):515-521

[61] Lin C, Tsai C-H, Chen P-Y, Wu C-Y, Chang Y-L, Yang Y-L, et al. (2018) Biological control of potato common scab by *Bacillus amyloliquefaciens* Ba01. PLoS ONE 13(4): e0196520. https://doi.org/10.1371/journal.pone.0196520

[62] https://www.frontiersin.org/articles/10.3389/fmicb.2018.02213/full

[63] Alori ET, Babalola OO. Microbial inoculants for improving crop quality and human health in Africa. Frontiers in microbiology. 2018;*9*:2213

[64] Aliye N, Fininsa C, Hiskias Y. Evaluation of rhizosphere bacterial antagonists for their potential to bioprotect potato (Solanum tuberosum) against bacterial wilt (Ralstonia solanacearum). Bio. Control. 2008;**47**(3):282-288

[65] Xue QY, Chen Y, Li SM, Chen LF, Ding GC, Guo DW, et al. Evaluation of the strains of Acinetobacter and Enterobacter as potential biocontrol agents against Ralstonia wilt of tomato. Bio. Control. 2009;**48**(3):252-258. DOI: 10.1016/j.biocontrol.2008.11.004

[66] Sarwar A, Latif Z, Zhang S, Hao J, Bechthold A. A potential biocontrol agent Streptomycesviolaceusniger AC12AB for managing potato common scab. Frontiers in microbiology. 2019;*10*:202

[67] Tariq M, Yasmin S, Hafeez FY. Biological control of potato black scurf by rhizosphere associated bacteria. Brazilian Journal of Microbiology. 2010;*41*(2):439-451

[68] Meng Q, Hanson LE, Douches D, Hao JJ. Managing scab diseases of potato and radish caused by Streptomyces spp. using Bacillus amyloliquefaciens BAC03 and other biomaterials. Biological Control. 2013;*67*(3):373-379

[69] Jahagirdar S, Hegde G, Krishnaraj PU, Kambrekar DN. Microbial consortia for plant disease management and sustainable productivity. In: *Emerging Trends in Plant Pathology*. Singapore: Springer; 2021. pp. 367-384

[70] Bradáčová K, Florea AS, Bar-Tal A, Minz D, Yermiyahu U, Shawahna R, et al. Microbial consortia versus single-strain inoculants: An advantage in PGPM-assisted tomato production? Agronomy. 2019;*9*(2):105

[71] Sarma BK, Yadav SK, Singh S, Singh HB. Microbial consortium-mediated plant defense against phytopathogens: Readdressing for enhancing efficacy. Soil Biology and Biochemistry. 2015;*87*:25-33

[72] Wang Z, Li Y, Zhuang L, Yu Y, Liu J, Zhang L, et al. A rhizosphere-derived consortium of Bacillus subtilis and Trichoderma harzianum suppresses common scab of potato and increases yield. Computational and structural biotechnology journal. 2019;*17*:645-653

[73] Maciag T, Krzyzanowska DM, Jafra S, Siwinska J, Czajkowski R. The great five—An artificial bacterial consortium with antagonistic activity towards Pectobacterium spp. and Dickeya spp.: Formulation, shelf life, and the ability to prevent soft rot of potato in storage. Applied microbiology and biotechnology. 2020;*104*(10):4547-4561

[74] Leksomboon C, Thaveechai N, Kositratana W. "Effect of Thai

Medicinal Plant Extracts on Growth of Phytopathogenic Bacteria," in Proceeding of the 36th Kasetsart University Annual Conference. Plant Section: Kasetsart University, Bangkok, Thailand; February 1998

[75] Leksomboon C, Thaveechai N, Kositratana W, Paisooksantivatana Y. Antiphytobacterial activity of medicinal plant extracts. Science. 2000;**54**:91-97

[76] Uppal AK, El Hadrami A, Adam LR, Tenuta M, Daayf F. Biological control of potato Verticillium wilt under controlled and field conditions using selected bacterial antagonists and plant extracts. Biological control. 2008;*44*(1):90-100

[77] Hassan MAE, Bereika MFF, Abo-Elnaga HIG, Sallam MAA. Direct antimicrobial activity and induction of systemic resistance in potato plants against bacterial wilt disease by plant extracts. The Plant Pathology Journal. 2009;*25*(4):352-360

[78] Fiers M, Edel-Hermann V, Chatot C, Le Hingrat Y, Alabouvette C, Steinberg C. Potato soil-borne diseases. A review. Agronomy for Sustainable Development. 2012;*32*(1):93-132

[79] Globally, plant diseases are major constraints to crop production (Villa et al., 2017).

[80] Curl EA. Control of plant diseases by crop rotation. The Botanical Review. 1963;*29*(4):413-479

[81] Koike, S. T., Gaskell, M., Fouche, C., Smith, R., & Mitchell, J. (2000). Plant Disease Management for Organic Crops.

[82] Larkin RP, Griffin TS. Control of soil-borne potato diseases using Brassica green manures. Crop protection. 2007;*26*(7):1067-1077

[83] Larkin RP, Halloran JM. Management effects of disease-suppressive rotation crops on potato yield and soil-borne disease and their economic implications in potato production. American Journal of Potato Research. 2014;*91*(5):429-439

[84] Larkin RP, Honeycutt CW. Crop rotation effects onRhizoctoniacankerand black scurf of potato in Central Maine, 1999 and 2000. Biological and cultural tests (online). Report 17, PT06.DOI:10.1094/BC17. The American Phytopathological society, St. Paul. In: MN. 2002

[85] Larkin RP, Honeycutt CW. Effects of different 3-yr cropping systems on soil microbial communities and soil-borne disease of potato. Phytopathology. 2006;**96**:68-79

[86] Abbas MF, Naz F, Irshad G. Important fungal diseases of potato and their management: A brief review. Mycopath. 2013;**11**(1):45-50

[87] Park TH, Vleeshouwers VGAA, Jacobsen E, van der Vossen E, Visser RGF. Molecular breeding for resistance to *Phytphthora infestans* (Mont.) de Bary in potato (*Solanum tuberosum* L.): A perspective of cisgenesis. Plant breeding. 2009;**128**:109-117

[88] Naess S, Bradeen J, Wielgus S, Haberlach G, McGrath JM, Helgeson J. Resistance to late blight in Solanum bulbocastanum is mapped to chromosome 8. Theor. Appl. Genet. 2000;**101**:697-704. DOI: 10.1007/s001220051533

[89] Tek AL, Stevenson WR, Helgeson JP, Jiang J. Transfer of tuber soft rot and early blight resistances from Solanum brevidens into cultivated potato. Theor. Appl. Genet. 2004;**109**:249-254. DOI: 10.1007/s00122-004-1638-4

[90] Quintanilla P, Brishammar S. Systemic induced resistance to late blight in potato by treatment with salicylic acid and *Phytophthora cryptogea*. Potato Research. 1998;**41**:135-142

[91] Nadia, G., El-Gamal, F., Abd-El-Kareem, Y. O., Fotouch and Nehal, S. El-Mougy (2007). Induction of systemic resistance in potato plants against late and early blight diseases using chemical inducers under greenhouse and field conditions. Res. *Journal of Agriculture and Bioscience* 3(2): 73 – 81.

[92] Liljeroth E, Bengtsson T, Wiik L, Andreasson E. Eratumto: Induced resistance in potato to Phytophthora infestans effects of BABA in greenhouse and field tests with different potato varieties. European Journal of Plant Pathology. 2010;**127**:303

[93] Andreu AB, Guevara MG, Wolski EA, Daleo GR, Caldizi DO. Enhancement of natural disease resistance in potatoes by chemicals. Pest Management Science. 2016;**62**(2):162-170

[94] Elazouni I, Abdel-Aziz S, Amira R. Microbial efficiency as biological agents for potato enrichment as well as biocontrols against wilt disease caused by *Ralstonia solanacearum*. World Journal of Microbiology and Biotechnology. 2019;**35**:30

[95] El-Naggar MA, Abouleid HZ, El-Deeb HM, Abd-El-Kareem F, Elshahawy IE. Biological control of potato late blight by means of induction of systemic resistance and antagonism. Research Journal of Pharmaceutical, Biological and Chemical Sciences. 2016;**7**(1):1338-1348

[96] Kumar S, Thind TS, Bela A, Gupta AK. Induced resistance in potato against *Phytophthora infestans* using chemicals and biological agents. Plant Disease Research. 2010;**25**(1):12-18

[97] Schoenherr AS, Rizzo E, Jackson N, Manosalva P, Gomez SK. Mycorrhiza-induced resistance in potato involves priming of defence responses against cabbage looper (Noctuidae: Lepidoptera). Environmental Entomology. 2019;**48**(2):370-381

[98] Schoenheer, A. P., Rizzo, E., Jason, N., Manosalva, P. and Gomez, S. K. (2019). Mycorrhiza-induced resistance in potato involves priming of defence responses against cabbage looper (Noctuidae: Lepidoptera). Environmental Entomology 48(2): 370 – 381

[99] Nasr-Eldin M, Messiha N, Othman B, Megahed A, Elhalag K. Induction of potato systemic resistance against the potato virus Y (PVY^{NTN}) using crude filtrates of *Streptomyces* spp. Under greenhouse conditions. *Egyptian Journal of Biological Pest Control*. 2019;**29**(62):1-11

[100] Webi EN, Daniel K, Johnson K, Anne N, Marc G, Eric M. Extreme resistance to late blight disease by transferring 3 *R* genes from wild relatives into African farmer-preferred potato varieties. African Journal of Biotechnology. 2019;**18**(29):845-856

[101] Cavatorta J, Perez KW, Gray SM, Eck JV, Yeam I, Jahn M. Engineering virus resistance using a modified potato gene. Plant Biotechnology Journal. 2011;**9**:1014-1021

[102] Jo KR, Kim C, Kim T, Bergervoet M, Jongsma MA, Visser RGF, et al. Development of light blight resistant potatoes by cisgene stacking. BMC Biotechnology. 2014;**14**(50):1-10

[103] Listanto, E., Riyanti, E. I. and Ambarwati, A. D. (2020). Transformation using RNAi technology for developing potato lines resistance to late blight (*Phytophthora infestans*). Conference Series: Earth and Environmental Sciences **482** 012030

[104] Majumdar R, Rajasekaran K, Cary JW, Cary JW. RNA interference (RNAi) as a potential tool for control of mycotoxin contamination in crop plants: Concepts and considerations. Frontiers in Plant Science. 2017;**8**

[105] Sakha BM, Rai GP, Dhital SP, Nepal RB. Disease-free pre-basic seed potato production through tissue culture in Nepal. Nepal Agric. Resource. 2007;**8**:1-8

[106] Shaheb MR, Begum MM, Ahmed KU, Nazrul MI, Wiersema SG. Challenges of seed potato (*Solanum tuberosum* L.) production and supply system in Bangladesh: A review. The Agriculturists. 2015;**13**(1):173-188

[107] Morais TP, Asmar SA, Silva FJ, Luz JMQ, Melo B. Application of tissue culture techniques in potato. Bioscience Journal. 2018;**34**(4):952-969

[108] Dango SD, Barakate A, Stephens J, Caliskan ME, Baklish A. Genome editing of potato using CRISPR technologies: Current development and future prospective. Plant Cell, Tissue and Organ Culture. 2019;**139**:403-416

[109] Pyott DE, Sheehan E, Molnar A. Engineering of CRISPR/Cas9-mediated potyvirus resistance in transgene-free Arabidopsis plants. Mol Plant Pathol. 2016;**17**:1276-1288. DOI: 10.1111/mpp.12417

[110] Zaidi SS-E-A, Tashkandi M, Mansoor S, Mahfouz MM. Engineering plant immunity: Using CRISPR/Cas9 to generate virus resistance. Front Plant Sci. 2016;7:1673. DOI: 10.3389/fpls.2016.01673

[111] Lima FSO, Mattos VS, Silva ES, Carvalho MAS, Teixeira RA, Silva JC, Correa VR. Nematodes affecting potato and sustainable practices for their management, potato. In: Yildiz, M, editor. From Incas to All Over the World, Intech Open, 2018; DOI: 10.5772/intechopen.73056. Retrieved from: https://www.intechopen.com/books/potato-from-incas-to-all-over-the-world/nematodes-affecting-potato-and-sustainable-practices-for-their-management.

[112] Westerdahl BB. UC IPM Pest management guidelines- agriculture: Potato Pest management guidelines: Nematodes. Retrieved from: Nematodes / Potato / Agriculture: Pest Management Guidelines / UC Statewide IPM Program (UC IPM) (ucanr.edu). 2019; Accessed: May 18, 2021.

[113] Grabau ZJ, Noling JW. Nematode management in potatoes (Irish or white). Entomology and Nematology Department. UF/IFAS extension. Retrieved from: ENY-029/NG029: Nematode Management in Potatoes (Irish or White) (ufl.edu). 2021; Accessed: May. 2021:18

[114] Hameed A, Mahmood MA, Shahid M, Fatima S, Khan A, Ali S. Prospects for potato genome editing to engineer resistance against viruses and cold-induced sweetening. GM Crop and Food. 2020;**11**:185-205

[115] Kieu NP, Lenman M, Wang ES, et al. Mutations introduced in susceptibility genes through CRISPR/Cas9 genome editing confer increased resistance in potatoes. Scientific Reports. 2021;**11**:4487

Chapter 12

Molecular Host-Nematode Interactions and Tuber Development

Refik Bozbuga and Selman Uluisik

Abstract

Potato, *Solanum tuberosum*, the most important non-grain food crop and essential crop globally, has been widely cultivated around the world for centuries. The significance of this plant is increasing due to high nutritional value of the tubers combined with the simplicity of its propagation. As a plant organ, tuber of potato, is mainly edible part of it and popular as nutrient for almost all nations. Tuberization in potato is a very complex biological occurrence affected by numerous ecological signals, genetics, plant nutrition and several different hormones. Many pests including nematodes limit potato tuber development that plant hormones play roles in nematode feeding cell formation. Parasitic nematodes, important pests which cause damage to plants, tubers, suck up nutrients from plants and weaken plant development and yield losses. Many genes involve in tuber development and plant response nematodes. The aim of this chapter is to demonstrate the new advances in the field of molecular host-nematode interactions and tuber development.

Keywords: Nematode, gall, tuber, potato, molecular, gene, interactions

1. Introduction

Potato (*Solanum tuberosum*) is one of the first domesticated vegetables with cultivation over 6000 years. It is the fourth most important staple food crop produced worldwide with continuously growing production capacity up to 370 million tonnes/year [1]. Potato tuber is rich in health-promoting carotenoids, anthocyanins, and antioxidants such as polyphenols, essential minerals, and amino acids [2].

The production of potatoes has been expeditiously increasing in the last forty years, especially in industrialising countries. However, the average amount of potatoes produced in developing countries is only half that of developed countries. The reasons for this are that modern agriculture is quite different between both developed and developing countries, and only limited contributions have been observed on potato yields revealed by modern breeding strategies in developing countries [3]. Because of these reasons, novel genes associated with yield, such as those related to flowering, tolerance to a/biotic stress conditions, and enhanced postharvest quality attributes should be characterised and introgressed into cultivated potato genotypes. The advances in different omic platforms (transcriptomic, metabolomic, and proteomic) not only reduces the costs but also provides expanded knowledge about diversity in crop genomes. The datasets provide an excellent resource for

IntechOpen

selecting new genetic resources (e.g., single nucleotide polymorphisms, SNPs arrays) for introducing agronomically important improved varieties. The better linkage maps, gene annotations and much easier deciphering the genes related to different quality parameters, such as tuberization have been provided by releasing of potato genome sequence [4]. For example, 185 clones that had previously been SNP genotyped by the Solanaceae Coordinated Agricultural Project (SolCAP) and detected 981 features which represent a mixture of metabolites, and hydrolysed fragments of abundant proteins were examined [5]. Therefore, with the help of new genetic technologies, the quicker screening of large populations which improve the identification of quality candidate traits and genes will be more accessible and chargeable [6].

Potato tuberisation (tuber formation) is a complex physiological phenomenon regulated by both exogenously (environmental factors) and endogenously (metabolic pathways, hormones and genes) [7, 8]. Contrary to most plants that develop from roots, potato tuber originates from an underground specialised stem or stolons, accumulates starch which results in enlargement in favourable conditions [9]. This complex development process can be examined in four stages in its simplest form, which are stolon initiation, enlargement of apical and subapical parts of the stolon, cell divisions and enlargement for tuber is triggered, and resource storage (starch accumulation) until tuber reaches its final mass [10]. The induction of tuberisation is favoured under conditions of long dark periods, cool temperatures, and low amount of nitrogen fertilisation, regulation of a graft-transmissible signal transported from leaves to stolon tips for tuber-inducing stimuli [11]. Initiation of tuberisation signalling and the transition from stolon to tuber is a very dynamic process at the molecular level. Identification of FLOWERING LOCUS T (FT)-like protein (StSP6A), CONSTANS (CO), POTATO HOMEOBOX 1 (POTH1), StBEL5 transcription factor, and microRNA156 and-172 revealed the governing the tuber formation process in potato [12–15]. In stolon tips, before the onset of tuber initiation, StBEL5/StKNOX complex coordinates hundreds of genes, including the genes involved in phytohormone synthesis [11]. Signalling and crosstalk of phytohormones, abscisic acid (ABA), auxins, cytokinins (CKs), gibberellins (GAs), ethylene, and strigolactones (SLs), and other compounds, such as carbohydrates and organic acids are known to play important key roles in regulating the morphological events of tuber development [16].

Several biotic stress factors effect negatively on potato plants that plant parasitic nematodes which are among them cause significant damage to potato growth and tuber development.

2. Plant parasitic nematodes and host-plant interactions

Plant-parasitic nematodes are significant crop pests and cause billions of dollars around the globe [17]. Plant-parasitic nematodes (PPNs) have more than 4,100 species in the world [18]. They infect many crops encompassing from the Solanaceae family to Fabaceae and Poaceae families [19]. Plant-parasitic nematodes may divide based on feeding behaviour as ectoparasites, semi-endoparasites, and endoparasites [19, 20]. Ectoparasites do not spend their life cycle within the plant. However, endoparasitic nematodes spend all their life cycle within plant hosts. Root-knot nematodes (RKNs) are best examples of endoparasitic nematodes that complete their life cycle within a plant after entering the root. The RKN (*Meloidogyne* species) and *Globodera rostochiensis* and *Globodera pallida* (cyst nematodes) are known as noteworthy sedentary endoparasitic nematodes. The RKNs cause a unique feeding structure termed feeding site by modification of cell wall molecular architecture. The host cell differentiation occurs in plant tissues following the infection of RKNs [21, 22]. After entering the root, cyst nematodes move intercellularly, Meloidogyne species move between the cells within the plant roots and cause galls. RKNs and cyst nematodes secrete nematode effectors to manipulate plant defence mechanisms and manipulate the cell wall for increasing the nematode parasitism [23].

During the pathogen attack, plants recognise pathogens with different pathogen recognition systems such as pathogen-associated recognition systems (PAMP) and damage-associated molecular patterns (DAMPs) [19, 24].

Many plant-parasitic nematode (PPNs) species cause damage in potatoes and decrease the tuber quality. There are many nematodes species are found in potato plants: *B. longicaudatus, H. pseudorobustus, H. galeatus, T. claytoni, Pratylenchus andinus, P. brachyurus, P. coffeae, P. crenatus, P. mediterraneus, P. minyus, P. neglectus, P. penetrans, P. scribneri, P. thornei, P. vulnus, P. zeae, N. dorsalis, D. dipsaci, Paratrichodorus spp., Trichodorus spp., Belonolaimus longicaudatus, Helicotylenchus pseudorobustus, Hoplolaimus galeatus, Tylenchorhynchus claytoni, Rotylenchulus reniform, Radopholus similis, Meloidogyne acronea, M. arenaria, M. incognita, M. fallax, M. hapla, M. javanica* and *Xiphinema spp* species [25]. Among those, some of them are major species: potato cyst nematodes (PCNs) *Globodera rostochiensis* and *G. pallida*, RKNs *Meloidogyne* spp., specifically *M. chitwoodi*, the root-lesion nematode *Pratylenchus* spp., *Ditylenchus destructor, Nacobbus aberrans* [25].

Among the plant-parasitic nematodes, RKNs are one of the most damaging nematode genera, particularly *Meloidogyne chitwoodi*, the most damaging species on tuber and decreases tuber quality (**Figure 1**). Therefore, this chapter mainly focuses on plant-root knot interactions.

Root-knot nematodes, which are found in the *Meloidogyne* genus, are economically significant PPNs in the world. They are obligate PPNs that cause damage to roots and tubers, resulting in a high amount of yield losses. This group of nematodes is mostly found in tropical and temperate zones around the globe. In addition to direct crop loss, RKNs have also quarantined organisms for many countries and need regulation [26]. There are many RKN species in the world. Meloidogyne chitwoodi is one of the most common and most damaging RKNs in potato areas among these species.

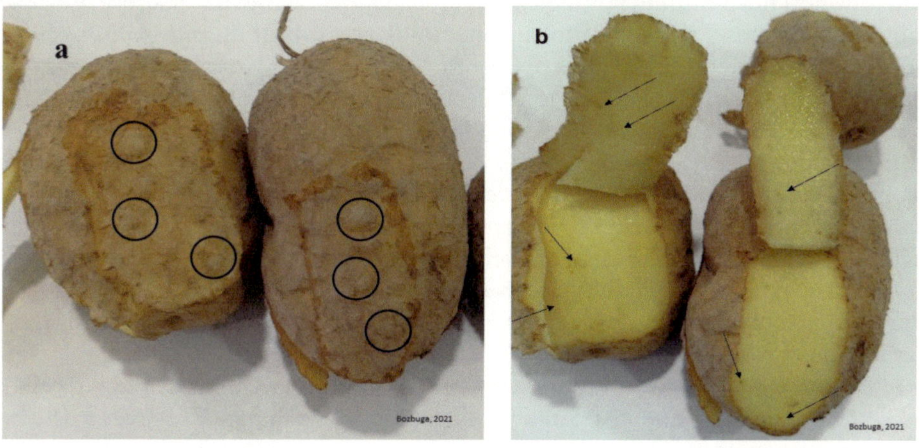

Figure 1.
Symptoms of an RKN species, Meloidogyne chitwoodi, *on potato tubers.* M. chitwoodi *Induced tuber deformations are shown (a). The nematode caused small swellings (pimple-like structures) on the tuber represent within circles (a). Damage caused from the nematode is seen when the potato tuber is peeled (b). Nematodes can be found in discoloured spots (indicated by arrow) in the potato tubers and feed there (b).*

Meloidogyne incognita, Meloidogyne arenaria, Meloidogyne luci and *Meloidogyne javanica*, are RKN species which are found in vegetable areas in Turkey [27]. Even *Meloidogyne* species has different races, for instance, *M. incognita* race 1, 2, 3, *M. javanica* race 3, and *M. arenaria* race1 and 3 [27]. Nematodes are densely found in many orchards where agriculture is carried out with cultivated plants belonging to the Solanaceae family, such as tomato, pepper, and eggplant [28]. *M. javanica, Meloidogyne acronea, M. fallax, M. chitwoodi, M. incognita, M. hapla, M. arenaria* are RKN species that cause damage to potatoes [25].

Potatoes are exposed to diseases and pests while growing. Nematodes that damage the tuber due to the propagation of potatoes by tubers constitute a serious problem in potato production. Nematode species such as potato cyst nematodes, RKNs (*Meloidogyne* spp.) are important potato pests [29]. Root-knot nematodes take the first place among plant-parasitic nematodes in terms of the level of economic damage they cause on plants [17].

In the second stage, juveniles and males of *M. chitwoodi* are thread-shaped, females are pear or lemon-shaped microscopic worms. The life cycle of *M. chitwoodi* takes place in approximately 3-4 weeks under favourable conditions. Although most reproduce parthenogenetically, sexual reproduction is also seen [30]. *M. chitwoodi* spends its first offspring on the potato roots, infects the tuber in the following generations, and develops there. A female can lay approximately 1000 eggs. The number of offspring per year varies, depending on the host plant condition and environmental conditions, especially temperature. *M. chitwoodi* needs 600-800 days to complete the first generation and 500-600 days to complete the next generations [31]. *M. chitwoodi* may infect many plants, but potato and tomato are good hosts and economically important nematode causing damage on potato [32]. It causes many pimple-like structures on tubers, and it is added to the quarantine list in Europe to prevent the distribution within this continent [32]. *Meloidogyne chitwoodi,* feeding in potato plant tubers, may occur in the form of spots caused by the colours visible on the bottom when the tuber is peeled off, causing quality problems (**Figure 1a**). Many necrotic spots are seen on the fleshy parts of the potato tuber. Therefore, the tuber's quality decreases, and it's caused by the nematode (**Figure 1b**).

The second stage is the juvenile root-knot nematodes, an infective stage that is found in free form in the soil which enters the root tip [33]. Chemotactic genes may be involved in host-finding strategies, e.g., *Meloidogyne incognita*. Sucrose, glucose, arabinose, galactose, and mannitol are chemo-attractants of *Meloidogyne incognita*, and signal transduction may involve Mi-odr-1, Mi-odr-3, Mi-tax-4 and Mi-tax-2genes [34–36]. Vanillic acid, lauric acid (signal transduction may require *Mi-odr-1, Mi-odr-3, Mi-tax-2 and Mi-tax-4* genes) [34–37], arginine, lysine [34–36] and calcium chloride [35, 38], *Mi-odr-3, Mi-tax-2, Mi-tax-4* genes are chemotactic genes involve in *Meloidogyne incognita* and predicted functions are membrane-bound guanylyl cyclase that produces secondary messenger, α protein that regulates cyclic nucleotide metabolism, subunits of cyclic nucleotide-gated cation channel involved in G-protein-mediated signalling, respectively [35, 36]. Carbon dioxide (CO2) is an important attractant released by roots for RKNs [39], and lauric acid controls the chemotaxis of root-knot nematodes [37].

In the second stage, juveniles move between the cells (without damaging cells) and reach the feeding site [40]. Sugar transporter genes: Sugars Will Eventually be Exported Transporter (SWEET), vacuolar glucose transporter (VGT), tonoplast monosaccharide transporter (TMT), and sucrose transporter (SUT/SUC) genes may be involved during early infection of *M. incognita* [41]. The host gene expression is manipulated by RKNs [42]. Nematodes secrete several effectors to enable parasitism that macrophage migration inhibitory factors (MIFs) are among them that MIF-like effector overwhelms the *Arabidopsis* immunity and enables *M. incognita* parasitism by cooperating with plant annexins [43]. Similarly, SlWRKY3 plays a role in plant resistance to *Meloidogyne javanica* by involving lipids and hormone activation [44]. Mi gene decreased ability for the nematode infection in tomato through the infection of *M. incognita* [45]. *Meloidogyne incognita* Profilin 3 (MiPFN3) effector results in the actin cytoskeleton of *Arabidopsis* [46].

During the feeding, the nematode creates a feeding tube where it inserts the stylet to release nematode secretions of glands to manipulate plant resistance and create a feeding site [47]. Karyokinesis occurs without cytokinesis in nematode feeding sites termed giant cells in plant tissues [48]. Several nuclei are found in giant cells, and giant cells are much larger than normal cells. The thickness of giant cell walls in the vascular cylinder is much higher than the thickness of neighbouring cell walls (CWs) induced by *M. incognita*. The thickness of giant cell walls may change depending on the host plant [49]. The thickness of giant CWs of Aduki bean is thicker than *Arabidopsis* and maize, and the giant cell walls are a minimum of 2.5 times thicker than neighbouring cell walls [49].

3. Formation of galls and plant- nematode molecular interactions

Nematodes cause damage to plants by influencing the phytohormone structure and modify plant development to establish feeding sites in plants [50]. Plant hormones such as auxin and cytokinin play an important role in forming a sedentary nematode (Cyst and RKN) feeding site [50]. Auxin, a plant hormone, is involved in the formation of galls after infection of RKN, *Meloidogyne javanica* in plant roots [51]. Auxin triggers the gall initiation; however, it is not needed for the later development of the galls [51]. Cytokinin and ethylene may be involved in plant gall formation processes [48]. Ethylene involves in RKN, *Meloidogyne javanica*, induced gall formation in tomato plants [52]. Some plant hormones (jasmonate acid and salicylate (SA)) are involved in plant defence; however, the nematode secretes chorismate mutase to decrease plant defence [50]. The increased level of Pathogenesis related 1 and Pathogenesis related 5 gene expressions are seen during the SA-induced *M. incognita* infection [53, 54]. Auxin performs a function in a cell division and development in host roots [55]. Auxin transport involves developing gall and expansion in the roots of *Arabidopsis thaliana* after the infection of *M. incognita* [56]. Modification of the auxin accumulation and distribution in the roots of plants is observed after infection of *M. javanica* [51]. Plant growth hormones, particularly cytokinin and auxin, play an important role in causing plant galls in pathogen-infected hosts [57].

Small RNAs are differentially expressed in the galls induced by *Meloidogyne javanica* in *Arabidopsis* [58]. Acting as vital mechanisms in gene expression, MicroRNAs are small non-coding RNAs, play an important role in plant nematode interactions. For example, miR159 and MYB33 play an essential role in establishing giant cells of *Arabidopsis* infected by RKN [59]. The specific gene expression patterns appear in nematode induced galls caused by the RKN [60]. Root-knot nematodes and cyst nematodes (CNs) are significant plant parasitic nematode genera of PPNs [17]. They cause hypertrophied and multinucleate feeding cells in the host plant to allow nutrient flow, and they are metabolically active with many organelles, dense cytoplasm, and modifying cell walls [49, 61, 62]. The second stage juveniles of RKNs choose few parenchyma cells and stimulate dedifferentiation into giant cells through succeeding mitosis deprived of cytokinesis [22, 63]. During the nematode infection, nematodes manipulate plant functions, plant defence, phytohormone [50], and cell wall modification [22]. Auxin and ethylene are involved in the transfer cells (TCs) in initial nematode feeding cells of *Arabidopsis* [64]. Auxin and cytokinin are involved in expansion of phloem in nematode induced feeding sites [65]. The atypical transcription factor, DP-E2F-like 1 (DEL1), suppresses salicylic acid (SA) gathering in *Meloidogyne incognita*-induced galls and increased the level of lignification in galls are found in the roots of *Arabidopsis thaliana* [66].

Pattern-triggered immunity (PTI) responses involve camalexin and glucosinolate biosynthesis that BAK1-dependent and -independent PTI are nematode recognition mechanisms in *Arabidopsis* [67]. Msp40 effector of RKN manipulates plant immunity to enable parasitism by suppressing PTI and/or ETI signals [68]. Nematode-associated molecular pattern (NAMP) plays an important role [69].

Microbes attaching to endoparasitic phytonematodes: PTI-responsive defence genes, particularly jasmonic acid-mediated PTI marker genes TFT1 and GRAS4.1, are up-regulated following microbe infections and *M. hapla* in suppressive soil, stimulating initial basal defences in plants by this way overwhelming nematode act in plant roots [70]. TIR-NB-LRR immune receptor DSC1 (DOMINANT SUPPRESSOR OF Camta 3 NUMBER 1) and TIR-NB-LRR-WRKY-MAPx protein WRKY19 adjust basal stages of immunity against *M. incognita* in *Arabidopsis* [71].

Nematodes may modify several plant hormones for successful parasitism. Furthermore, each defined hormone co-ordinately stimulates (IAA, CKs, ABA, and JA) or suppresses (GAs) the formation of tuberization. Numerous researches have reported the importance of the hormones and the genes to play key roles in the synthesis for tuberization. In this part of the chapter, recent studies will be discussed by bringing together the genes related to hormones that are involved in the formation of potato tubers.

4. Hormonal regulation of tuberisation

With respect to the involvement of hormones, gibberellic acid (GA) has been described as one of the most important regulators for tuber development [72, 73]. It is the required hormone for the elongation of stolon meristems during the initiation of tuberisation [74]. Copalyl pyrophosphate synthase (CPS), ent-kaurene synthase (KS), ent-kaurene oxidase (KO), GA-20 oxidase (GA20ox), and GA-2 oxidase (GA2ox) are described as the key enzymes involved in the synthesis of GAs. CPS is the first key enzyme of the gibberellin biosynthesis pathway, which can be stopped by mutating the CPS. However, there is no study that reveals the functioning mechanism of the *CPS* gene, its expression increases during potato elongation [75]. GA20ox and GA3ox catalyse the last two steps of active GA biosynthesis; the former is directly related to the photoperiod of short/long days [76]. Knocking down the expression of the potato *GA20ox-1* gene, resulted in reduced stem elongation and increased tuberisation and yield of tubers [77]. While over-expression of *StGA3ox2* slightly delayed tuberisation phenotype, down-regulation of it did not change the time point of tuber initiation with a smaller average tuber weight [78]. Higher expression of *StGA2ox1* was observed during the early stages of potato tuber development, increased and decreased levels of the gene expression resulted in earlier and delayed tuberisation, respectively [79]. In a recent study, potato plants transformed with the AtGA20-oxidase or *AtGA2-oxidase* genes, the former promotes biosynthesis of bioactive gibberellins (GAs) and the latter acting oppositely, respectively. While tuber formation was increased in plants transformed with *AtGA2-oxidase*, the potato productivity was reduced in plants transformed with AtGA20-oxidase, which promotes active GA synthesis [80]. Overall, GAs levels are quite high at the stolon tips of potato plants and go down intensely when the stolon tip starts swelling and remains at a low-level during tuber formation [81]. These previous and recent studies confirm that GAs are the main tuberisation inhibitors.

Auxin is an exceptional plant hormone. It plays pronounced roles in many plant developmental processes, including tuber initiation, which is crosstalk with gibberellin and strigolactone. In other words, at the initiation of tuber development, the number of GA decreases, whereas that of auxin increases in the stolon subapical region which results in a swollen stolon [82]. The roles of auxin hormone in various biosynthesis metabolisms have been explained in detail [83]. The amount of endogenous auxin positively correlated to tuber growth rate [84]. If it is zoomed at molecular studies, changes in the expression of auxin transport (PIN gene family), auxin response factors (ARF), and Aux/IAA genes during the tuber initiation have been shown [85, 86]. Auxin transcription factor6 (ARF6) decreased its expression several-fold during the transition from longitudinal to transverse cell division at swelling stolon tips [85]. In transgenic potato plants, tuber formation was stimulated by an additional auxin biosynthesis gene (*tms1*) under the control of the tuber-specific B33 promoter [80]. *StARF1/2a* expression was relatively high in stolons, which might have contributions to the swelling of stolons [87]. Indole-3-acetic acid (IAA), one of the most abundant natural forms of auxin, was found extensively across the plants. The role of *StIAA* genes in tuber development was assessed, and 12 genes highly expressed in stolon organs and during the tuberisation stages. Therefore, *Aux/IAA* genes could be used as novel potential candidate genes to improve tuber development of potatoes. With the advent of bioinformatic analysis, it was observed that the gene regulatory network and tuberisation pathway controlled by mobile RNAs (StBEL5 and POTH1) and proteins (StPTB1/6 and StSP6A) have appeared to be conserved among storage root crops like potato, carrot and radish. In this way, StBEL5 targeting genes involved in auxin biosynthesis was unveiled and may prove to be one of the key factors involved in the initiation of potato tuberisation [88]. The *PIN* genes have a central role in polar auxin transport and subsequently mediate the growth of different plant tissues, and 10 of PIN genes were identified in potatoes [73]. *StPIN2* and *StPIN4* genes are highly homologous with *Arabidopsis thaliana* PINs, displayed a role for auxin in tuber development [86]. Although it is insufficient to examine the auxin hormone alone, and the exact role of this hormone is still controversial, a moderate organ-specific increase in auxin level may be suggested as an encouraging approach for improvement of potato productivity by biotechnological methods.

Abscisic acid (ABA) is also well characterised and has been shown to have a supportive effect on tuber development when applied exogenously and to act antagonistically towards GAs, auxins and cytokinins [89]. However, the main role for ABA was determined as dormancy induction and maintenance by different working groups [90]. Genes encoding most enzymes of the ABA synthesis pathway have been identified and cloned from different species [91]. Over-expression of the ABA synthetic gene *StNCED2* promotes tuber yield due to the increase of single potato tuber weight, not the tuber number. ABA signalling transcription factor (TF) *StABF1* and GA metabolism gene *StGA2ox1* were up-regulated while GA synthetic genes *StGA3ox2, StGA20ox1*, and GA signalling TF GAMYB were down-regulated in stolon and tubers of over-expression lines, suggesting there might be a direct interaction between ABA and GA. Ectopic expression of *Arabidopsis ABF4* or *ABF2* (ABRE-binding factor) proteins are transcription factors involved in ABA and stress signalling, which positively regulate potato tuber induction. Increasing of ABA also resulted in decreased expression of GA metabolism genes, which shows ABA-GA signalling crosstalk during tuberisation [92].

Among the phytohormones, it has long been known that cytokinins (CKs) function as universal regulators of storage-organ formation in plants. It was previously shown that CKs have a stimulating effect on tuber formation [93, 94]. CKs are an agronomically and commercially important trait, as CK application before tuber formation can increase tuber yield [95]. However, although there are many effects of CKs on tuber development, tuber development regulated by CKs has not been fully elucidated at molecular level. The role of CKs for tuberisation is closely related to differential expression level of the genes, which can directly reflect the changes of related protein levels and metabolism regulation. Over-expression of *AtCKX1* from *A. thaliana* in soil-grown potato (*Solanum tuberosum* L.) displayed a severely altered phenotype, including reduced tuber yield and morphology. *AtCKX1*-over-expression negatively affected tuber number and tuber size per plant, proving that cytokinin deficiency had significant effects on tuber induction and tuber initiation/growth [96]. In another study, introducing of the *ipt* gene related to bacterial cytokinin biosynthesis, under control of a chalcone synthase promoter (PCHS) generated potato plants with more tubers but reduced tuber weight and nitrogen content [97].

Strigolactones (SLs), carotenoid-derived plant metabolites, have emerged as an important new plant hormone, making it more attractive than other endogenous plant hormones. They mainly regulate various aspects of plant architecture, including the inhibition of shoot branching [98]. Because SLs is a new hormone class, knowledge about SLs related genes in tuberisation and their regulation is much less compared to other hormones. Transgenic potato plants generated by down-regulating *CAROTENOID CLEAVAGE DIOXYGENASE8* (*CCD8*) gene, key in the SL biosynthetic pathway, resulted in changes in potato tuber morphology [99]. Therefore, interestingly, stolons of the StCCD8 RNAi lines tend to emerge from the soil and form aerial shoots. The transgenic lines also provided a higher number of tubers but smaller in size. As it has just been mentioned, SLs is quite a new plant hormone. Therefore, more genes on the SLs synthesis pathway should be functionally characterised in potato tuber development.

5. Postharvest

Potato tubers are generally consumed fresh, but they can also be consumed throughout the year. Therefore, it might be necessary to store them under favourable conditions for an extended period like from one growing season to another one. After the potato has completed its maturation process, they transit to the dormancy period, in which reserves of starch and protein are kept for future sprouts [100]. A major commercial issue is dormancy breakage following sprouting, resulting in quality losses and reduced tuber marketability. CIPC ([isopropyl-N-(3-chlorophenyl) carbamate) is particularly important as a sprout suppressant for potatoes during storage. However, CIPC has been proven not to be safe for humans and the environment in recent years [101]. Therefore, alternate sprout suppressant approaches, for example constant ethylene supplement, could be used to suppress post-harvest sprouting [102]. Storing potato tubers which were treated with/without ethylene binding inhibitor 1-methylcyclopropene (1-MCP at 1 μL L – 1 for 24 h), in air or air enhanced with constant ethylene (10 μL L – 1) [103], revealed extended ecodormancy in the potato samples treated with grouping of ethylene plus 1-MCP, while the inhibited sprout elongation in exogenous ethylene treated samples. Moreover, at the molecular level, continuous ethylene application activated two genes coding 1-aminocyclopropane-1-carboxylate oxidase (ACO) and parenchymatic ABA catabolism via *CYP707A*, encoding ABA 8´hydroxylase upregulation. Consequently, this novel study provided information on how exogenous ethylene and/or 1-MCP elicited their results on the tuber quality of potato.

Another technology that can be used instead of CIPC is in order to grow new potato cultivars with reduced rate of sprout development and/or a long dormancy period. For example, down- and up-regulation of StCEN accompanying to an enhanced and decreased of sprout development than controls, respectively [103, 104], showed that there is no link between exogenous ethylene and *StCEN* expression. This result supports [104] as endogenous ethylene production from transgenic *StCEN* tubers, generally, is not meaningfully dissimilar from controls. The dormancy period of tubers is regulated by both internal/external factors and plant hormones, genetic factors, post-harvest storage conditions, and particular signalling molecules, such as nitric oxide (NO) and gibberellins [105, 106]. Although there are many studies specific to the mentioned factors, this part will try to concentrate on main molecular studies related to the post-harvest condition of potato tuber.

The plant cell wall composed of mainly pectin, is a complex and dynamic network of polysaccharides. Cell wall compositions function in plant development, stress responses, shelf life and plant growth. Basically, the primary cell wall (CW) consists of cellulosic (1,4-β-D-glucan), hemicellulosic polysaccharides for example xyloglucan (XG), and pectic polysaccharides for example homogalacturonan (HG) and rhamnogalacturonans I-II, which are all explained very well in different studies [107]. The recent vision of the plant cell wall (PCW) suggests that the relationship of cellulose–pectin is more extensive and makes more important contributions to wall biomechanical properties than was previously thought [108]. The CWs of tuber tissues are constitute of cellulose and hemicellulose which hold together a large amount of pectic polysaccharides [109]. The texture of plant products is highly affected by the cell wall structure, and modifications of this part of the cell are the biggest contributors to texture. Generally, during fruit maturation, enzyme activities of hemicelluloses (HCL), celluloses (Cel), β-galactosidases (β-Gal), polygalacturonase (PG) and increase to lessen the intercellular associations and accomplish cell seperation, ensuing in modifications in fruit roughness and softening [110, 111]. Potato tuber texture is one of the most important quality characteristics of cooked potato and an obviously dominant trait that influences consumer preference, as mainly affecting the taste, aroma, and mouthfeel of the storage roots in potato [112]. Two types of potatoes that differ in terms of texture represented an extreme variant in textural properties. The expression levels of the genes encoding two important cell wall degrading enzymes, pectin acetylesterase and xyloglucan endotransglycosylase, were significantly higher in Phureja, an accession that greatly reduced cooking time compared to Tuberosum accession [113]. In another recent study, the correlation between the texture of cooked potato and β-amylase activity shows the negative correlation between the enzyme activity and firmness in cooked sweet potato [114].

Moreover, various studies have been conducted to elucidate the cell wall mechanism and texture changes in potato tuber. For example, two potato varieties showing significant differences in texture (Yushu No 10 with soft texture, Mianfen No 1 with firm texture) have been recently characterised in terms of the cell wall composition content and cell wall-related enzyme activities [60]. The 'Yushu No 10' have more than twice soluble pectin content than 'Mianfen No 1', but the unsolvable pectin ingredient was lower than that of 'Mianfen No 1'. It has been an important correlation of gumminess and chewiness between hemicellulose activity of 'Yushu No 10', and 'Mianfen No 1' having an unimportant correlation with Cel, PG, HCL, and β-Gal enzymes [115].

Potato is a highly heterozygous crop. Therefore, genetic advance of this crop using conventional breeding is labour-intensive and time-consuming work. For this reason, genetic engineering offers an opportunity to progress a limited genetic gain whilst retaining the well-known advantages of traditional varieties. The genetically modified potatoes show developments in quality traits that benefit farmers [116], consumers [117], and for the land in terms of sustainability [118]. In recent years, with the increasingly aggravated global warming conditions, the research concentrated more on generation potato crops tolerant against extreme conditions such as salinity and drought [119]. However, due to the concept of this chapter, we try to cover the transgenic studies using cell wall related enzymes. Transgenic potato made by the introduction of the gene encoding rhamnogalacturonan lyase (RGL) from *Aspergillus aculeatus* had a surface with a wrinkled appearance [120]. The expression of a β-galactosidase (β-Gal) gene from *Cicer arietinum* introduced into the potato and resulted in the removal of the galactan side chains from RG-I [121]. In a more recent study, genes encoding β-Gal or RG-I lyase were introduced to wild-type potato Karnico. The mutant lines of β-Gal contained 54% less galactose, representing shorter galactan side chains. Over-expression RG-I lyase potato lines contained more galacturonic acid and less galactose, which was due to the removal of galactan-rich RG-I branches [122]. Over-expression of endo-1,5-α-arabinanase of *A. aculeatus* caused no modified phenotype comparison the wild type but reduced galactan sidechains of RGI and increased the number of uronic acids [123].

High-throughput RNA sequencing (RNA-Seq) is a powerful tool for revelaing the variability of gene expression levels between different samples. An RNA-Seq was performed to investigate the potato tuber dormancy release process, and 5912 and 3885 DEGs (differentially expressed genes) from dormancy tuber (DT) vs. dormancy release tuber (DRT) and DRT vs. sprouting tuber (ST), respectively [124]. In another study carried out by iTRAQ labelling strategy, a total of 1752 proteins associated with tuber dormancy release in DT, DRT, and ST were identified. lncRNAs generally have structural features of mRNA, with exceptional roles in DNA methylation, histone modification, chromatin remodelling, and other biological processes. Moreover, lncRNAs regulated the expression of target genes by interacting with DNA, RNA, and proteins [125]. In a recent study, 235 potato miRNAs out of 386 lncRNAs differentially expressed during sprouting were identified as putative targets. The results provided lncRNAs were involved in the potato tuber sprouting process and identified their possible functions in dormancy and sprouting [126]. Based on these results, it can be said that tuber dormancy release is a complex process, and the genes upregulated during this period suggest the activation of multiple mechanisms enabling the tuber dormancy release.

Enzymatic browning is a serious problem for both producers and the industry as the tubers can be affected during storage and distribution. This problem is usually overcomed by applying chemical and/or physical agents or storing the potato in controlled storage conditions [127]. However, keeping harvested potato tubers at low temperatures causes physiological changes, such as photosynthetic capacity, electrolyte leakage, and respiration rates [128]. Transcriptomic and proteomic analysis were carried out in potato tubers stored at 15°C, 4°C, and 0°C to examine the mechanism of cold responses during post-harvest storage. The results showed that sugar accumulation increased at low temperatures.

Moreover, fifteen heat shock proteins (Hsps) were upregulated by low temperatures, which may act to prevent damage from cold stress [129]. Application of the CRISPR/Cas9 to induce mutations in the StPPO2 gene in the tetraploid variety 'Desiree' reduced up to 69% in tuber PPO (Polyphenol oxidase) activity and 73% in enzymatic browning in transgenic lines compared to control [130]. This result demonstrated that the CRISPR/Cas9 system has been successfully used to generate new potato varieties that reduce enzymatic browning through specific regulation of a single member of the StPPO gene family.

6. Conclusion

Plant hormones are involved in the gall formation and tuber development of potato plants. Numerous nematode species infects potato and cause an adverse effect on plant development and crop quality. Specifically, CNs (*Globodera rostochiensis* and *G. pallida*) and RKNs (*Meloidogyne chitwoodi*) cause severe damage to potato plants and RKNs cause gall formations in the roots of plants. The gall formation has not been fully understood yet. *M. chitwoodi* causes damage to tubers, too. In addition to direct damage of these nematodes, some nematode species, such as root-knot nematodes, are also quarantined organisms that cause restriction of trade. For this reason, they are extremely important organisms for potato production due to causing crop losses. Many genes are involved in nematode parasitism and plant defence mechanism. Plant-parasitic nematodes cause damage in potato tuber by manipulating plant hormones to create feeding sites in potatoes. Understanding the molecular mechanisms of plant nematode interactions (gall formation) and molecular mechanism of tuber development may share some similarities, leading to the researcher creating better potato crops against biotic stress as a future aspect. In this way, using improved pest management strategies and new insights in genetic breeding against nematode may lead to producing healthier crops and high-quality tubers using new molecular and genetic methods.

To improve future potato tuber quality, it should be worked with industry and academic groups to meet producer and consumer preferences. With molecular and improved phenotyping techniques, knowledge about the mechanisms affecting potato tuber development, texture and post-harvest storage conditions will be increased for potato tuber quality. Furthermore, this combined information will profit the improvement of new cultivars by enlarging sustainable agricultural practices and storing approaches. Therefore, the combination of novel molecular techniques (gene-editing technologies) and pre/post-harvest applications will help the improvement, protection and viability of upcoming tuber quality.

Conflict of interest

The authors declare no conflict of interest.

Author details

Refik Bozbuga[1*] and Selman Uluisik[2]

1 Biological Control Research Institute, Adana, Turkey

2 Burdur Food Agriculture and Livestock Vocational School, Burdur Mehmet Akif Ersoy University, Burdur, Turkey

*Address all correspondence to: refikbozbuga@gmail.com

References

[1] FAO, Food and Agriculture Organization of the United Nations, http://www.fao.org/home/en/ (Date of access: 21.03.2021).

[2] Fogelman E, Oren-Shamir M, Hirschberg J, Mandolino G, Parisi B, Ovadia R, Tanami Z, Faigenboim A, Ginzberg I. Nutritional value of potato (*Solanum tuberosum*) in hot climates: anthocyanins, carotenoids, and steroidal glycoalkaloids. Planta. 2019; 249: 1143-1155. DOI: https://doi.org/10.1007/s00425-018-03078-y

[3] Ghislain M., Douches DS. The Genes and Genomes of the Potato. In: Campos H., Ortiz O. (eds) The Potato Crop. Springer, Cham. 2020. https://doi.org/10.1007/978-3-030-28683-5_5

[4] The Potato Genome Sequencing Consortium. Genome sequence and analysis of the tuber crop potato. Nature. 2011; 475: 89-195. DOI: https://doi.org/10.1038/nature10158

[5] Levina A V, Hoekenga O, Gordin M, Broeckling C, De Jong WS. 2020. Genetic analysis of potato tuber metabolite composition: Genome-wide association studies applied to a nontargeted metabolome. Crop Science. 2020; 61 (1): 591-603. DOI: 10.1002/csc2.20398

[6] Khlestkin VK, Erst TV, Rozanova IV, Efimov VM, Khlestkina EK. 2020. Genetic loci determining potato starch yield and granule morphology revealed by genome-wide association study (GWAS). Plant Biology. 2020; 8:e10286.

[7] Paradiso R, Arena C, Rouphael Y, d'Aquino L, Makris K, Vitaglione P, De Pascale S. Growth, photosynthetic activity and tuber quality of two potato cultivars in controlled environment as affected by light source. Plant Biosystems - An International Journal Dealing with all Aspects of Plant Biology. 2019; 153(5): 725-735. https://doi.org/10.1080/11263504.2018.1549603

[8] Cheng L, Wang D, Wang Y, Xue H, Zhang F. 2020. An integrative overview of physiological and proteomic changes of cytokinin-induced potato (*Solanum tuberosum* L.) tuber development in vitro. Plant Physiology. 2020; 168(3): 675-693. DOI: 10.1111/ppl.13014

[9] Fernie AR, Willmitzer L. Molecular and biochemical triggers of potato tuber development. Plant Physiology. 2001; 127: 1459-1465. DOI: https://doi.org/10.1104/pp.010764

[10] Viola R, Roberts AG, Haupt S, Gazzani S, Hancock RD, Marmiroli N, Machray GC, Oparka KJ. Tuberization in potato involves a switch from apoplastic to symplastic phloem unloading. Plant Cell, 2001; 13: 385-398. DOI: 10.1105/tpc.13.2.385

[11] Hannapel DJ, Sharma P, Lin T, Banerjee AK. The multiple signals that control tuber formation. Plant Physiology, 2017; 174: 845-856. DOI: https://doi.org/10.1104/pp.17.00272

[12] Martin A, Adam H, Díaz-Mendoza M, Zurczak M, González-Schain ND, Suarez-Lopez P. Graft-transmissible induction of potato tuberization by the microRNA miR172. Development. 2009; 136 (17): 2873-2881. DOI: https://doi.org/10.1242/dev.031658

[13] Navarro C, Abelenda, JA, Cruz-Oró E, Cuéllar CA, Tamaki S, Silva J, Shimamoto K, Prat S. Control of flowering and storage organ formation in potato by FLOWERING LOCUS T. Nature. 2011; 478(7367):119-22. DOI: 10.1038/nature10431.

[14] González-Schain ND, Díaz-Mendoza M, Żurczak M, Suárez-López P. Potato CONSTANS is involved

in photoperiodic tuberization in a graft-transmissible manner. The Plant Journal. 2012; 70(4): 678-690. https://doi.org/10.1111/j.1365-313X.2012.04909.x

[15] Sharma P, Lin T, Grandellis C, Yu M, Hannapel DJ. The BEL1-like family of transcription factors in potato. Journal of Experimental Botany. 2014; 65: 709-723. https://doi.org/10.1093/jxb/ert432

[16] Alamar MC, Tosetti R, Landahl S, Bermejo A, Terry LA. Assuring Potato Tuber Quality during Storage: A Future Perspective. Frontiers in Plant Science. 2017; 8: 2034. doi: 10.3389/fpls.2017.02034

[17] Jones JT, Haegeman A, Danchin EG, Gaur, HS, Helder J, Jones MG., et al. Top 10 plant-parasitic nematodes in molecular plant pathology. Mol. Plant Pathol. 2013; 14 (9): 946-961. doi: 10.1111/mpp.12057

[18] Decraemer W, Hunt D. Structure and classification, in Plant Nematology, eds Perry RN, Moens M. (Wallingford: CAB International. 2006; 3-32.

[19] Sato K, Kadota Y, Shirasu K. Plant Immune Responses to Parasitic Nematodes. Front. Plant Sci. 2019; 10:1165. doi: 10.3389/fpls.2019.01165

[20] Decraemer W, Hunt, DJ. Structure and classification, in Plant Nematology, 2 ed. Eds. Perry RN, Moens M. Wallingford, UK: CABI Publishing. 2013; 3-39. doi: 10.1079/9781780641515.0003

[21] Perry RN, Moens M, Starr JL. Root-knot Nematodes Root-Knot Nematodes. 2010. Oxfordshire, UK: CAB International.

[22] Bozbuga R, Lilley CJ, Knox JP, Urwin PE. Host-specific signatures of the cell wall changes induced by the plant parasitic nematode, *Meloidogyne incognita*. Sci Rep. 2018; 8, 17302 https://doi.org/10.1038/s41598-018-35529-7

[23] Gheysen G, Mitchum MG. How nematodes manipulate plant development pathways for infection. Curr. Opin. Plant Biol. 2011; 14 (4), 415-421. doi: 10.1016/j.pbi.2011.03.012

[24] Jones JD, Dangl JL. The plant immune system. Nature. 2006; 444 (7117), 323-329. doi: 10.1038/nature05286

[25] Lima FSO, Vanessa SM, Edvar SS, Maria A.S. Carvalho, Teixeira RA, Silva JC, Correa VR. Nematodes Affecting Potato and Sustainable Practices for Their Management, Potato - From Incas to All Over the World, Mustafa Yildiz, IntechOpen 2018; DOI: 10.5772/intechopen.73056.

[26] Moens M, Perry RN, Starr JL .*Meloidogyne* species: A diverse group of novel and important plant parasites. In: Perry RN, Moens M, Starr JL (eds) Root-Knot Nematodes. CABI Publishing, Wallingford. 2009; 1-17

[27] Gürkan B, Çetintaş R, Gürkan T. Gaziantep ve Osmaniye Sebze Alanlarında Bulunan Kök-ur Nematodu Türleri (*Meloidogyne* spp.)'nin Teşhisi ile Bazı Nematod Popülasyon Irklarının Belirlenmesi. KSÜ Tarım ve Doğa Derg. 2019; 22(Ek Sayı 1): 113-124. DOI: 10.18016/ksutarimdoga.v22i49073.551240

[28] Yağcı M, Kaşkavalcı G. Distribution and identification of root-knot nematode (*Meloidogyne* spp) species in Peach growing areas of Aegean Region, Ege Univ., Ziraat Fakültesi dergisi. 2018; 55, (3): 305-310.

[29] Avcı GK, Pırlak U, Toktay H, Özarslandan A, İmren M, Canhilal R. Tohumluk Patates Yumrularinda Kök-Ur Nematodu (*Meloidogyne chitwoodi*) Zarari Ve Mücadele Yöntemleri . Uluslararası Katılımlı Türkiye 6.

Tohumculuk Kongresi Bildiri kitabı. 2018; 78-83.

[30] Van der Beek JG, Karssen G. Interspecific hybridization of meiotic parthenogenetic *Meloidogyne chitwoodi* and *M. fallax.* Phytopathology. 1997; 87:1061-1066 . 10.1094/PHYTO. 1997.87.10.1061

[31] Pinkerton JN, Santo GS, Mojtahedi H. Population dynamics of *Meloidogyne chitwoodi* on Russet Burbank potatoes in relation to degree-day accumulation. Journal of Nematology. 1991; 23, 283-290.

[32] Hunt D, Handoo Z. Taxonomy, Identification and Principal Species, 55-88. In Root- Knot Nematodes, (Eds. R. N. Perry, M. Moens & J. L. Starr). CABI, London. 2009; 488.

[33] Bleve-Zacheo T, Melillo MT, Castagnone-Sereno P. The contribution of biotechnology to root-knot nematode control in tomato plants. Pest Technology, Global Science Books. 2007; 1: 1-16.

[34] Fleming TR, Maule AG, Fleming CC. Chemosensory Responses of Plant Parasitic Nematodes to Selected Phytochemicals Reveal Long-Term Habituation Traits. J. Nematol. 2017; 49 (4), 462-471.

[35] Tsai AY, Oota M, Sawa S. Chemotactic Host-Finding Strategies of Plant Endoparasites and Endophytes. Front Plant Sci. 2020; 11:1167. doi: 10.3389/fpls.2020.01167.

[36] Shivakumara TN, Dutta TK, Chaudhary S, von Reuss SH, Williamson VM, Rao U. Homologs of *Caenorhabditis elegans* Chemosensory Genes Have Roles in Behavior and Chemotaxis in the Root-Knot Nematode *Meloidogyne incognita*. Mol. Plant Microbe Interact. 2019; 32 (7), 876-887. 10.1094/MPMI-08-18-0226-R

[37] Dong L, Li X, Huang L, Gao Y, Zhong L, Zheng Y, et al. Lauric acid in crown daisy root exudate potently regulates root-knot nematode chemotaxis and disrupts Mi-flp-18 expression to block infection. J. Exp. Bot. 2014; 65 (1), 131-141. 10.1093/jxb/ert356

[38] Wang C, Masler EP, Rogers ST. Responses of *Heterodera glycines* and Meloidogyne incognita Infective Juveniles to Root Tissues, Root Exudates, and Root Extracts from Three Plant Species. Plant Dis. 2018; 102 (9), 1733-1740. 10.1094/PDIS-09-17-1445-RE

[39] Rasmann S, Ali JG, Helder J, van der Putten WH. Ecology and evolution of soil nematode chemotaxis. J. Chem. Ecol. 2012; 38, 615-628. 10.1007/s10886-012-0118-6

[40] Wyss U, Grundler FMW. Feeding-behavior of sedentary plant parasitic nematodes. Netherlands Journal of Plant Pathology. 1992; 98, 165-173.

[41] Zhao D, You Y, Fan H, Zhu X, Wang Y, Duan Y, Xuan Y, Chen L. The Role of Sugar Transporter Genes during Early Infection by Root-Knot Nematodes. Int J Mol Sci. 2018; 19(1):302. doi: 10.3390/ijms19010302.

[42] Bird DM. Manipulation of host gene expression by root-knot nematodes. J Parasitol. 1996; 82(6):881-8.

[43] Zhao J, Li L, Liu Q, Liu P, Li S, Yang D, Chen Y, Pagnotta S, Favery B, Abad P, Jian H. A MIF-like effector suppresses plant immunity and facilitates nematode parasitism by interacting with plant annexins. J Exp Bot. 2019; 70(20):5943-5958. doi: 10.1093/jxb/erz348.

[44] Chinnapandi B, Bucki P, Fitoussi N, Kolomiets M, Borrego E, Braun Miyara S. Tomato SlWRKY3 acts as a positive regulator for resistance against

the root-knot nematode *Meloidogyne javanica* by activating lipids and hormone-mediated defense-signaling pathways. Plant Signal Behav. 2019; 14(6):1601951. doi: 10.1080/15592324.2019.1601951.

[45] Bozbuga R, Dasgan HY, Akhoundnejad Y, Imren M, Günay O, Toktay H. Effect of *Mi* gene and nematode resistance on tomato genotypes using molecular and screening assay. Cytology and Genetics. 2020; 54:154-164. https://doi.org/10.3103/S0095452720020048

[46] Leelarasamee N, Zhang L, Gleason C. The root-knot nematode effector MiPFN3 disrupts plant actin filaments and promotes parasitism. PLoS Pathog. 2018; 14(3):e1006947. doi: 10.1371/journal.ppat.1006947.

[47] Hussey RS, Mims CW. Ultrastructure of feeding tubes formed in gient-cells induced by root-knot nematode *Meloidogyne incognita*. Protoplasma. 1991; 162: 99-107. https://doi.org/10.1007/BF02562553

[48] Karssen G, Moens M. Root-knot nematodes. In: Perry RN, Moens M. (eds) Plant Nematology. 2006; 59-90.

[49] Bozbuga R. Characterisation of cell walls at the feeding site of *Meloidogyne incognita*, PhD thesis, University of Leeds, Leeds. 2017.

[50] Gheysen G, Mitchum MG. Phytoparasitic nematode control of plant hormone pathways. Plant Physiol. 2019; 179, 1212-1226. doi: 10.1104/pp.18.01067

[51] Hutangura P, Mathesius U, Jones MGK, Rolfe BG. Auxin induction is a trigger for root gall formation caused by root-knot nematodes in white clover and is associated with the activation of the flavonoid pathway. Australian Journal of Plant Physiology. 1999; 26, 221-231. https://doi.org/10.1071/PP98157

[52] Glazer, I., Orion, D. & Apelbaum, A. 1983. Interrelationships between ethylene production, gall formation, and root-knot nematode development in tomato plants infected with *Meloidogyne javanica*. Journal of Nematology, 15, 539-544.

[53] Bozbuga R. Effect of Submerging *Solanum lycopersicum* Roots in Salicylic Acid (SA) Solution for Different Durations on Nematode Infection and Expressions of SlPR5 Gene. Horticultural Studies. 2020; 37: 144-149.

[54] Bozbuga R. Expressions of pathogenesis related 1 (PR1) gene in *Solanum lycopersicum* and Influence of salicylic acid exposures on host-*Meloidogyne incognita* interactions. Doklady Biochemistry Biophysics. 2020; 494:266-269. https://doi.org/10.1134/S1607672920050038

[55] De Smet I, Lau S, Voss U, Vanneste S, Benjamins R, Rademacher EH, Schlereth A, De Rybel B, Vassileva V, Grunewald W, Naudts M, Levesque MP, Ehrismann JS, Inze D, Luschnig C, Benfey PN, Weijers D, Van Montagu MCE, Bennett MJ, Juergens G, Beeckman T. Bimodular auxin response controls organogenesis in Arabidopsis. Proceedings of the National Academy of Sciences of the United States of America. 2010; 107, 2705-2710. https://doi.org/10.1073/pnas.0915001107

[56] Kyndt T, Goverse A, Haegeman A, Warmerdam S, Wanjau C, Jahani M, Engler G, De Almeida Engler J, Gheysen G. Redirection of auxin flow in *Arabidopsis thaliana* roots after infection by root-knot nematodes. Journal of Experimental Botany. 2016; 67, 4559-4570. DOI: 10.1093/jxb/erw230

[57] Kant U, Ramani V. Insect induced plant galls in tissue-culture. Proceedings

of the Indian Academy of Sciences-Animal Sciences. 1990; 99, 257- 265.

[58] Cabrera J, Barcala M, García A, Rio-Machín A, Medina C, Jaubert-Possamai S, Favery B, Maizel A, Ruiz-Ferrer V, Fenoll C, et al. Differentially expressed small RNAs in Arabidopsis galls formed by Meloidogyne javanica: A functional role for miR390 and its TAS3-derived tasiRNAs. New Phytol. 2016; 209: 1625-1640. DOI: 10.1111/nph.13735

[59] Jaubert-Possamai S, Noureddine Y, Favery B. MicroRNAs, New Players in the Plant-Nematode Interaction. Front Plant Sci. 2019; 10:1180. doi: 10.3389/fpls.2019.01180.

[60] Hammes UZ, Schachtman DP, Berg RH, Nielsen E, Koch W, Mcintyre LM, Taylor CG. Nematode-induced changes of transporter gene expression in Arabidopsis roots. Molecular Plant-Microbe Interactions. 2005; 18, 1247-1257. DOI: 10.1094/MPMI-18-1247

[61] Grundler FMW, Sobczak M, Golinowski W. Formation of wall openings in root cells of *Arabidopsis thaliana* following infection by the plant- parasitic nematode *Heterodera schachtii*. Eur. J. Plant Pathol. 1998; 104, 545-551. doi: 10.1023/A:1008692022279

[62] Favery B, Quentin M, Jaubert-Possamai S, Abad P. Gall-forming root-knot nematodes hijack key plant cellular functions to induce multinucleate and hypertrophied feeding cells. J. Insect Physiol. 2016; 84, 60-69. doi: 10.1016/j.jinsphys.2015.07.013

[63] Caillaud M-C, Dubreuil G, Quentin M, Perfus-Barbeoch L, Lecomte P, de Almeida Engler J, et al. Root-knot nematodes manipulate plant cell functions during a compatible interaction. J. Plant Physiol. 2008; 165, 104-113. doi: 10.1016/j.jplph.2007.05.007

[64] Cabrera J, Barcala M, Fenoll C, Escobar C. Transcriptomic signatures of transfer cells in early developing nematode feeding cells of Arabidopsis focused on auxin and ethylene signaling. Front Plant Sci. 2014; 5:107. doi: 10.3389/fpls.2014.00107.

[65] Absmanner B, Stadler R, Hammes UZ. Phloem development in nematode-induced feeding sites: the implications of auxin and cytokinin. Front Plant Sci. 2013; 4:241. doi: 10.3389/fpls.2013.00241.

[66] Nakagami S, Saeki K, Toda K, Ishida T, Sawa S. The atypical E2F transcription factor DEL1 modulates growth-defense tradeoffs of host plants during root-knot nematode infection. Sci Rep. 2020; 10(1):8836. doi: 10.1038/s41598-020-65733-3.

[67] Teixeira MA, Wei L, Kaloshian I. Root-knot nematodes induce pattern-triggered immunity in *Arabidopsis thaliana* roots. New Phytol. 2016;211(1): 276-87. doi: 10.1111/nph.13893.

[68] Niu J, Liu P, Liu Q, Chen C, Guo Q, Yin J, Yang G, Jian H. Msp40 effector of root-knot nematode manipulates plant immunity to facilitate parasitism. Sci Rep. 2016 ;6:19443. doi: 10.1038/srep19443.

[69] Kaloshian I, Teixeira M. Advances in Plant-Nematode Interactions with Emphasis on the Notorious Nematode Genus *Meloidogyne*. Phytopathology. 2019;109(12):1988-1996. doi: 10.1094/PHYTO-05-19-0163-IA.

[70] Topalović O, Bredenbruch S, Schleker ASS, Heuer H. Microbes Attaching to Endoparasitic Phytonematodes in Soil Trigger Plant Defense Upon Root Penetration by the Nematode. Front Plant Sci. 2020;11:138. doi: 10.3389/fpls.2020.00138.

[71] Warmerdam S, Sterken MG, Sukarta OCA, van Schaik CC,

Oortwijn MEP, Lozano-Torres JL, Bakker J, Smant G, Goverse A. The TIR-NB-LRR pair DSC1 and WRKY19 contributes to basal immunity of Arabidopsis to the root-knot nematode Meloidogyne incognita. BMC Plant Biol. 2020;20(1):73. doi: 10.1186/s12870-020-2285-x

[72] Ewing EE, Struik PC. Tuber formation in potato: Induction, initiation, and growth, horticultural reviews, Horticultural Reviews, John Wiley & Sons, Inc. 1992;14. https://doi.org/10.1002/9780470650523.ch3

[73] Roumeliotis E, Kloosterman B, Oortwijn M, Visser RGF, Bachem CWB. The PIN family of proteins in potato and their putative role in tuberization. Frontiers in Plant Science 2013; 4: 524. 10.3389/fpls.2013.00524

[74] Dutt S, Manjul AS, Raigond P, Singh B, Siddappa S, Bhardwaj V, Kawar PG, Patil V U, Kardile HB. Key players associated with tuberization in potato: potential candidates for genetic engineering. Critical Reviews in Biotechnology. 2017; 37 (7): 942-957. DOI: 10.1080/07388551.2016.1274876

[75] Bin Shi, J, Ji Ye G, Zhi Yang Y, Wang F, Zhou Y, Hong Sun H, Wang AJ. The Differential Gene Expression of Key Enzyme in The Gibberellin Pathway In The Potato *Solanum tuberosum*) Mutant M4p-9. Pakistan Journal of Botany. 2016; 48(3): 1107-1117. DOI: 10.13560/j.cnki.biotech.bull.1985.2016.01.021

[76] Hedden P, Phillips AL. Gibberellin metabolism: new insights revealed by the genes. Trends in Plant Science. 2000; 5: 523-530. DOI: 10.1016/s1360-1385(00)01790-8.

[77] Carrera E, Bou J, García-Martínez JL, Prat S. Changes in GA 20-oxidase gene expression strongly affect stem length, tuber induction and tuber yield of potato plants. The Plant Journal, 2000; 22(3): 247-256. DOI: 10.1046/j.1365-313x.2000.00736.x.

[78] Roumeliotis, E, Kloosterman B, Oortwijn M, Lange T, Visser RGF, Bachem CWB. Down regulation of StGA3ox genes in potato results in altered GA content and affect plant and tuber growth characteristics. Journal of Plant Physiology. 2013; 170: 1228-1234. DOI: 10.1016/j.jplph.2013.04.003.

[79] Kloosterman B, Navarro C, Bijsterbosch G. Lange T, Prat S, Visser RGF, Bachem CWB. StGA2ox1 is induced prior to stolon swelling and controls GA levels during potato tuber development. The Plant Journal, 2007; 52: 362-373. 10.1111/j.1365-313X.2007.03245.x

[80] Kolachevskaya OO, Lomin SN, Kojima M, Getman IA, Sergeeva IA, Sakakibara H, Romanov GA. Tuber-Specific Expression of Two Gibberellin Oxidase Transgenes from Arabidopsis Regulates over Wide Ranges the Potato Tuber Formation. Russian Journal of Plant Physiology. 2019; 66: 984-991. https://doi.org/10.1134/S1021443720010094

[81] Ševčíková H, Mašková P, Tarkowská D, Mašek T, Lipavská H. Carbohydrates and gibberellins relationship in potato tuberization. Journal of Plant Physiology. 2017; 214; 53-63. DOI: 10.1016/j.jplph.2017.04.003.

[82] Roumeliotis E, Visser RG, Bachem CW. A crosstalk of auxin and GA during tuber development. Plant Signaling & Behavior. 2012; 7: 1360-1363.

[83] Kondhare KR, Patil AB, Giri AP. Auxin: An emerging regulator of tuber and storage root development. Plant Science. 2021; 306: 110854. DOI: https://doi.org/10.1016/j.plantsci.2021.110854

[84] Puzina TI, Kirillova IG, Yakushkina NI. Dynamics of

indole-3-acetic acid in organs of potato at different stages of ontogenesis and its role in regulation of tuber growth (in Russian). Izvestiya Akademii Nauk SSSR. Seriya Biologicheskaya. 2000; 2: 170-177. DOI: 10.1007/s00299-019-02395-0.

[85] Faivre-Rampant O, Cardle L, Marshall D, Viola R, Taylor MA. Changes in gene expression during meristem activation processes in *Solanum tuberosum* with a focus on the regulation of an auxin response factor gene. Journal of Experimental Botany. 2004; 55: 613-622. https://doi.org/10.1093/jxb/erh075

[86] Kloosterman B, De Koeyer D, Griffiths R, Flinn B, Steuernagel B, Scholz U, Sonnewald S, Sonnewald U, Bryan GJ. Prat S, Banfalvi Z, Hammond JP, Geigenberger P, Nielsen KL, Visser RG, Bachem CW. Genes driving potato tuber initiation and growth: identification based on transcriptional changes using the POCI array. Functional & Integrative Genomics. 2008; 8: 329-340. https://doi.org/10.1007/s10142-008-0083-x

[87] Song S, Hao L, Zhao P, Xu Y, Zhong N, Zhang H, Liu N. Genome-wide Identification, Expression Profiling and Evolutionary Analysis of Auxin Response Factor Gene Family in Potato (*Solanum tuberosum* Group Phureja). Scientific Reports. 2019; 9: 1755. DOI: https://doi.org/10.1038/s41598-018-37923-7

[88] Natarajan B, Kondhare KR, Hannapel DJ, Banerjee AK. Mobile RNAs and Proteins: Prospects in storage organ development of tuber and root crops. Plant Science. 2019; 284, 73-81. DOI:10.1016/j.plantsci.2019.03.019

[89] Xu X, van Lammeren AA, Vermeer E, Vreugdenhil D. The role of gibberellin, abscisic acid, and sucrose in the regulation of potato tuber formation in vitro. Plant Physiology. 1998a; 117: 575-584. DOI: https://doi.org/10.1104/pp.117.2.575

[90] Mani F, Bettaieb T, Doudech N, Hannachi C. Physiological mechanisms for potato dormancy release and sprouting: a review. African Crop Science Journal. 2014; 22: 155-174.

[91] Nambara E, Marion-Poll A. Abscisic acid biosynthesis and catabolism. Annual Review of Plant Biology. 2005; 56: 165-185. DOI: https://doi.org/10.1146/annurev.arplant.56.032604.144046

[92] Muñiz García MN, Stritzler M, Capiati DA. Heterologous expression of Arabidopsis ABF4 gene in potato enhances tuberization through ABA-GA crosstalk regulation, Planta. 2013; 239(3): 615-631. DOI: 10.1007/s00425-013-2001-2.

[93] Sarkar D, Pandey SK, Sharma S. Cytokinins antagonize the jasmonates action on the regulation of potato (*Solanum tuberosum*) tuber formation in vitro. Plant Cell, Tissue and Organ Culture. 2006; 87: 285-295. https://doi.org/10.1007/s11240-006-9166-3

[94] Malkawi A, Jensen BL, Langille AR. Plant hormones isolated from "Katahdin" potato plant tissues and the influence of photoperiod and temperature on their levels in relation to tuber induction. Journal of Plant Growth Regulation. 2007; 26: 308-317. DOI: https://doi.org/10.1007/s00344-007-9010-y

[95] Pavlista AD. Thidiazuron increased yield of potato cultivars. PGRSA Quart. 2001; 29: 72-80.

[96] Raspor M, Motyka V, Ninkovic S, Dobrev PI, Malbeck J, Cosic T, Cingel A, Savic J, Tadic V, Dragicevic IC. Endogenous levels of cytokinins, indole-3-acetic acid and abscisic acid in

in vitro grown potato: A contribution to potato hormonomics. Scientific Reports. 2020; 10: 3437. DOI: https://doi.org/10.1038/s41598-020-60412-9

[97] Tao G, Letham DS, Yong JWH, Zhang K, John PCL, Schwartz O, Wong SC, Farquhar GD. Promotion of shoot development and tuberisation in potato by expression of a chimaeric cytokinin synthesis gene at normal and elevated CO2 levels Functional Plant Biology. 2010; 37(1): 43-54. DOI: https://doi.org/10.1071/FP07032

[98] Zwanenburg B, Blanco-Ania D. Strigolactones: new plant hormones in the spotlight. Journal of Experimental Botany. 2018; 69: 2205-2218.DOI: https://doi.org/10.1093/jxb/erx487

[99] Pasare SA, Ducreux LJM, Morris WL, Campbell R, Sharma SK, Roumelitotis E, Kohlen W, van der Krol S, Bramley PM, Roberts AG, Fraser PD, Taylor MA. The role of the potato (*Solanum tuberosum*) CCD8 gene in stolon and tuber development. The New Phytologist. 2013; 198(4): 1108-1120. DOI: 10.1111/nph.12217

[100] Aksenova NP, Sergeeva LI, Konstantinova TN, Golyanovskaya SA, Kolachevskaya OO, Romanov GA. Regulation of potato tuber dormancy and sprouting. Russian Journal of Plant Physiology. 2013; 60: 301-312. DOI: 10.1007/s00425-013-1968-z

[101] Meng Liu J, Shan Wang S, Zheng X, Jin N, Lu J, Tao Huang Y, Fan B, Zhong Wang F. Antimicrobial Activity Against Phytopathogens and Inhibitory Activity on Solanine in Potatoes of the Endophytic Bacteria Isolated from Potato Tubers. Frontiers in Microbiology. 2020; 11. DOI: https://doi.org/10.3389/fmicb.2020.570926

[102] Foukaraki SG, Cools K, Chope GA, Terry LA. Impact of ethylene and 1-MCP on sprouting and sugar accumulation in stored potatoes. Postharvest Biology and Technology. 2016; 114: 95-103. DOI: https://doi.org/10.1016/j.postharvbio.2015.11.013

[103] Tosetti R, Waters A, Chope G. A, Cools K, Alamar MC, McWilliam S, Thompson AJ, Terry LA. New insights into the effects of ethylene on ABA catabolism, sweetening and dormancy in stored potato tubers. Postharvest Biology and Technology. 2021; 173: 111420. DOI: https://doi.org/10.1016/j.postharvbio.2020.111420

[104] Morris WL, Alamar MC, Lopez-Cobollo RM, Cañete JC, Bennett M, Van der Kaay J, Stevens J, Sharma SK, McLean K, Thompson AJ, Terry AJ, Turnbull CGN, Bryan GJ, Taylor MA. A member of the TERMINAL FLOWER 1/ CENTRORADIALIS gene family controls sprout growth in potato tubers. Journal of Experimental Botany. 2018; 70(3): 835-843. DOI: https://doi.org/10.1093/jxb/ery387

[105] Wang Z, Ma R, Zhao M, Wang F, Zhang N, Si H. NO and ABA Interaction Regulates Tuber Dormancy and Sprouting in Potato. Frontiers in Plant Science. 2020; 11: 311. DOI: 10.3389/fpls.2020.00311

[106] Deligios PA, Rapposelli E, Mameli MG, Baghino L, Mallica GM, Ledda L. Effects of Physical, Mechanical and Hormonal Treatments of Seed-Tubers on Bud Dormancy and Plant Productivity. Agronomy. 2020; 10(1): 33. https://doi.org/10.3390/agronomy10010033

[107] Pogorelko G, Lionetti V, Bellincampi D, Zabotina O. Cell wall integrity. Plant Signal Behaviour. 2013; 8(9): e25435. DOI: 10.4161/psb.25435

[108] Wang D, Yeats TH, Uluisik S, Rose JKC, Seymour GB. Fruit Softening:

Revisiting the Role of Pectin. Trends Plant Science. 2018; 23(4): 302-310. DOI:https://doi.org/10.1016/j.tplants.2018.01.006

[109] Vincken JP, Borkhardt B, Bush M, Doeswijk-Voragen C, Dopico B, Labrador E, Lange L, McCann M, Morvan C, Muños F, Oomen R, Peugnet I, Rudolph B, Schols H, Sørensen S, Ulvskov P, Voragen A, Visser R. Proceedings of the Second European Conference on Plant Biotechnology. ed. / Gert E de Vries; Karin Metzlaff. Elsevier. 2006; p: 245-256

[110] Poles L, Gentile A, Giuffrida A, Valentini L, Endrizzi I, Aprea E, Gasperi F, Distefano G, Artioli G, Malfa SL, Costa F, Lovatti L, Di Guardo M. Role of fruit flesh cell morphology and MdPG1 allelotype in influencing juiciness and texture properties in apple. Postharvest Biology and Technology. 2020; 164: 111161. DOI: https://doi.org/10.1016/j.postharvbio.2020.111161

[111] Gao Y, Guo Y, Su Z, Yu Y, Zhu Z, Gao P, Wang X. Transcriptome analysis of genes related to fruit texture in watermelon. Scientia Horticulturae. 2020; 262: 109075. DOI: https://doi.org/10.1016/j.scienta.2019.109075

[112] García-Segovia P, Andrés-Bello A, Martínez-Monzó J. Textural properties of potatoes (*Solanum tuberosum* L., cv. Monalisa) as affected by different cooking processes. Journal of Food Engineering. 2008; 88: 28-35. DOI: https://doi.org/10.1016/j.jfoodeng.2007.12.001

[113] Ducreux LJM, Morris WL, Prosser IM, Morris JA, Beale MH, Wright F, Shepherd T, Bryan GJ, Hedley PE, Taylor MA. Expression profiling of potato germplasm differentiated in quality traits leads to the identification of candidate flavour and texture genes. Journal of Experimental Botany. 2008; 59: 4205-4219. DOI: 10.1093/jxb/ern264.

[114] Banda L, Kyallo M, Domelevo Entfellner JB, Moyo M, Swanckaert J, Mwanga ROM, Onyongo A, Magiri E, Gemenet DC, Yao N, Pelle R, Muzhingi T. Analysis of β-amylase gene (Amyβ) variation reveals allele association with low enzyme activity and increased firmness in cooked sweetpotato (*Ipomoea batatas*) from East Africa. Journal of Agriculture and Food Research. 2021; 4: 100121. DOI: https://doi.org/10.1016/j.jafr.2021.100121

[115] Dong W, Li L, Cao R, Xu S, Cheng L, Yu M, Lv Z, Lu G. Changes in cell wall components and polysaccharide-degrading enzymes in relation to differences in texture during sweet potato storage root growth. Journal of Plant Physiology. 2020; 254: 153282. DOI: https://doi.org/10.1016/j.jplph.2020.153282

[116] Che Y, Zhang N, Zhu X, Li S, Wang S, Si H. Enhanced tolerance of the transgenic potato plants overexpressing Cu/Zn superoxide dismutase to low temperature. Scientia horticulturae. 2020; 261: 108949. https://doi.org/10.1016/j.scienta.2019.108949

[117] Upadhyaya CP, Bagri DS. Biotechnological Approaches for Nutritional Improvement in Potato (*Solanum tuberosum* L.). Genome Engineering for Crop Improvement. 2021 John Wiley & Sons Ltd. DOI: https://doi.org/10.1002/9781119672425.ch15

[118] Upadhyaya DC, Bagri DS, Upadhyaya CP, Kumar A, Thiruvengadam M, Jain SK. Genetic engineering of potato (*Solanum tuberosum* L.) for enhanced

α-tocopherols and abiotic stress tolerance. Physiologia Plantarum. 2020. https://doi.org/10.1111/ppl.13252

[119] Li S, Zhang N, Zhu X, Ma R, Yang J, Tang X, Si H. Enhanced drought tolerance with artificial microRNA-mediated StProDH1 gene silencing in potato. Crop Science. 2020; 60(3): 1462-1471. DOI: https://doi.org/10.1002/csc2.20064

[120] Oomen RJFJ, Doeswijk-Voragen CHL, Bush MS, Vincken JP, Borkhardt B, van den Broek LAM, Corsar J, Ulvskov P, Voragen AGJ, McCann MC, Visser RGF. In muro fragmentation of the rhamno galacturonan I backbone in potato (*Solanum tuberosum* L.) results in a reduction and altered location of the galactan and arabinan side-chains and abnormal periderm development. The Plant Journal. 2002; 30:403-413. DOI: 10.1046/j.1365-313x.2002.01296.x

[121] Martín I, Dopico B, Munoz FC, Esteban R, Oomen RJFJ, Driouich A, Vincken JP, Visser R, Labrador E. In vivo Expression of a *Cicer arietinum* β-galactosidase in Potato Tubers Leads to a Reduction of the Galactan Side-chains in Cell Wall Pectin. Plant Cell Physiology. 2005; 46(10): 1613-1622.

[122] Hong Huang J, Kortstee A, Dess DCT, Trindade LM, Schols HA, Gruppen H. Modification of potato cell wall pectin by the introduction of rhamnogalacturonan lyase and β-galactosidase transgenes and their side effects. Carbohydrate Polymers. 2016; 144:9-16. DOI: 10.1016/j.carbpol.2016.02.037

[123] Oxenboll Sorensen S, Pauly M, Bush M, Skjot M, McCann MC, Borkhardt B, Ulvskov P. Pectin engineering: modification of potato pectin by in vivo expression of an endo-1,4-beta-D-galactanase. PNAS. 2000; 97:7639-7644. DOI: https://doi.org/10.1073/pnas.130568297

[124] Liu B, Zhang N, Wen Y, Jin X, Yang J, Si H. Transcriptomic changes during tuber dormancy release process revealed by RNA sequencing in potato. Journal of Biotechnology. 2015; 198, 17-30. DOI: 10.1016/j.jbiotec.2015.01.019

[125] Chekanova JA. Long non-coding RNAs and their functions in plants. Current Opinion in Plant Biology. 2015; 27, 207-216. DOI: 10.1016/j.pbi.2015.08.003

[126] Hou X, Du Y, Liu X, Zhang H, Liu Y, Yan N, Zhang Z. Genome-Wide Analysis of Long Non-Coding RNAs in Potato and Their Potential Role in Tuber Sprouting Process. International Journal of Molecular Science. 2018; 9(1):101; DOI: https://doi.org/10.3390/ijms19010101

[127] Zhang Z, Wang J, Zhang X, Shi Q, Xin L, Fu H, Wang Y. Effects of radio frequency assisted blanching on polyphenol oxidase, weight loss, texture, color and microstructure of potato. Food Chemistry. 2018b; 248:173-182. DOI: 10.1016/j.foodchem.2017.12.065.

[128] Khan TA, Yusuf M, Ahmad A, Bashir Z, Saeed T, Fariduddin Q, Hayat S, Mock HP, Wu T. Proteomic and physiological assessment of stress sensitive and tolerant variety of tomato treated with brassinosteroids and hydrogen peroxide under low-temperature stress. Food Chemistry. 2019; 15: 289:500-511. DOI: 10.1016/j.foodchem.2019.03.029.

[129] Lin Q, Xie Y, Guan W, Duan Y, Wang Z, Sun C. Combined transcriptomic and proteomic analysis of cold stress induced sugar accumulation and heat shock proteins expression during postharvest potato tuber storage. Food Chemistry. 2019;

297, 124991. DOI: https://doi.org/10.1016/j.foodchem.2019.124991

[130] Nicolás González M, Massa GA, Andersson M, Turesson H, Olsson N, Sofie Fält A, Storani L, Décima Oneto CA, Hofvander P, Feingold SE. Reduced Enzymatic Browning in Potato Tubers by Specific Editing of a Polyphenol Oxidase Gene via Ribonucleoprotein Complexes Delivery of the CRISPR/Cas9 System. Frontiers in Plant Science. 2020. DOI. https://doi.org/10.3389/fpls.2019.01649

Chapter 13

Gamma Radiation Effect on *Agrobacterium tumefaciens*-Mediated Gene Transfer in Potato (*Solanum tuberosum* L.)

Murat Aycan, Muhammet Cagri Oguz, Yasin Ozgen, Burak Onol and Mustafa Yildiz

Abstract

Potato (*Solanum tuberosum* L.) is one of the major crops of the world. Significant improvements can be achieved in terms of yield and quality by the determination of efficient transformation methods. On the other hand, low transformation frequency seriously limits the application of molecular techniques in obtaining transgenic crops. In the present study, the effect of gamma radiation on *Agrobacterium tumefaciens*-mediated transformation to the potato was firstly investigated. Sterile seedlings of potato cv. 'Marabel', which was grown on Gamborg's B5 medium in Magenta vessels, were irradiated with different gamma radiation doses (0-control, 40, 80, 120 Gy ^{60}Co). Stem parts having axillary meristems were excised from irradiated seedlings and inoculated by *A. tumefaciens* (GV2260), which harbors the binary plasmid p35S GUS-INT contains and *GUS* (β-glucuronidase) gene controlled by 35S promoter (CaMV) and *nptII* (neomycin phosphotransferase II) gene driven by NOS (nopaline synthase) promoter). Inoculated stem parts having axillary meristems explants were then directly transported to a selection medium containing duocid (500 mg l^{-1}), and kanamycin (100 mg l^{-1}), 4 mg l^{-1} gibberellic acid, 1 mg l^{-1} BAP and 0.1 mg l^{-1} NAA. The adult transgenic plants were detected by polymerase chain reaction (PCR) analysis. According to the number of transgenic plants determined by PCR analysis, results obtained from explants treated with 40 Gy gamma gave the best results compared to the control (0 Gy) application. The doses over 40 Gy were also found statistically significant compared to the control (0 Gy). It is expected that the protocol described in this study make the transformation studies easier by skipping the stages of 'co-cultivation', 'culturing explants on selection medium' and 'recovery of transgenic shoots on selection medium' not only for potato but also for other crop plants. This study was supported by a grant from the Scientific and Technological Research Council of Turkey (TUBİTAK) (Grant number 113O280 to Prof. Dr. Mustafa YILDIZ).

Keywords: transformation efficiency, gamma radiation, potato, *A. tumefaciens*

1. Introduction

Genetic transformation technologies developed with recombinant DNA technology and *in vitro* regeneration methods are successfully used to overcome the species differences and taxonomic obstacles encountered in traditional breeding programs.

Agrobacterium-mediated transformation method is the most proficient and commonly used plant transformation method among different gene transfer techniques. However, there are a number of variables that affect the success of *Agrobacterium*-mediated genetic transformation [1]. The success of genetic transfer efficiency with *A. tumefaciens* depends on, *Agrobacterium* strain used, bacteria concentration, antibiotic types and concentration used for *in vitro* selection, inoculation time, and temperature [2]. Besides, the type of the target plant, plant explants, hormone combinations used in *in vitro* regeneration, pH, etc., are among the factors affecting the recovery of transgenic plants [3]. Increasing the transformation efficiency by reducing the limiting factors in genetic transformation studies will significantly contribute to the success rate [4].

Gamma radiation treatments are innovative biotechnological interventions used to increase yield and quality. Gamma radiation technique is successfully applied in plant breeding programs to increase genetic diversity and biotic/abiotic stress tolerance [5]. With the use of this method, thousands of mutant varieties have been obtained from approximately 200 plant species [6]. Gamma radiation is considered a physical mutagen that has significant effects on cytological, biochemical, molecular, physiological and morphological processes in plants [7–9]. The biological effect of gamma radiation is due to its interaction with atoms and water molecules in the cell [10]. As a result of the interaction of gamma rays with atoms, free radicals are produced at the cellular level. These radicals affect physiological and metabolic activities in plants [11].

Low doses of gamma have positive effects on cell proliferation; cell and tissue growth, germination percentage, enzyme activity, chlorophyll content, biotic and abiotic stress tolerance and crop yield [12–15]. On the other hand, high doses of gamma particles cause damage to protein synthesis, enzyme activity hormone balance, water exchange and leaf gas exchange [16].

The adverse effects of gamma rays are divided into two as direct and indirect. While its direct impact is realized by the interaction between radiation and target living molecules, its indirect effect arises from the formation of free radicals [17]. These free radicals are called radiation hormones and inhibit the growth of the plant [18]. Besides, the effect of gamma-ray on the plant depends on the source and dose of gamma radiation, exposure time, target plant species and variety, plant tissue and the plant's growth period [19].

New approaches are needed to increase the low success rate in genetic transformation studies. The positive effect of a reduced dose of gamma radiation on genetic transformation efficiency has been investigated in a few studies [4, 20]. Determining the effect of gamma radiation on genetic transformation frequency in different plant species will contribute to genetic transformation and molecular assisted cultivar development studies. This study was aimed to determine the effect of gamma radiation to increase the genetic transformation frequency in potato and to create a repeatable successful protocol.

2. Materials and methods

2.1 Plant material

Potato (*S. tuberosum* L.) tubers of cv. 'Marabel' were used in the study.

2.2 Explant material

Stem parts having axillary meristems isolated from irradiated sterile seedlings were used as explant (**Figure 1**).

2.3 Radiation source

0.8 kGy h^{-1} of ^{60}Co γ ray source at the Sarayköy Nuclear Research and Training Center, Turkish Atomic Energy Authority, Sarayköy, Ankara.

2.4 *A. tumefaciens* strain

The GV2260 including p35S GUS-INT plasmid of *A. tumefaciens* strain was utilized for inoculation. The characteristic of p35S GUS-INT binary plasmid described Yildiz et al. [20]. GV2260 strain (OD = 0.6) was incubated overnight in a liquid medium (Nutrient Broth) including rifampicin (50 mg l^{-1}) and kanamycin (50 mg l^{-1}) in an incubator (rotary shaker) at 180 rpm under 28°C and used for transformation studies.

2.5 Irradiation of seedlings

One-month-old sterile seedlings were irradiated with different doses (0-control, 40, 80 and 120 Gy) of ^{60}Co γ source. Fricke and alanine dosimeters were used for dose mapping and determination of dose rates of gamma source. Seedlings were irradiated along with a dosimeter for each dose to be sure that ionization was uniform.

2.6 Culture conditions

Seedlings were grown on the Gamborg's B5 medium containing the mineral salts and vitamins, sucrose (3%, w/v) and agar (0.7%, w/v). The pH of the medium was adjusted to 5.8 before autoclaving. All cultures were grown at 25 ± 1°C under cool white fluorescent light (27 μmol m^{-2} s^{-1}) with a 16/8 h day/night photoperiod in the growth chamber.

2.7 Transformation procedure

A. tumefaciens GV2260 carrying p35S GUS-INT plasmid was incubated overnight and diluted with a liquid medium to $1X10^8$ cell/ml. Stem parts having

Figure 1.
Stem parts having axillary meristems excised from irradiated sterile seedlings.

axillary meristems excised from irradiated sterile seedlings were inoculated in a liquid regeneration medium containing 4 mg l^{-1} gibberellic acid, 1 mg l^{-1} BAP and 0.1 mg l^{-1} NAA for 20 min. After inoculation, stem parts having axillary meristems were directly transferred to selection medium bypassing co-culture stage in Magenta vessels containing 4 mg l^{-1} gibberellic acid, 1 mg l^{-1} BAP and 0.1 mg l^{-1} NAA, supplemented with 100 mg l^{-1} kanamycin and 500 mg l^{-1} duocid for 2 weeks.

2.8 Recovery of putative transgenic plants

Seedlings were planted to pots having commercial soil in a growth chamber for 3 weeks where temperature (24 ± 1°C), light (27 μmol m^{-2} s^{-1}) and humidity were controlled. To keep the humidity high, the pots were covered with a thin nylon transparent bag and placed in the growth chamber. The humidity was gradually reduced by making small holes in the bags every 2-3 days. After 10 days, the bags were completely removed. By this way, humidity was reduced gradually from 100–40%. Candidate transgenic plants were irrigated with 50 ml water including kanamycin (100 mg l^{-1}) at 2 day-intervals during 14 days for further selection.

2.9 gDNA (genomic DNA) extraction

The gDNA was extracted from fresh leaves of putative transgenic plants and from control (non-transformed) plants with slight modification of the protocol described by [21].

2.10 Polymerase chain reaction (PCR)

PCR amplification was performed to detect the *npt-II* gene with the following designed specific primer sets Forward: 5′-TTGCTCCTGCCGAGAAAG-3′ and Reverse: 5′-GAAGGCGATAGAAGGCGA-3′. PCR amplification of the chromosomal virulence gene (*chv*) was carried with the following primer sets Forward: 5′-CGAACCGCTGTTCGGCCTGTGG-3′ and Reverse: 5′-GTTCAGGAGGCCGGCATCCTGG-3′ for determine of *A. tumefaciens* contamination in putative transgenic plants.

The PCR was conducted in 2 μL containing 100 ng of DNA, 10 pmol of each forward and reverse primers, 0.25 μM dNTP, 2 mM $MgCl_2$, 1× PCR buffer, and 0.625 U of DreamTaq DNA polymerase enzyme (Thermo Scientific, Waltham, Massachusetts, USA). The PCR was run with an initial denaturation of the DNA template at 95°C for 5 min followed by 36 cycles, each consisting of 95°C for 1 min, 58°C for 1 min and 72°C for 1 min, and final extension at 72°C for 5 min in a Prime G Gradient Thermal Cycler (Techne, Staffordshire, UK). Amplified PCR products were electrophoresed on a 1% agarose in TAE (tris-acetate EDTA) buffer. The bands were stained with ethidium bromide staining and visualized with UV light.

2.11 Observations

Number of explants cultured on selection medium, number of plants growing on selection medium, number of putative transgenic plants transferred to soil, number of PCR positive (+) plants, number of PCR (+) plants after *chv* gene analysis and transformation efficiency were determined.

2.12 Statistical analysis

Five replicates of rooted plants in the pots were tested and considered the units of replication. One-way Analysis of Variance (ANOVA) was used to test the effect of gamma radiation on gene transformation efficiency. All experiments were repeated two times. Data were statistically analyzed by "IBM SPSS Statistics 22" computer program. Duncan's multiple range test was used to compare the means [22].

3. Results and discussion

Results of gene transformation to stem parts having axillary meristems of irradiated seedlings were given in **Table 1**. In the current study, inoculation was performed to 30 stem parts having axillary meristems at each of the gamma doses (0-control, 40, 80 and 120 Gy). Only 14 plants were grown in control treatment where gamma radiation was not applied. As the result of PCR analysis, band of *npt-II* gene was detected in only 5 putative transgenic plants. However, after *chv* gene analysis, it was determined that none of 5 putative transgenic plants was real transgenic which meant band of *npt-II* gene detected in 5 putative transgenic plants came from bacteria being on the plants in control treatment. In all gamma treatments, increases were observed in the number of plants growing on selection medium. In all the parameters examined, the highest values were recorded in plants grown from stem parts having axillary meristems to which 40 Gy gamma dose was applied. At 40 Gy gamma dose, 28 out of 30 inoculated stem parts having axillary meristems of irradiated seedlings were successfully grown in soil. Thirty three out of 28 putative transgenic plants were found PCR(+) (**Table 1**, **Figure 2**). The presence of *chv* gene was checked in 33 putative transgenic plants, and consequently, 28 plants were confirmed as real transgenic without bacterial contamination. Transformation efficiency was calculated as 100% (**Table 1**, **Figure 3**).

Results showed positive effects of gamma radiation on transformation at 40 Gy as compared to control. Higher gamma doses over 40 Gy, transformation hindered significantly. PCR analysis confirmed that 28 plants out of 33 were transgenic at 40 Gy gamma treatment (**Table 1**).

A. tumefaciens, a plant pathogen, is commonly used as a vector for genetic transformation to plants [23, 24]. The genetic transformation prosperity of *A. tumefaciens* method is limited in plant species largely, because the mechanism of plant's resistance will be active when pathogen attacks. That is why, genetic manipulations of the plant, physical conditions and bacteria have been applied to increase the virulence of bacteria and to increase the transformation efficiency [25, 26].

Before inoculation, pre-culturing explants [25, 27], alteration of temperature [25, 28] and medium pH [28, 29], addition chemicals to inoculation medium such as acetosyringone [25, 26, 28, 30–32], altering bacterial density and co-cultivation time [27, 29, 31] and vacuum infiltration [33–35] have been reported to increase transformation.

Possible molecular effects of gamma radiation in plants include activations of RNA and protein synthesis, acceleration of cell division, and direct or indirect activation of genes [36, 37]. Ionizing radiation causes a single strand break and replication inhibition at high doses, while at low doses it causes only minor replication blockade [38]. Gamma radiation cause chromosome strand breaks and consequently integration of genes transferred from extracellular to DNA. Köhler et al. [39] reported an increase in the frequency of transgenic plants regenerated from protoplasts exposed to gamma radiation. It has been reported that this occurred as a

Gamma radiation dose (Gy)	Number of stem parts cultured on selection medium	Number of plants growing on selection medium	Number of putative transgenic plants transferred to soil	Number of PCR (+) plants	Number of PCR(+) plants after *chv* gene analysis	Transformation efficiency (%)*
0	30.00	14.00 b	14.00 b	5.00 c	0.00 c	0.00 d
40	30.00	28.00 a	28.00 a	33.00 a	28.00 a	100.00 a
80	30.00	26.00 ab	26.00 ab	15.00 b	14.00 b	53.85 c
120	30.00	25.00 ab	25.00 ab	17.00 b	17.00 b	68.00 b

Values in a column followed by different letters are significantly different at the 0.01 level.
**Transformation efficiency = (Number of PCR (+) plants after chv gene analysis/Number of putative transgenic plants transferred to soil) x 100.*
The reason why number of PCR (+) plants is higher than number of putative transgenic plants transferred to soil at 40 Gy gamma treatment is the development of new shoots from buds on tuber.

Table 1.
Results of PCR analysis in plants grown from stem parts having axillary meristems of irradiated seedlings inoculated by Agrobacterium tumefaciens.

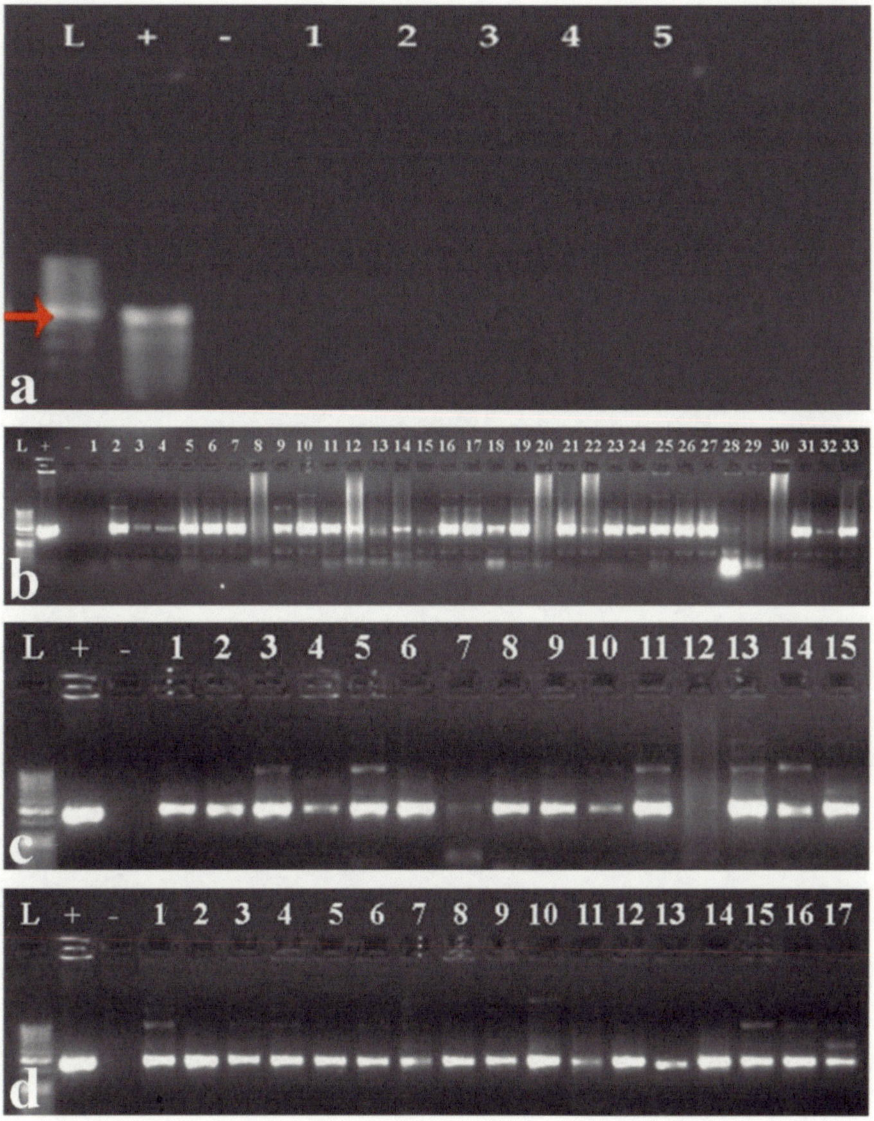

Figure 2.
PCR analysis of genomic DNA of putative transgenic plants grown from stem parts having axillary meristems of irradiated seedlings for the amplification of npt-II gene. L-DNA ladder, + positive control, – negative control. a. 0 Gy, b. 40 Gy, c. 80 Gy, d. 120 Gy.

result of the increased recombination mechanism in the irradiated cells, resulting in an increased number of transformed colonies with high integration rates. Similarly, in our study, the positive effect of low dose gamma dose on potato transformation efficiency was determined. Another possible effect of gamma radiation on genetic transformation efficiency may be related to the process of radiation of the target plant. It was reported that protoplast radiation one hour before transformation increased the success rate, whereas radiation performed one hour after transformation had no effect on the transformation efficiency [39]. In our study, gamma irradiation was applied before the gene transfer stage. The results obtained from our study coincide with the results stated above.

From the results of the current study, in the gene transformation to potato stem parts having axillary meristems by *A. tumefaciens*, it was observed that 40 Gy

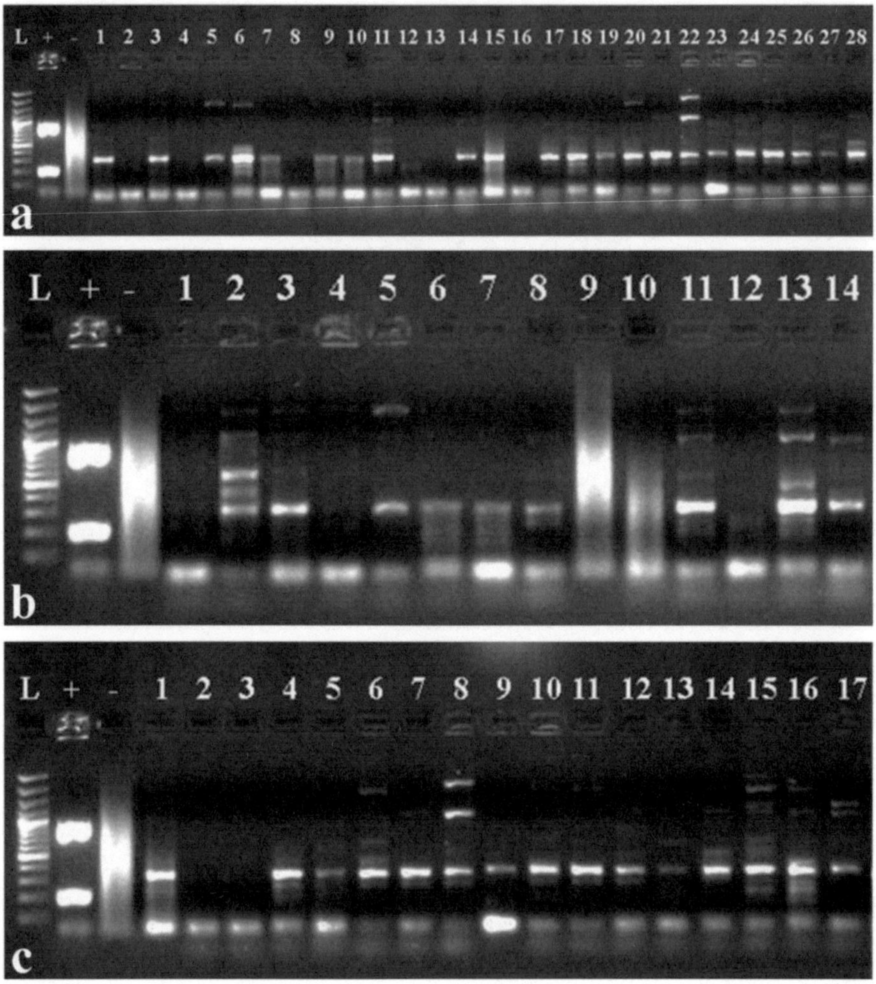

Figure 3.
PCR analysis of genomic DNA of putative transgenic plants grown from stem parts having axillary meristems of irradiated seedlings for the amplification of chv gene. L-DNA ladder, + positive control, – negative control. a. 40 Gy, b. 80 Gy, c. 120 Gy.

gamma dose significantly increased the transgenic plant frequency compared to control in which no gamma was used. To our knowledge, this was the first study revealing gene transformation to stem parts having axillary meristems via *A. tumefaciens* in potato.

Author details

Murat Aycan[1], Muhammet Cagri Oguz[2], Yasin Ozgen[3], Burak Onol[3] and Mustafa Yildiz[3*]

1 Laboratory of Biochemistry, Faculty of Agriculture, Niigata University, Niigata, Japan

2 Department of Field Crops, Graduate School of Natural and Applied Sciences, Ankara University, Ankara, Turkey

3 Department of Field Crops, Faculty of Agriculture, Ankara University, Ankara, Turkey

*Address all correspondence to: myildiz@ankara.edu.tr

References

[1] Gürel S, Oğuz MÇ, Turan F, Kazan K, Özmen CY, Gürel E, Ergül A. Utilization of sucrose during cocultivation positively affects Agrobacterium-mediated transformation efficiency in sugar beet (*Beta vulgaris* L.). Turk. J. Agric. For. 2019;43(6):509-517. DOI:10.3906/tar-1812-90

[2] Pathi KM, Tula S, Tuteja N. High frequency regeneration via direct somatic embryogenesis and efficient *Agrobacterium*-mediated genetic transformation of tobacco. Plant Signal. Behav. 2013;8(6):e24354. DOI:10.4161/psb.24354

[3] Nofouzi F, Oğuz MÇ, Khabbazi SD, Ergül A. Improvement of the *in vitro* regeneration and *Agrobacterium*-mediated genetic transformation of *Medicago sativa* L. Turk. J. Agric. For. 2019;43(1):96-104. DOI: 10.3906/tar-1804-52

[4] Yildiz M, Beyaz R. The effect of gamma radiation on *Agrobacterium tumefaciens*-mediated gene transfer in durum wheat (*Triticum durum* Desf.). Fresenius Environ. Bull. 2019;28:488-494

[5] Aly AA, Eliwa NE, Maraei RW. Physiological and molecular studies on ISSR in two wheat cultivars after exposing to gamma radiation. Science Asia. 2019;45:436-445. DOI:10.2306/1513-1874.2019.45.436

[6] Sen A, Ozturk I, Yaycili O, Alikamanoglu S. Drought tolerance in irradiated wheat mutants studied by genetic and biochemical markers. J Plant Growth Regul. 2017;36:669-679. DOI:10.1007/s0034 4-017-9668-8

[7] Ashraf M, Cheema AA, Rashid M, Qamar Z. Effect of gamma rays on M1 generation in Basmati rice. Pak. J. Bot. 2003;35:791-795

[8] Melki M, Dahmani T. Gamma irradiation effects on durum wheat (*Triticum durum* Desf.) under various conditions. Pak. J. Biol. Sci. 2009;12:1531-1534

[9] Heidarieh M, Mirvaghefi A, Akbari M, Farahmand H, Sheikhzadeh N, Shahbazfar A, Behgar M. Effect of dietary Ergosan on growth performance, digestive enzymes, intestinal histology, hematological parameters and body composition of rainbow trout (*Oncorhynchus mykiss*). Fish Physiol. Biochem. 2012;38:1169-1174. DOI:10.1007/s10695-012-9602-8

[10] Kovács E, Keresztes Á. Effect of gamma and UV-B/C radiation on plant cells. Micron. 2002;33(2):199-210. DOI: 10.1016/S0968-4328(01)00012-9

[11] Borzouei A, Kafi M, Khazaei H, Naseriyan B, Majdabadi A. Effects of gamma radiation on germination and physiological aspects of wheat (*Triticum aestivum* L.) seedlings. Pak. J. Bot. 2010;42:2281-2290.

[12] Qi W, Zhang L, Wang L, Xu H, Jin Q, Jiao Z. Pretreatment with low-dose gamma irradiation enhances tolerance to the stress of cadmium and lead in Arabidopsis thaliana seedlings. Ecotoxicol. Environ. Saf. 2015;115:243-249. DOI: 10.1016/j.ecoenv.2015.02.026.

[13] Wang X, Ma R, Cui D, Cao Q, Shan Z, Jiao Z. Physio-biochemical and molecular mechanism underlying the enhanced heavy metal tolerance in highland barley seedlings pre-treated with low-dose gamma irradiation. Sci. Rep. 2017;7(1):1-14. DOI: DOI:10.1038/s41598-017-14601-8.

[14] Gordeeva EI, Shoeva OY, Yudina RS, Kukoeva TV, Khlestkina EK. Effect of seed pre-sowing gamma-irradiation treatment in bread wheat

lines differing by anthocyanin pigmentation. Cereal Res. Commun. 2018;46(1):41-53. DOI: 10.1556/0806.45.2017.059.

[15] Beyaz R. Impact of gamma irradiation pretreatment on the growth of common vetch (Vicia sativa L.) seedlings grown under salt and drought stress. Int. J. Radiat. Biol. 2020;96(2):257-266. DOI: 10.1080/09553002.2020.1688885.

[16] Hameed A, Mahmud TS, Atta BM, Haq MA, Sayed H. Gamma irradiation effects on seed germination and growth, protein content, peroxidase and protease activity, lipid peroxidation in desi and Kabuli chickpea. Pak. J. Bot. 2008;40:1033-1041.

[17] Esnault AM, Legue F, Chenal C. Ionizing radiation: advances in plant response. Environ. Exp. Bot. 2010;68:231-237. DOI: 10.1016/j.envexpbot.2010.01.007.

[18] Macovei A, Garg B, Raikwar S, Balestrazzi A, Carbonera D, Buttafava A, Tuteja N. Synergistic exposure of rice seeds to different doses of γ -ray and salinity stress resulted in increased antioxidant enzyme activities and gene-specific modulation of TC-NER pathway. BioMed Res. Int. 2014;676934,15. DOI:10.1155/2014/676934.

[19] Arena C, De Micco V, Macaeva E, Quintens R. Space radiation effects on plant and mammalian cells and mammalian cells. Acta Astronaut. 2014;104:419-431. DOI:10.1016/j.actaastro.2014.05.005

[20] Aycan M, Beyaz R, Bahadir A, Yildiz M. The effect of magnetic field strength on shoot regeneration and *Agrobacterium tumefaciens*-mediated gene transfer in flax (*Linum usitatissimum* L.). Czech J. Genet. Plant Breed., 2019;55:20-27. DOI:10.17221/195/2017-CJGPB

[21] Dellaporta SL, Wood J, Hicks JB. 1983 A plant DNA miniprepration: version II. Plant Mol. Biol. Rep. 1983;4:9-21.

[22] Snedecor GW, Cochran WG. Statistical Methods. The Iowa State University Press, Iowa, USA, 1967.

[23] Zupan J, Muth TR, Draper O, Zambryski P. The transfer of DNA from Agrobacterium tumefaciens into plants: a feast of fundamental insights. Plant J. 2000;23:11-28. DOI: 10.1046/j.1365-313x.2000.00808.

[24] Joubert P, Beaupe`re D, Lelie`vre P, Wadouachi A, Sangwan RS, Sangwan-Norreel BS. Effects of phenolic compounds on Agrobacterium vir genes and gene transfer induction. A plausible molecular mechanism of phenol binding protein activation. Plant Sci. 2002;162:733-743. DOI:10.1016/S0168-9452(02)00012-2.

[25] Chakrabarty R, Viswakarma N, Bhat SR, Kirti PB, Singh BD, Chopra VL. Agrobacterium-mediated transformation of cauliflower: optimization of protocol and development of Bt-transgenic cauliflower. J Biosci. 2002;27:495-502. DOI:10.1007/bf02705046.

[26] Henzi MX, Christey MC, McNeil DL. Factors that influence Agrobacterium rhizogenesmediated transformation of broccoli (Brassica oleracea L. var. italica). Plant Cell Rep. 2000;19:994-999. DOI:10.1007/s002990000221.

[27] Amoah BK, Wu H, Sparks C, Jones HD. Factors influencing Agrobacterium-mediated transient expression of uidA in wheat inflorescence tissue. J. Exp. Bot. 2001;52:1135-1142. DOI:10.1093/jexbot/52.358.1135

[28] De Clercq J, Zambre M, Van Montagu M, Dillen W, Angenon G.

An optimized Agrobacterium-mediated transformation procedure for *Phaseolus acutifolius* A. Gray. Plant Cell Rep. 2002;21:333-334. DOI:10.1007/s00299-002-0518-0.

[29] Mondal TK, Bhattacharya A, Ahuja PS, Chand PK. Transgenic tea (*Camellia sinensis* [L.] O. Kuntze cv. Kangra Jat) plants obtained by Agrobacterium-mediated transformation of somatic embryos. Plant Cell Rep. 2001;20:712-720. DOI:10.1007/s002990100382.

[30] Le VQ, Belles-Isles J, Dusabenyagasani M, Tremblay FM. An improved procedure for production of white spruce (*Picea glauca*) transgenic plants using Agrobacterium tumefaciens. J. Exp. Bot. 2001;52:2089-2095. DOI:10.1093/jexbot/52.364.2089.

[31] Lopez SJ, Kumar RR, Pius PK, Muraleedharan N. *Agrobacterium tumefaciens*mediated genetic transformation in tea (*Camellia sinensis* [L.] O. Kuntze). Plant Mol. Biol. Rep. 2004;22:201-202. DOI:10.1007/BF02772730.

[32] Somleva MN, Tomaszewski Z, Conger BV. Agrobacterium-mediated transformation of switchgrass. Crop Sci. 2002;42:2080-2087. DOI:10.2135/cropsci2002.2080

[33] Mahmoudian M, Yucel M, Öktem HA. Transformation of lentil (*Lens culinaris* M.) cotyledonary nodes by vacuum infiltration of Agrobacterium tumefaciens. Plant Mol. Biol. Rep. 2002;20:251-257. DOI:10.1007/BF02782460

[34] Wang WC, Menon G, Hansen G. Development of a novel Agrobacterium-mediated transformation method to recover transgenic Brassica napus plants. Plant Cell Rep. 2003;22:274-281. DOI:10.1007/s00299-003-0691-9

[35] Spokevicius AV, Van Beveren KS, Bossinger G. Agrobacterium-mediated in vitro transformation of wood-producing stem segments in eucalypts. Plant Cell Rep. 2005;23:617-624. DOI:10.1007/s00299-004-0856-1

[36] Kuzin AM, Vagabova ME, Revin AF. Molecular mechanisms of the stimulating action of ionizing radiation on seeds part 2 activation of protein and high molecular rna synthesis. Radiobiology. 1976;16:259-261

[37] Kovalchuk I, Molinier J, Yao Y, Arkhipov A, Kovalchuk O. Transcriptome analysis reveals fundamental diferences in plant response to acute and chronic exposure to ionizing radiation. Mutat. Res. 2007;624:101-113. DOI: 10.1016/j.mrfmmm.2007.04.009

[38] Lavin M, Schroeder AL. Damage-resistant DNA synthesis in eukaryotes. Mutation Res. 1988;193:193-206

[39] Köhler F, Cardon G, Pöhlman M, Gill R, Schieder O. Enhancement of transformation rates in higher plants by low-dose irradiation: Are DNA repair systems involved in the incorporation of exogenous DNA into the plant genome?. Plant Mol. Biol. 1989;12(2):189-199.

Chapter 14

Biotechnological Strategies for a Resilient Potato Crop

Elena Rakosy-Tican and Imola Molnar

Abstract

The aim of this chapter is to describe in a synthetic manner the most efficient biotechnological techniques which can be applied in potato breeding with emphasis on multiple resistance traits. To this end, most important results of all biotechnological techniques will be pointed out including new biotechnological tools of genome editing. The somatic hybridization will be the core of the presentation as the only non-GMO strategy with good results in transferring multiple resistances into potato gene pool. The chapter is presenting all data in a synthesized form and made comparisons between the existing techniques and their possible adoption in breeding in different parts of the world, depending on regulations and consumer choice. Moreover, the recently discovered value of potato as a healthy food and its possible applications in cancer treatment will be also discussed with new data on both potato and some of its wild relatives.

Keywords: advantages, genetic transformation, multiple resistance traits, new biotechnological techniques, potato breeding, somatic hybridization

1. Introduction

As a major food staple, the potato is contributing to the UN Millennium Development Goals of food security and poverty eradication. Today, potato is the most widly grown non-cereal crop [1] and important vegetable for human consumption [2]. The wide climatic adaptability and short growing time of potato facilitated its spread across diverse geographical regions. To date more than three thousand potato cultivars are cultivated in 165 countries with a production exceeding 350 million tonnes per year, particularly under temperate, subtropical and tropical regions, covering a major economic share in the global agricultural market [2]. For the last two decades, potato cultivation and utilization have also been notably increased in developing countries such as China, India and Bangladesh [3]. Although, classical breeding has developed thousands of new cultivars, potato is still sensitive to countless diseases and pests, which lead to 44.9% yield losses in every year [4]. Diseases such as late blight produced by the oomycete *Phytophthora infestans* (*Pi*), viruses like potato virus Y (PVY) and pests as Colorado potato beetle (CPB) are able to completely destroy a potato field if left uncontrolled. Even today the main way to combat diseases and pests is massive application of pesticides. Pesticides increase pollution of the environment, are toxic for non-target organisms including humans and exert selection pressure on the diseases and pests, which develop resistance. New sustainable and effective ways to combat diseases and pests of potato are required and biotechnological approaches have been lately developed

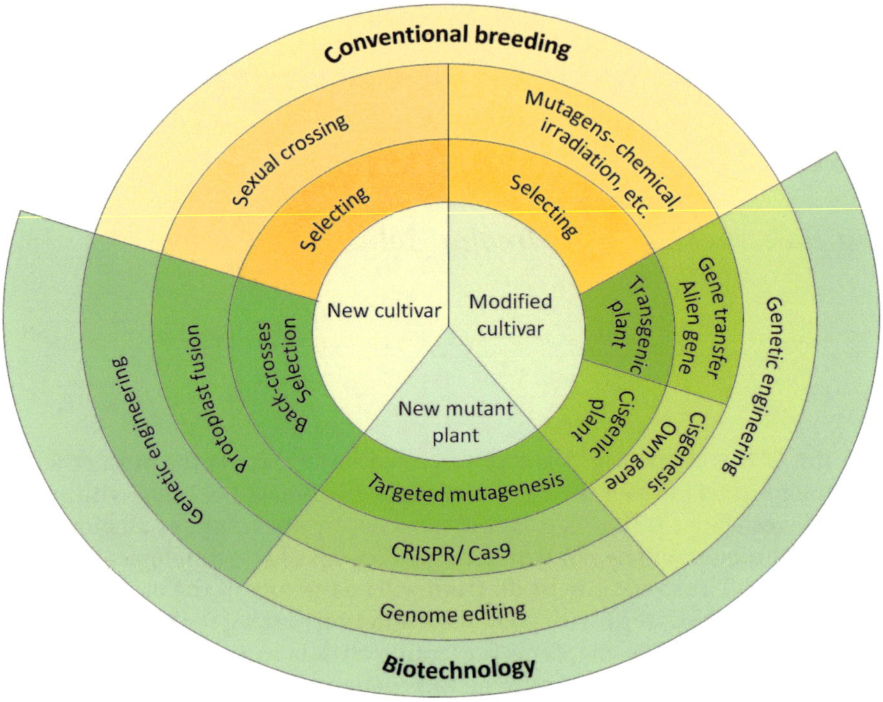

Figure 1.
Overview of classical breeding tools, as well as biotechnology and their applications for improving crops in general and potato resilience, in particular.

also to address this challenging issue (**Figure 1**). Moreover, climate change has challenged potato production worldwide in the last decades and new strategies to develop resilient potato to drought, high temperature, salt and other abiotic stresses or multiple stresses are an urgent need for potato cultivation. To achieve these goals, both classical breeding and biotechnology are aware of the resources of resistance genes in the crop wild relatives, as for example the project of International Potato Centre (CIP). There are published several books and reviews dealing with potato biotechnology and breeding [1, 2, 5, 6], but in this chapter we are going to overview, synthetize and point out those techniques that are included in potato genetic improvement for a resilient potato crop in order to develop a sustainable agriculture and reduce poverty.

2. Genetic engineering sustainability for a resilient potato crop

Modern biotechnology is defined as the technology which use living cells, microorganisms, or functional parts, such as enzymes, proteins, DNA or RNA molecules to develop basic research and deploy new useful products [7]. Genetic engineering, as part of plant biotechnology, covers techniques which change the genome of plants. In its larger sense, plant genetic engineering includes: (i) somaclonal variation, (ii) cell fusion and regeneration of somatic hybrid plants, (iii) gene transfer and (iv) genome editing. Since somaclonal variation has already been presented in detail and its results are currently not widely used in potato breeding [8], in this chapter we are presenting the other genetic engineering techniques and obtained results in developing resilient potato crop. Potato crop requires considerable inputs of: nutrients, pesticides, and water to maintain yield, tuber quality, and protection from its pathogens, pests and extreme climate conditions. Genetic variations for the most important traits is low in commercial cultivars, but related wild relatives contain many unique, valuable traits missing from cultivars, which represent a rich genetic source for potato improvement [9]. Potato breeding efforts have historically focused primarily on yield, fresh market and processing quality, storability as well as disease resistance. Only after developing genetic transformation and/or other biotechnological approaches, a faster transfer of valuable traits like quality of tuber composition and resistance to biotic and abiotic stresses became possible. Moreover, with using classical breeding one new cultivar can be produced in 10 to 15 years from the initial cross to cultivar release, while with biotechnology, particularly gene transfer, shorter time is required, from some months (6–12 months) to a few years, ignoring the long regulatory clearances [6]. There are many attempts and results on the transfer and integration of economically important genes in potato crop and some previous reviews have presented the state of art in plants or in this tuberous crop [6, 8, 10].

2.1 Gene transfer to develop resilient potato to biotic and abiotic stresses

Genetic transformation of potato was first achieved in 1988 [11, 12], potato being the third plant to be successfully transformed. This technology uses *Agrobacterium tumefaciens* - mediated gene transfer, which is reported as the most efficient for potato crop and some of potato wild relatives [13]. The first commercially grown potato was introduced by Monsanto as New Leaf™ in 1995, the first released genetically modified crop of the company. Besides gene transfer from bacteria, fungi, animal or other plant species commonly called transgenesis, more recently wild species are considered as a rich reservoir of resistance genes. The transfer of genes from the same genus, i.e. from related species that can be crossed, is called cisgenesis. Because the genes can be also integrated into the recipient plant genome by classical breeding, cisgenesis was thought to be exempted from GMO low in Europe. Plant own genes can be also transferred in order to increase their expression, and this technique is called intragenesis [14, 15]. *Solanum* wild species, that evolved to resist in diverse climates in South and North America, are indeed a rich reservoir of genes which can be introgressed in potato genome. It is estimated that around 190 wild tuber-bearing relatives of potato, in the section Petota of the genus *Solanum*, are available for resistance breeding [16, 17]. Moreover, besides their rich genetic resources, potato and its wild relatives benefit from a good amenability to *in vitro* tissue and protoplast culture, making it possible to exploit this diversity through genetic engineering [8].

2.1.1 Single or multiple resistance gene transfer to improve pathogen and pest resistance

Genetic engineering has the potential to transfer single genes to increase disease or pest resistance, if the selectable marker gene, which is necessary for transgenic plant selection is not considered. Such single genes can be introgressed in potato elite varieties to improve one resistance trait. The frequently used marker gene during potato gene transfer is *npt*II (bacterial neomycin phosphotransferase II gene), which renders transgenic cells resistant to aminoglycoside antibiotics, including kanamycin and G418 [18]. Selection based on kanamycin has been proven to generate escapes in potato crop [13]. In this study both genes: *npt*II and reporter *gfp* (green fluorescent protein), have been used to reveal the transgene transfer efficiency, which allowed to evaluate the escape events. In order to transfer single genes that increase host plant resistance to pathogens and pests, the researchers have to identify and clone the genes of interest (GOI). At this stage, a good knowledge of mechanisms of host plant– pathogen interaction and gene characterization is necessary. In the last decades new insights into the complex molecular race between pathogens and/or pests and crop hosts were advanced and many genes are characterized and some cloned [19, 20]. With the advent of Potato Genome Sequencing Consortium [21] and completion of the first reference genome of potato [17], and later the release of genome data for some of its wild relatives i.e. *S. commersoni* [22], and *S. chacoense* [23], potato breeding and biotechnology entered into the genomic-based improvement era. Gene transfer is already taking advantage of genome sequencing data in first instance through the transfer of potato own resistance genes and secondly utilization of potato wild relative (PWR) genes. In **Table 1**, examples of the latest year's single and multiple gene transfer for improving potato resilience to biotic and abiotic stresses are given, as well as some results on insect resistance. Potato wild relatives have evolved defense mechanisms against pathogens and pests at multilayer level (**Figure 2**). The interaction between host potato species and its pathogens involves the following mechanisms: (1) physical and physiological barriers that prevent the pathogens to enter into the plant cells; (2) plasma membrane-bound and intracellular immune receptors that initiate defense responses upon the perception of pathogens; (3) interference RNA (RNAi) used by plants to detect invading viruses and fragment their RNA [20]. Pathogens as bacteria and fungi, respond to potato defense through: (1) production and release of cell-wall-degrading enzymes; (2) production and delivery into host cytoplasm of effector proteins, some of which suppress host defense and promote susceptibility; (3) viruses produce suppressors of host plant RNAi and/or hijack host RNAi to silence host genes and promote viral pathogenicity [20]. On the other hand, the interaction between herbivorous insect pests and plants also involves various mechanisms: (1) non-glandular and especially glandular trichomes that act as physical and physiological barrier to insect feeding; (2) toxins such as glycoalkaloids, which are well characterised in the *Solanum* genera; (3) enzyme inhibitors such as protease inhibitors; (4) use of bacterial insecticidal genes [61] (references herein) (**Figure 2**). All genes involved in host plant resistance to pathogens and pests as well as pathogenesis susceptibility genes can be transferred to produce resistant potato crop.

For instance, genes for pattern recognition receptors (PRRs), from other species can recognize pathogen associated molecular patters (PAMPs) and activate defense responses, as was demonstrated in *Arabidopsis thaliana* lectin receptor kinase LecRK1.9 transferred into potato that increased resistance to *Phytophthora infestans* (*Pi*) (**Table 1**) [31]. This first level of defense is known as pathogen targeted immunity (PTI). It is likely that there are different type of PRRs in potato but one was identified as ELR protein, which was capable to recognize the INFI elicitin from *Pi* [62]. Others are known from tomato and other species [6]. The tomato PRR *Ve1*, which recognize the Ave1 protein from *Verticillium dahliae*, when was expressed in potato was conferring resistance to this disease [63]. Gene transfer gave good results when R genes could be isolated and cloned. R proteins represent the second level of defense recognizing specific effector proteins of the pathogen, called effector targeted immunity (ETI) (**Figure 2**) [6]. Compared to PRR system, effectors use a similar defense response in the host plant, but effectors coupled with R genes elicit a stronger response which activates hypersensitive reaction (HR) in resistant plants. HR imply cell death surrounding the pathogen attack and represent a barrier for further pathogen spread. Pathogen effectors have high diversity but R genes have two conserved domains: nucleotide binding (NB) and leucine reach repeat (LRR), which makes their identification easier [6]. In the last two decades many R genes were cloned from potato wild relatives that induce resistance to *Pi* and transferred into potato varieties, either as single or multiple genes (**Table 1**). Some examples

Trait	Gene/s	Result/resistance to:	References
Resistance to bacteria	*5-UGT*	*Erwinia carotovora*	[24]
	ScSN1	*Erwinia carotovora Rhizoctonia solani*	[25]
	Overexpression of peptides with anti-fungal properties	*Rhizoctonia solani*	[26]
Resistance to late blight (*Pi*)	*Rpi-vnt1.1*	*Pi* in field trials	[27]
	Rpi-vnt1.1, Rpi-sto1	*Pi* cisgenic marker-free	[28]
	RB (*Rpi-blb*1)	Tolerance to *Pi*	[29]
	Rpi-vnt1.1, Rpi-sto1, Rpi-blb3	*Pi*, stacking three cisgenes	[30]
	LecRK1.9	*Pi*	[31]
	AtROP1	*Pi*	[32]
	hp-PiGPB1	Pi – (HIGS)	[33]
	Rpi-blb2, Rpi-blb1, Rpi-vnt1.1	Pi, stacking 3 Rpi genes in African varieties	[34, 35]
Resistance to diseases	*MsrA2*	Broad-spectrum fungi and bacteria	[36]
	MsrA3 with tissue specific promoter	Mitigates biotic and abiotic responses	[37]
	hLf	bacteria and fungi	[38]
	Accumulation of cystatin	Diseases, insects and fungi	[39]
Nematode resistance	*Gpa2*	*Globodera pallida*	[40]
	Gro1–4	*Globodera rostochinensis*	[41]
	Peptide disrupting chemoreception of nematodes	*Globodera pallida*–no off target side effects	[42]

Trait	Gene/s	Result/resistance to:	References
Virus resistances	dsRNA PVY coat protein (CP)	RNAi - PVY	[43]
	shRNA with *ipt* gene	PVY^{NTN} in a marker-free system	[44]
	CP gene	PVY in the field	[45]
	shRNA with I	PVY^{NTN} in a marker-free system	[46]
	shRNA	PVY	[47]
	eIF4E-1 variant *Eva1* (*S. chacoense*)	PVY	[48]
	eIF4E	PVY	[49]
	RNAi	PVY	[50]
	CRISPR-Cas13a	Durable resistance all strains PVY	[51]
	Overexpression of *StSAR1A*	PVY and PVA	[52]
Insect resistance	Hybrid *Bt* endotoxin	Both coleopteran and lepidopteran pests	[53]
	cry1Ac9	Tuber moth	[54]
	Cysteine *Pls*	Western flower thrips	[55]
	RNAi incapsulated in bacteria (*E. coli*)	CPB	[56]
	DsRNA for CPB control	CPB	[57]
	ACT-dsRNA expressed in chloroplasts	CPB	[58]
	RNAi of molting-associated EcR gene of CPB	CPB	[59]
	CRISPR-Cas9 site directed editing of *vest* gene in CPB	CPB	[60]

ACT – gene for βactin; AtROP1 – Arabidopsis thaliana gene for protein at the plasma membrane; CPB – Colorado potato beetle; hLf – human lactoferrin gene; MsrA2 – gene for frog antimicrobial peptide; LecRK1.9 - Arabidopsis lectin receptor kinase; Pi- Phytophthora infestans; Pls – protease inhibitors; ScSN1 - Snakin-1, a cysteine-rich peptide from Solanum chacoense; 5-UGT - anthocyanin 5-O-Glucosyltransferase; SAR1a - secretion-associated RAS super family 1 gene; vest – vestigial gene involved in wing development.

Table 1.
Synthesis of transgenesis and cisgenesis results presenting the transfer of single or multiple resistance genes in order to improve biotic stress resistance in potato.

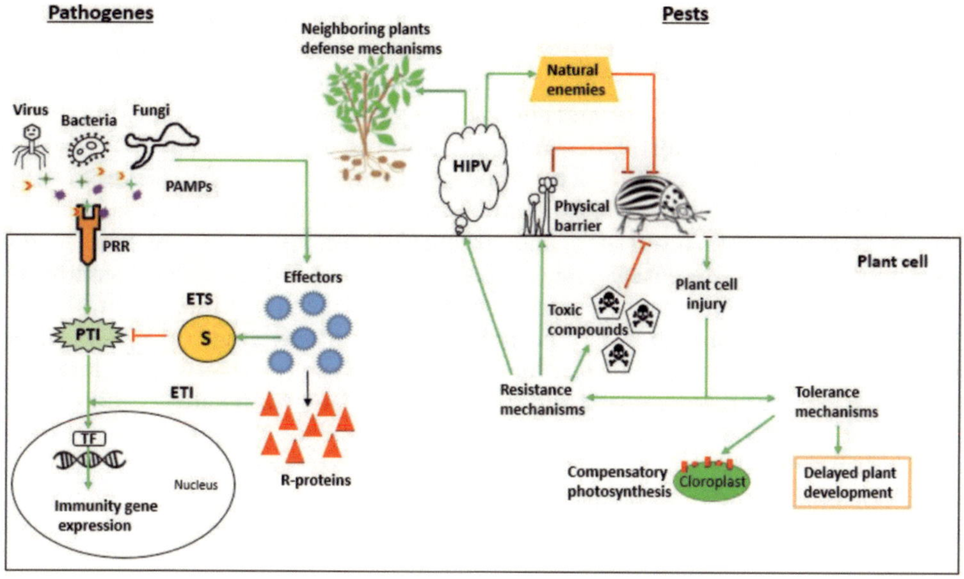

Figure 2.
The principal mechanisms of interaction between pathogens (bacteria, fungi and viruses) on the left and insect pests on the right with the potato host: the pathogens trigger two responses PTI (pathogen triggered immunity) and ETI (effector triggered immunity); in PTI the membrane proteins PRR recognize pathogen molecular patterns (PAMPs) and induce transcription factors (TFs) which activate immunity genes; in ETI effector molecules interact with specific resistance genes (R), but they can also interact with sensitivity genes (S) to inhibit PTI; insect pests interaction with its host is less understood but at first the pest interacts with leaf trichomes, glandular and/or non-glandular, mainly acting as a physical barrier; after wounding the leaf cells are inducing either tolerance responses like compensatory photosynthesis and delayed plant development, or resistance responses through synthesis of toxins like glycoalkaloids. Resistance mechanisms can activate HIPV (herbivore induced plant volatiles).

of R genes are: *R1*, *R2* and *R3a*, *R3b*, originally identified in *S. demissum*; *Rpi-blb1* (RB), *Rpi-blb2*, *Rpi-blb3* from *S. bulbocastanum*; *Rpi-vnt1.1* and *Rpi-vnt1.2* from *S. venturii*; *Rpi-mcq 1from S. mochiquense* [6, 64], etc. R genes were also delivered into potato varieties as gene stacks. In Europe BASF Company petitioned for the release of potato Fortuna resistant to late blight (*Pi*) after stacking of two R genes: *Rpi-blb1* and *Rpi-blb2*, obtained after a long effort of breeding, but unfortunately, this cultivar was never marketed [6]. The Simplot's second generation Innate® potato which besides reduced browning and bruising, also carries R genes and hence is resistant to late blight (*Pi*), was approved for cultivation in USA [65], and for cultivation and consumption in Canada [66]. One important research project was developed in Netherland between 2006 and 2015 on Durable Resistance in potato against *Phytophthora* (DuRPh) at Wageningen University and Research Centre [64]. The aim of this project was to identify and clone new durable resistance genes from potato wild relatives and transfer them as single or stalked genes into varieties by cisgenesis using marker assisted selection (MAS). Through this project a great deal of data has been accumulated and cisgenic varieties resistant to late blight were produced but these will require some more backcrosses to be released as resistant and productive varieties [64]. Still cisgenesis is considered as GM in Europe. A successful cisgenic approach was applied in Africa, where highland varieties were transformed with an efficiency of 75% using three Rpi genes: *Rpi-blb1*, *Rpi-blb2* and *Rpi-vnt1.1* (**Table 1**) [34]. R genes that improve resistance to other pathogens were also discovered: *Rx1* and *Rx2* (from *S. tuberosum* ssp. *andigena* and *S. acaule*, respectively), that confer resistance to potato virus X (PVX) [67]; *Gro1–4* from *S. spegazzinii*, confer resistance to root cyst nematode *Globodera rostochinensis* (**Table 1**) [41]. Another strategy for resistance to a broad spectrum of pathogens is overexpression of a single gene located upstream in signalling cascades and thus regulates large number of defense-responsive genes. There are many examples of successful engineered plants using different constructs to overexpress trans- and endogenous genes in crops, including potato. Overexpression of these upstream signalling genes and defense-related genes can lead to a constitutive expression of resistance phenotype. In plant disease resistance, a vital role is played by small G-proteins and subsequent cellular responses to pathogens such as bacteria, fungi and viruses [52]. A number of G-proteins have been transferred to different plant species including potato where stable overexpression of *AtRop1* (DN-*AtRop1*) increased resistance to *Pi* infection (**Table 1**) [32]. An important breakthrough is the continuous research identifying new molecular markers linked to resistance genes or more recently QTLs (quantitative trait loci) such are: AFLP, RFLP, SSR, RAPD and their maps available for potato breeding [68]. At International Potato Center a continuous effort, as mentioned above, aims to store genetic diversity and improve it for the benefit of the next generations and efficient alleviation of underdeveloped nations' poverty. Several other genes were also cloned and transferred into potato crop for improvement of resistance to: PVY (*eIF4E-1* variant *Eva1*) and *Pi* – host induced gene silencing (HIGS) (**Table 1**) [33, 48]. The aim of the latest strategy is to achieve more durable resistance than R genes, but this also uses gene constructs that fall under GM rules [6].

2.1.2 Insect resistant potato crop

Insects are also a plague for potato production but the most difficult to control is the voracious Colorado potato beetle (CPB). It is estimated that 75% of potato production can be lost by pests if left uncontrolled [69]. CPB develop on potato crop, larvae and adults eat leaves and are able to completely skeletonize the plants. During development, the three stages of instar larvae consume around 40 cm^2 of potato leaves [70]. Plant breeding and biotechnology were not able to release a variety resistant to CPB without GM technology. Wild potato relatives are a reservoir of resistance traits as it was discussed for pathogens. Two natural host plant resistances are known: glandular trichomes and specific glycoalkaloids, the leptines I and II [71]. Detailed knowledge on the interaction between potato and resistant relatives with the voracious beetle are still scarce (**Figure 2**). Another interesting mechanism of resistance was discovered [72], the hypersensitive reaction of plants to CPB egg masses and egg drop. Any breakthrough into the physical, physiological and molecular mechanisms of resistance will fasten the progress of resistance breeding using biotechnology. The main strategy of genetic engineering to induce resistance to CPB was based on bacterial toxin from *Bacillus thuringiensis* (*Bt*), a bacterium also used in integrated pest management by spraying bacterial suspensions in the field. The technology is very specific for a certain species of pest, because *Bt* not only has a large repertoire of the *cry* genes that produce the protoxins involved in pest induced mortality, but the toxin is formed only in the gut of feeding pests and would not affect non-targeted beneficial insects [71]. The first success was introducing by gene transfer the *cry3a* gene into potato cv. Russet Burbank to protect it from CPB attack [73]. The GM variety with resistance to CPB was approved for human consumption and was commercially available in USA between 1996 until 2001, proving to control the beetle in the field without any unwanted effects on the cultivar [74]. NewLeaf™ potato, developed by Monsanto, containing *cry3a* proved to supress CPB populations at greater extent as insecticides or sprays based on formulations from *Bt* bacteria containing CRY3A protein [71]. In the next years, after the first success with cry3a, other *cry* genes have been optimize and transferred into potato: *cry3Ca1*, *cry1*, *cry3Bb1* [71] (**Table 1**).

Coombs *et al* [75] combined leptines, glycoalkaloids considered as toxic to CPB, with glandular trichomes and *Bt-cry3a* to obtain transgenic potato host plants resistant to CPB. In that way the main problem of *Bt* potato, the development of resistance, could be also managed [75]. To date, there are no *Bt* potato on the market, as discussed in public acceptance of GM potatoes. Recent studies have been focusing on RNAi technology, including direct spraying of dsRNA in the field [71]. The first success with dsRNA used in a transgenic approach [76], lead to long or short double stranded RNA used to target a specific gene at posttranscriptional level determining mRNA fragmentation and hence silencing the gene. This proof of concept brought about a growing interest for the use of RNAi technology for controlling the CPB pest [57]. Moreover, non-transgenic alternatives were developed including dsRNA spraying on the plants [59, 77], but in this year (2021), resistance development in CPB populations after dsRNA foliar-delivery in potato has been already observed [78]. Sequence of CPB transcriptome can assist in the identification of new target genes for RNAi that can be used to control this pest [79]. To date, 24 target genes with important roles in cellular functions were silenced using RNAi, as reviewed by Balaško *et al* [71]. Knockdown of those genes affect insect morbidity and mortality. There were also different delivery methods of dsRNA into CPB, like the use of bacteria, liposomes and nanocarriers, all of them able to protect and deliver dsRNA [77]. Moreover, other improvements for CPB control were the xenobiotic transcription factor Cap 'n' collar isoform C (CncC) that regulates the expression of multiple cytochrome P450 genes, and plays crucial roles in CPB insecticide resistance. The suppression of CncC by RNAi reduced imidacloprid resistance of CPB [80]. Ochoa-Campuzano *et al* [81] identified prohibitin, an essential protein for CPB viability, as Cry3Aa binding protein. Combination of feeding prohibitin dsRNA and treatment with Cry3Aa enhanced the toxic effect by threefold and CPB was killed faster with 100% mortality in five days. The molecular mechanisms of synergism between prohibitin, RNAi and Cry3Aa toxin are not understood, but this study proposes an interesting method, combining toxins derived from bacteria or other organisms with RNAi in order to improve efficiency of dsRNA in pest control. Moreover, recently targeted mutagenesis using CRISPR-Cas9 technology in CPB was demonstrated [60], a technology which holds great promise for the future.

2.1.3 Gene transfer for resilience to abiotic stress

Abiotic stresses such as drought, salt, high temperatures and extreme weather also limit potato yield around the world. With global climate change, abiotic stress is expected to be less predictable in the years to come and also affect pathogen attacks and pest effects on potato and other crops. The response of the plants to abiotic stresses involve generally the expression of inducible resistance genes. In particular, transcription factors (TFs) that control resistance genes are a key in gene regulatory networks that control the expression of many genes involved in stress responses [82]. Transgenesis uses genes for such TFs like WRKY, MYB or DREB, the last also used in potato crop (**Table 2**). Other genes that were engineered in potato are related to response of the plants to abiotic stress, like *StProDH1,* which is a key player in potato response to drought stress [93]. Through the manipulation of abscisic acid signal transduction after loss of function of cap-binding protein (CBP) [71], in cv. Désirée, a higher tolerance to drought was reported [92]. Through transgenic approach, potato lines with increased betaine aldehyde dehydrogenase, an enzyme for glycine betaine biosynthesis, which has important role in drought stress, has been able to induce drought tolerance in potato [88]. Transcriptome analysis, comparing control with drought stressed potato plants, has indicated many genes that are overexpressed

Trait	Gene/s	Result/tolerance to:	Reference
Heat tolerance	*CaPF1*	High temperature	[83]
	AtCBF3	Heat tolerance	[84]
	Allelic variant *HSc70*	Moderately high temperatures in cv. Désirée	[85]
Freezing tolerance	*Atrd29A::DREB1A*	Freezing	[86]
Drought tolerance	*ScTPS*	Studies on water content and photosynthesis	[87]
	Glycin betaine aldehyde dehydro-genase	Drought	[88]
	TPS1	Drought - increased threhalose	[89]
	PaSOD	Increased photosynthesis under drought	[90]
	AtDREB1/CBF	Drought	[91]
	CPB80	Drought	[92]
	amiRNA silencing of *StProDH1*	Drought	[93]
Salt tolerance	*Δ1-pyrroline-5-carboxylate synthetase*	Salt - increased proline	[94]
	HvNHX2	Salt	[95]
	Overexpression of *AtHKT1*	Salt	[96]
	StCYS1	Salt	[97]
Two stresses	*BADH*	Drought and salt	[88]
	*StEREBP*1	Cold and salt	[98]
	SOD, APX	Oxidative stress and high temperature	[99]
	At DREB1B	Drought and freezing tolerance	[100]
	StDREB1, StDREB2	Salt or drought tolerance	[101, 102]
	ggpPS	Drought and salt/ tuber increased glucosyl - glycerol	[103]
	GB	Salt and cold	[104]
	StWRKY1	Resistance to Pi and improved tolerance to drought	[105]
	AtABF4	Salt and drought, increased yield and tuber quality	[106]
	AtHXK1 and *SP6A*	Drought and heat	[107]
Multiple stresses	*CodA*/chloroplast	Oxidative, salt, and drought stresses	[108]
	SOD, APX, CodA/ chloroplast	Oxidative, salt, and drought stresses	[109]
	*StnsLTP*1	Multiple tolerance to heat, salt and drought	[110]
	IbOr	Multiple tolerance to drought, oxidative stress and high salinity, increased carotenoid contents	[111]

amiRNA – artificial miRNA; APX - ascorbate peroxidase; BADH - betaine aldehyde dehydrogenase; CaPF1- pepper transcription factor belonging to the family of TFs ERF/AP2; CBF - C-repeat Binding Factor;DREB - dehydration responsive element binding protein; CodA - choline oxidase; GB – Glycinebetaine; HSc70 – heat shock cognate 70 gene; HvNHX2 – Hordeum vulgare vacuolar Na+/H+ antiporter; IbOr – Ipomeaea batata orange gene; ScTPS-Saccharomyces cerevisiae trehalose-6-phosphate synthase; SOD - superoxide dismutase; StEREBP1 – S. tuberosum ethylene responsive element binding protein 1; StnsLTP1 – S. tuberosum nonspecific lipid transfer protein 1; StProDH1 – S. tuberosum proline dehydrogenase 1; TPS1- yeast trehalose-6-phosphate synthase 1.

Table 2.
Examples of single or multiple resistance genes transfer to improve abiotic stress tolerance in potato.

or underexpressed during drought stress, with genes involved in processes like: intracellular water and ion homeostasis, membrane structural stability, and reconstruction of primary and secondary metabolism, and stress regulatory genes, as calcium ions, TFs and receptor protein kinases that are involved in stress response through signal transduction and metabolic pathways [112].

Salt stress caused by soil salinization is an increasing threat to agriculture worldwide [113]. Different factors lead to the continuous salinization of the soil, mainly different agricultural practices such as irrigation and some fertilization procedures. The mechanisms that are involved in salt stress response are cellular and physiological: e.g. different cellular signalling, various ion transport, water management and specific gene expression which are involved in growth, development and survival [113]. Researchers are working on halophytes, plants that are adapted to salty soil, to get new insights on plant responses to salt stress. In the case of potato, as presented in **Table 2**, there are transgenic strategies which proved their utility in obtaining salt tolerance, either alone or in combination with other stress factors. Potato plants adapt to salinity stress through different mechanisms like osmotic adjustment by accumulating compatible solutes in the cytosol, decrease leaf water potential leading to reduced cell turgidity and growth retardation and tuber yield loss. One of the most important compatible solute is proline, which was accumulated in cv. Désirée 3.5 fold and 11 fold at 100 and 200 mM NaCl, respectively [114]. However the proline effects on salt tolerance need additional studies because foliar application of proline has no effect on salt tolerance of plants [115]. Potato is adapted to cool weather mostly preferring temperate zone. The vegetative part of plants grow properly at 20–25°C temperature, while tubers develop better at 15–20°C. The response of potato plants to high temperature varies across the cultivars, one example being the commercial cv. Russet Burbank, which exhibit maximum rates of photosynthesis at 24 to 30°C and a reduction of photosynthetic activity only at or above 35°C [116]. Global warming and drought are expected to drastically reduce the potato productivity, but with biotechnology heat tolerant potato was successfully obtained (**Table 2**). Plants exhibit different strategies to cope with high temperature stress involving physiological, morphological and molecular levels. At molecular level heat stress increase the activity of heat stress TFs (HSFs), which trigger the accumulation of heat shock proteins (HSPs). HSPs are known to govern heat stress response (HSR) and acquired thermo-tolerance through their role as molecular chaperones [117]. In a genome wide study 27 *StHSFs* in the *Solanum tuberosum* genome were identified [118], which have diverse regulatory functions during stress. Underlining the molecular mechanism of how heat stress induces HSFs trimerization, their activation and synthesis of HSPs is still underway. Elucidation of the mechanisms of heat stress response may offer new insights that will be useful in breeding new heat resilient cultivars with sustained or even enhanced potato crop productivity and quality in response to climate change.

2.1.4 Multiple stress factors

In nature, generally multiple stresses act on crops at the same time and all of them contribute to noticeable losses in production. Nowadays, there is knowledge about various genes that contribute to both biotic and abiotic stress response and resistance/tolerance. The effects of abiotic stress on potato crop under climate change is detailed in a recent review [117]. Molecular and genomic analysis revealed transcriptionally regulatory pathways involved in modulation of stress responsive genes. As mentioned above TFs are playing a crucial role, particularly in multiple stress response of plants [119]. Examples of TFs that activate stress responsive genes are AP2/ERF, containing AP2/ERF binding domain, a large superfamily that divides in AP2, ERF and RAV [120]. This family of genes participate in developmental processes. AP2 family is involved in regulation of development, together with ERF protein family. Based on the differences in DNA box-binding ability of the single AP2/ERF domain, the ERF family is divided in ERF and CBF/DREB (C-repeat Binding Factor/Dehydration Responsive Element-Binding) (**Table 2**). ERF proteins are mainly involved in inducing disease resistance in a negative or positive mode of action. Gangadhar *et al* [121] have identified 95 genes involved in heat tolerance in potato, eleven of them being associated with multiple stress tolerance, like drought, salt and heat. Prolamins are a group of plant storage proteins that represent useful factors implicated in controlling both abiotic and biotic stress-response in plants. The plant non-specific lipid transfer protein, nsLTP, is involved in phospholipid transfer but also various other biological functions as seed storage, lipid mobilization, cuticle synthesis, somatic embryogenesis and pollen tube adhesion [110]. Transgenic potato lines over-expressing *StnsLTP1* acquired improved tolerance to multiple abiotic stresses through enhanced activation of antioxidative defense mechanisms via cyclic scavenging of ROS and regulated expression of stress-related genes (**Table 2**) [110]. Another example is the use of TF *StWRKY1*, which successfully induced resistance to *Pi* and improved tolerance to water scarcity. This experiments prove the role of TFs and in particular WRKY in regulating both biotic and abiotic stress resistance thereby modulating plant basal defense networks and thus playing a significant role for potato crop improvement.

3. Use of cell fusion between potato crop and its wild relatives for resilient potato

Over the past fifty years the introgression of new traits from wild *Solanum* species have mainly achieved by using classical breeding methods. The number of wild species that could be integrated into potato breeding is quite limited because of sexual incompatibility and endosperm balance number (EBN), although there are techniques other than sexual crosses, such as manipulations of ploidy levels [122], breeding 2n gametes or using bridging species to integrate genes from wild *Solanum* species into modern cultivars [123]. Through sexual crosses the main source of resistance genes is still *S. demissum*, more than half of the modern cultivars contain introgressions from this species [123]. The main limitations of the potato classical breeding are tetraploidy and heterozygosity, which make breeding very complex and time-consuming [124]. Moreover, when genes from an incompatible wild species have to be exploited, as was in the case of *S. bulbocastanum*, the use of a bridging species was applied to produce new cultivars which took 49 years and then only one resistance gene (*Rpi-blb2*) against late blight was integrated into potato gene pool (cvs. Bionica and Toluca) [125]. Nowadays, somatic hybridization through protoplast fusion is a well refined and routinely used method in order to create *Solanum* hybrids with different useful properties [126, 127]. Plant protoplasts are naked somatic cells from which the cell wall has been removed by enzymatic digestion, therefore these cells can be used for gene transfer, somatic hybridization [128], and more recently for targeted mutagenesis and genome research. Protoplasts are still totipotent and they are able to regenerate new cell wall, divide to form new cell colonies, microcalluses, calluses and finally new plants. This protoplast technology proved to be very efficient in potato crop and is a reliable and useful way to regenerate large numbers of somatic hybrids (SHs) with distinct genetic backgrounds [129–131]. Among the agronomical important crops, potato was the first used in protoplast culture and somatic hybridization [132, 133], which opened the way for free gene transfer from potato wild relatives into potato crop [134]. Leaf mesophyll cells of *in vitro*-grown plants were used to isolate protoplasts [135], then the obtained fused products were cultured in VKM medium [136], followed by shoot development on the MS13K medium [137]. Recently, selection of SHs (*S. tuberosum* + *S. chacoense*) based on callus growth tagged with *gfp* has been also observed [138]. Different methods are available for protoplast fusion, but only two are generally used: electrofusion and PEG (polyethylene glycol) induced fusion [128]. Electrofusion is the most widely used method since its discovery in 1979 [139], and it consists in first instance of protoplast agglutination induced by the use of an alternating current (AC) field, the so-called dielectrophoresis [140]. In the second phase the agglutinated aligned protoplasts are induced to fuse by using direct current (DC) square wave pulses with a high intensity (2000 V cm^{-1}) and very short duration (10–100 μs) [141]. PEG-induced fusion generally has a similar efficiency as electrofusion, especially after applying calcium solution washing step [128]. Immediately after fusion or after the plants have been regenerated, the obtained SHs are subject to different analysis, such as cytological (flow cytometry, chromosome counts, chloroplasts counts in guard cells, FISH - fluorescence in situ hybridization and GISH - genomic *in situ* hybridization), molecular: isozyme, molecular markers (e.g. RAPD, RFLP, ISSR - inter simple sequence repeat, SSR-simple sequence repeat, AFLP - amplified fragment length polymorphism, and DArT-diversity array technology) [8, 129], phenotypic changes (e.g. foliage, stem, leaf, flower and tuber traits) and pollen fertility. Due to their stability and universality SSR markers are the most widely used [129, 130]. Recently, the application of DaRT made it possible to find out the composition of the SHs genome between potato and *S. x michoacanum*, which demonstrated the presence of both parents genome in hybrid plants, and provided evidence for late blight resistance trait transfer from wild relatives into SHs [142]. SHs are also analysed for cytoplasm types (haplotype of chloroplast/mitochondria: W/α, T/β, W/γ, W/δ and S/ε) [143], based on organelle segregation after fusion and organellar genome-specific markers as described by Lössl *et al* [144]. Finally, SHs are examined for the presence of target traits under field or controlled conditions eventually being tested for phenotype and tuber qualities in the field [8, 145]. Somatic hybridization through protoplasts fusion, which circumvents pre- and post-zygotic crossing barriers, can be successfully used to insert resistance into potato (**Table 3**) [143]. It has a greater potential for self-generating biodiversity in numerous nuclear and cytoplasmic genome combinations than sexual hybridization [184]. It also provides an opportunity for initiating recombination events between parental genomes. Moreover, homeologous recombinations (recombination between similar but not identical DNA molecules), can also be increased, that might increase the integration of valuable traits, by inducing a DNA repair deficiency, for instance, mismatch repair deficiency (MMR) [145, 175, 185]. MMR was successfully induced by *Agrobacterium*-mediated transfer of *AtMSH2* gene in antisense orientation or a dominant negative gene into *S. chacoense* [185], followed by somatic hybridization with potato tetraploid variety Delikat through electrofusion. Resistance to Colorado potato beetle (CPB) was more common in MMR deficient somatic hybrid plants [175]; MMR was also responsible for greater diversity and a novel trait tolerance to drought stress [180]. Since 1980s, different wild *Solanum* species have been hybridized with potato using protoplast fusion, and many of them express various valuable traits, including resistance to viruses [186], bacteria [187], fungi [188], insect pests [175] or tolerance to abiotic stresses (**Table 3**) [181]. Furthermore, multiple resistance can be also transferred from wild relatives into the potato gene pool [130] and even SHs with multiple parent lines can be produced, as in the case of the tri-species somatic hybrids [178].

Traits of interest	Somatic hybrid St + wild relative:	Tools for characterization and/or selection	Reference
Biotic factors			
Resistance to bacterial diseases			
Clavibacter	*S. acaule*	Glycoalkaloid aglicones	[146]
Erwinia carotovora	*S. brevidens* (*S. palustrae*)	RFLP, GISH, FISH	[147]
Ralstonia solanacearum	*S. chacoense*	SSR, cytoplasm type, MAS, BC_1	[148, 149]
	S. melongena	SSR, *smPGH1* gene	[150]
	S. stenotonum	Isoenzymes, SSR, PEPC/RUBISCO ratio	[151]
Streptomyces spp.	*S. brevidens* (*S. palustrae*)	Laboratory and field resistance tests	[152]
Resistance to fungal diseases			
Alternaria tomatophila	*S. brevidens* (*S. palustrae*)	RFLP, GISH, FISH	[147]
Phytophthora erythroseptica	*S. berthaultii* (+) *S. etuberosum*	NS	[153]
Phytophthora infestans	*S. bulbocastanum*	MAS for RB gene (Rpi-blb1) GISH, cytoplasmic DNA	[154, 155]
		SSR, cytogenetics, *Rpi-blb1*; *Rpi-blb3* gene	[131, 145]
	S. cardiophyllum	RAPD	[156]
		SSR, AFLP, MFLP, ploidy	[130]
		RAPD, SSR, ISSR, AFLP, cytoplasmic type molecular markers, FC	[157]
	S. circaeifolium	Morphology, RAPD, chromosomes	[158]
	S. chacoense	RAPD, morphology	[156]
	S. x michoacanum	Ploidy, RAPD	[159]
		DaRT	[142, 160]
	S. nigrum	Morphology, ploidy, RAPD	[161]
	S. pinnatisectum	RAPD, morphology	[156]
		Ploidy, cytoplasm type	[162]
		RAPD, SSR, cytoplasm type, FC	[163, 164]
		ISSR, BC_1 characterization, *Rpi-blb2* gene, field resistance tests	[165, 166]
	S. tarnii	SSR, AFLP	[129]
	S. verrucosum	RAPD	[167]
	S. villosum	RAPD, GISH, ROS	[168]

Traits of interest	Somatic hybrid St + wild relative:	Tools for characterization and/or selection	Reference
Pythium spp.	*S. berthaultii* (+) *S. etuberosum*	NS	[153]
	S. tuberosum cvs. Aminca (+) Cardinal Cardinal (+) Nicola	Isoenzymes, SSR, ISSR	[169]
Verticillium spp.	*S. commersonii*	Southern analysis of organelles	[170]
Resistance to viral diseases			
PRLV	*S. etuberosum*	Characterization of BC populations	[171]
	S. tuberosum x *S. berthaultii* (+) *S. etuberosum*	NS	[172]
PVX	*S. tuberosum* x *S. berthaultii* (+) *S. etuberosum*	NS	[172]
PVY	*S. cardiophyllum*	SSR, AFLP, MFLP, ploidy	[130]
	S. etuberosum	RAPD, SSR, GISH, cytoplasm type	[173]
		Cytoplasm type, FC, RAPD, SSR	[174]
	S. tuberosum x *S. berthaultii* (+) *S. etuberosum*	NS	[172]
	S. tarnii	SSR, AFLP	[129]
	S. tuberosum cvs. Aminca (+) Cardinal Cardinal (+) Nicola	Isoenzymes, SSR, ISSR	[169]
Resistance to insects			
Colorado potato beetle	*S. tuberosum* (+) *S. cardiophyllum*	RAPD	[156]
		SSR, AFLP, MFLP, ploidy	[130]
	S. tuberosum (+) *S. chacoense*	MMR deficiency, SSR, RAPD marker for leptines	[175]
	S. tuberosum x *S. berthaultii* (+) *S. etuberosum*	NS	[172]
	S. tuberosum (+) *S. pinnatisectum*	RAPD, morphology	[156]
Meloidogyne chitwoodi	*S. tuberosum* (+) *S. bulbocastanum*	Laboratory and field resistance tests	[176]
		MAS *RMc1(blb)*	[177]
Green peach and potato aphids, wireworm	*S. tuberosum* x *S. berthaultii* (+) *S. etuberosum*	NS	[172, 178]

Traits of interest	Somatic hybrid St + wild relative:	Tools for characterization and/or selection	Reference
Abiotic factors			
Drought tolerance	*S. tuberosum cvs.* Aminca (+) Cardinal Cardinal (+) Nicola	Greenhouse tolerance test	[179]
	S. chacoense	Laboratory and phenotyping	[180], Molnar *et al.* (under publication)
Frost tolerance	*S. malmeanum*	SSR, ploidy, BC_1 characterization, laboratory tolerance tests	[181]
Salt tolerance	*S. berthaultii*	ISSR, cytoplasmic DNA, FC	[182]
		Oxidative stress responses	[183]

NS – not specified.

Table 3.
The most important somatic hybrids with proved resistance to pathogens, pests and tolerant to abiotic stresses and the methods applied for their analysis.

One of the most economically valuable SH was obtained by fusion between the incompatible *S. bulbocastanum* species and cultivated tetraploid potato [189], which highlighted the advantages of somatic hybridization in potato genome improvement, because the SHs were highly resistance to *Pi* in the laboratory and a field under intense disease pressure. After back-crossing of these SHs with potato cultivars the resistance to this disease was not lost. Subsequently, RB gene involved in durable resistance was isolated, which is located on chromosome VIII [190]. Transgenic plants with RB, were regenerated after *Agrobacterium*–mediated gene transfer and proved durable resistant [191]. Since then, *S. bulbocastanum* demonstrated several times its value as a resource of durable resistance genes against late blight, therefore it has been an increasing interest in transferring the resistance traits of this species to cultivated potato [154, 192]. RB gene was the first durable resistance gene described for late blight, but soon many other genes were discovered both in *S. bulbocastanum* and other wild species. To date, there are four characterized resistance genes in *S. bulbocastanum*: *Rpi-blb1* (formerly RB), *Rpiblb2*, *Rpi-blb3* and *Rpi-bt1* [190, 193–196]. In addition, late blight resistance from other sources was also accessed by generation of interspecific SHs with the wild species *S. pinnatisectum* [163], *S. tarnii* [129], *S. cardiophyllum* [130] and more recently *S. x microachanum*, a wild diploid derived from a spontaneous cross between *S. bulbocastanum* and *S. pinnatisectum* [160]. These newly produced SHs were also tested in the field and were resistant after two or three years of assessment, therefore they are suitable for introducing in breeding. *S. stenotonum* is an exquisite source of resistance to bacterial wilt caused by *Ralstonia solanacearum*, and all of the SHs obtained by fusion of potato protoplasts with this wild species were as resistant as the wild parent line [197]. Similarly, *S. chacoense* was explored for molecular markers associated with bacterial wilt resistance, and for introgression of resistance into the potato gene pool [148]. A very successful approach involved the transgenic induction of MMR deficiency in a high leptine-producing accession of *S. chacoense*, followed by somatic hybridization, because large number of generated plants exhibited both antixenosis and antibiosis against CPB [175]. Recently, by using gene specific markers four *Pi* resistance genes: *Rpi-blb1, Rpi-blb3, R3a* and *R3b* were identified in *S. bulbocastanum* and derived SHs with potato cvs. Delikat and Rasant. The genes were present also in BC1 and BC2 progenies and resistance to late blight was maintained along good tuber traits [146]. The resistance gene pool of wild *Solanum* species can also be used to combat abiotic stresses like salt, drought and frost. For example SHs originating from fusion between potato and *S. bertaultii* are tolerant to salt stress [182]. Freezing is another abiotic factor, which decrease the yield of potato and SHs of *S. tuberosum* (+) *S. malmeanum* proved to be tolerant to frost [181]. Furthermore, SHs of potato and *S. chacoense* show different level of drought and salt tolerance beside resistance to CPB [184]. Interspecific somatic hybridization gave good results but the intraspecific somatic hybridization proved to be also suitable for potato improvement. Starting in the 1990s, somatic hybridization was used to study different dihaploid lines of potato generated by crossing with *S. phureja* [198] or pollen and anther *in vitro* culture. The results of the protoplast fusion of two dihaploid potato lines were at first not very promising, but the restoration of tetraploids from two dihaploid lines with valuable yield and resistance traits soon proved to be a valuable approach for potato breeding. Furthermore, resistance to nematodes, viruses (PVY) and *Phytium* bacterial diseases were achieved by intraspecific protoplast fusion [169, 199]. The intraspecific hybridization has a finite repository, but as long as this area is not exploited, it is worth considering. During interspecific somatic hybridization two obstacles may occur: (1) transfer of too much exotic, wild genetic material along with the desirable gene(s) from the wild species; and (2) genetic imbalance which lead to somatic incompatibility. These limitations result either in abnormal growth and development of the SHs, and/or regeneration of infertile plants. In order to reduce the wild imprint, the introgressive hybridization is followed by one or multiple back-crosses of the somatic hybrids with cultivars. The purpose of these cross-hybridization processes is on the one hand to eliminate the undesirable part of the wild genome, on the other to retain the target traits inherited from the wild parents and to restore the agronomic valuable cultivars, with high yield and adequate tuber quality [129, 130, 139]. Several experiments proved that, the above mentioned disadvantages could be eliminated through multiple back-crosses. Somatic hybrids of cultivated potato and *S. tarnii* were resistant to late blight and PVY, and these valuable traits were successfully transferred to BC1 progenies, which also presented good tuber yield and quality [129]. Multiple years of field evaluations of *S. etuberosum* + *S. tuberosum* and progenies showed stable transmission and expression of PLRV and PVY resistances in three BC1, BC2 and BC3 and two BC1 and BC2 generations, respectively [171]. Furthermore, late blight resistance can be transferred through breeding from tetraploid somatic hybrids (*S.* × *michoacanum* + *S. tuberosum* and autofused *S.* × *michoacanum*) to common varieties [142]. Bacterial wilt resistance was transferred to advanced progenies of somatic hybrids between *S. commersonii* and cultivated potato, and three highly resistant clones (BC1 and BC2) were selected as breeding materials [170]. In the case of potato there are many reports of symmetric interspecific somatic hybridization between diploid wild species and potato dihaploid lines [127]. The main problem with the majority of these hybrids was the infertility, which made difficult the restoration of valuable cultivar. For this reason symmetric somatic hybridization between tetraploid potato cultivars and diploid wild species became more popular [200]. The expected results after tetraploid with diploid protoplast fusion are hexaploid SHs, but among them aneuploid or mixoploid hybrids are often regenerated [131]. Genetically, the hybrids may be unstable and usually eliminate chromosomes from the wild species during the next stage of tissue culture, as occurred in the case of potato and *S. bulbocastanum* hybrids, but, after two back-crosses with cultivated potato, many of them re-stabilize at tetraploid level [131, 145]. Theoretically hexaploid or near hexaploid SHs of potato will tend to eliminate the wild species chromosomes and maintain only a few alien chromosomes or introgress some genes from the wild parent. Chromosome elimination in some interspecific somatic hybrids of potato largely depends on the phylogenetic relationship, type of genome: A, B, C, D and P [201], cell cycle synchronization after fusion and the two parent chromosomes interaction during mitosis [202]. Asymmetric somatic hybrids can be a result of the ordinary symmetric fusion or can be induced by fragmenting the donor species DNA by using the donor-recipient method [203]. Production of asymmetric somatic hybrid plants aroused interest of breeders, because with controlled chromosome transfer the restoration process of cultivars is faster and easier [204]. Usually, the donor protoplasts are treated with sub-lethal doses of ionizing irradiation, such as gamma, X rays [205, 206] or UV irradiation [207], in order to induce double-strand breaks and hence partial genome elimination [208]. In addition to irradiation, chemical agents can be used to induce chromosome elimination, such as restriction endonucleases, spindle toxin or chromosome condensation agents [209]. With applying these methods, asymmetric potato hybrids with some wild *Solanum* species [210] and intergeneric somatic hybrids were successfully produced [211, 212]. Another possible limitation of somatic hybridization is the production of somatic hybrids with resistant traits, but with decreased tuber yield and/or quality (mis-shaped tubers). Various solutions exists to overcome these disadvantages: use of haploidization and intra-specific hybridization of dihaploid potato lines [198], or the use of somatic fusion in which tetraploid potato cultivars are fused with sexually incompatible diploid wild species, when the resulted hexaploids are most of the time fertile and are crossable with other tetraploid cultivars [129–131]. Somatic hybridization produced a large number of somatic hybrids in potato some of them being integrated into pre-breeding and then breeding programs. The advantage of somatic hybridization is the transfer of multiple resistance genes, although it is difficult to control the genes transferred into the crop from its wild relative. It was thought that asymmetric fusion will allow better control on the genetic material to be transferred but soon it was demonstrated that only a low amount of donor DNA is eliminated and there is no correlation between the dose of radiation and DNA fragmentation. Nowadays, new strategies can be applied to better control the genetic fate of the SHs. Molecular markers can be used to select the traits or genes of interest [145], selection pressure like pathotoxins can be applied to increase the number of resistant SHs to a certain pathogen, *etc*. Moreover, there is a new opportunity to use all the genomic tools to get more insights into the complexity of the SHs and better understand the complex interaction between six genome forced together by artificial fusion in one cell. The main advantages of this biotechnological tool is its status as non-transgenic in Europe (directive 2001/18/EC, annex 1B) and its acceptance by consumers.

4. The new biotechnological techniques (NBT) and their success in improving potato crop

In the last decade, new plant breeding technologies (NPBT) have been developed to address plant breeding for important traits of current days. Those technologies were refinements of transgenesis and ended up with such advancements as leaving no foreign DNA in the new modified plants. In a review published by Lusser *et al* [213] the techniques used for NPBT were zinc finger nucleases (ZFN) and transcription activator-like effector nucleases (TALENs). In the same year a new NPBT was discovered and became the preferred alternative for plant genome editing, the clustered regulatory interspaced palindromic repeats or CRISPR [214]. CRISPR, a natural system used by bacteria and archaea to fight bacteriophages and foreign genetic fragments, has emerged as one of the most powerful and promising genome editing techniques shaping the future of biotechnology [215]. CRISPR-Cas method is based on a short single-guide RNA (sgRNA), with a 20 bp guide sequence complementary to a target region in recipient genome, a promoter and a sgRNA scaffold, which in combination with a Cas9 nuclease [214], can induce mutations in a target region of choice. Cas9 or an alternative nuclease induce double strand breaks (DSBs), that are repaired by the cell's own repair mechanism, either through non-homologous end joining (NHEJ) or homologous recombination (HR) [214]. NHEJ is an error prone and often leads to random-sized inserts or deletions (indels), which may cause a knockout of gene function. In potato the first results using CRISPR-Cas9 have shown mutations by using *Agrobacterium*-mediated stable transformation. The targeted genes were: gene encoding an Aux/IAA protein, the *StIAA2*, in a double haploid potato cultivar [216] and the *ALS* gene in both diploid and tetraploid potato [217]. More recently, TALEN and CRISPR-Cas9 were stably introduced targeting *ALS* and using a geminivirus-mediated guide, to facilitate designed mutations [218]. Because of its simplicity and cost efficiency CRISPR-Cas9 was adopted for many plant species [219], including potato as a tetraploid where targeted multialleles mutagenesis was achieved [220]. Traits such as: improved resistance to cold-induced sweetening, herbicide tolerance, processing efficiency, modified starch quality and self-incompatibility have been targeted in potato using CRISPR/Cas9 and TALEN editing technologies in diploid and tetraploid clones [221]. Potato varieties with knockout mutations in all alleles of the *VInv* (vacuolar invertase gene) through precise genome engineering were also produced [222]. This was accomplished by transiently expressing transcription activator-like effector nucleases (TALENs) designed to bind and cleave specific DNA sequences in the *VInv* locus. The double-stranded breaks (DSBs) created by the TALENs were repaired by NHEJ, which introduced indel (insertion/deletion) mutations that compromised *VInv* gene function. Due to the high levels of heterozygosity in the potato genome, the task of simultaneously targeting multiple alleles required careful TALEN design and optimization [223]. In contrast to previous RNAi work, TALENs achieved complete knockout lines without incorporating foreign DNA. As a result, the new potato lines have significantly lower levels of reducing sugars and acrylamide in heat-processed products [224]. In another attempt CRISPR-Cas9 was successfully applied to reduce browning of potato silencing *PPO* gene [225]. Increase resistance to late blight was obtained by mutating S (sensitivity gene) genes *StDND1* and *StCHL1* [226]. CRISPR-Cas13a was used to increase resistance to PVY, and it induced resistance to all strains of the virus [51], while RNAi confronted with many drawbacks because of the virus genetic evolution (**Table 1**) [50]. Although, the successfully edited plants by using CRISPR-Cas are deposited in Plant Genome Editing Database (PGED) [227], to date (2021-04-31) there is no registry for potato.

5. Acceptance by consumer and combinatorial biotechnology

Potato biotechnology has developed potato varieties with one or multiple genes, which resist one or multiple biotic and/or abiotic stresses. Unfortunately, the continuous debate and consumer lack of trust affect the GM cultivation in the field and specifically the adoption of GM plants in the food chain. There were success stories about genetically engineered potato crop, some of them were deregulated and had a short time of field cultivation. One of the examples that presents the fate of GM potato is the Monsanto potato story. Monsanto has developed GM potatoes with insect resistance (IR) and virus resistance (VR). In 1995, Monsanto received US government approval for Cry3A *Bt* potato, resistant to CPB. 600 ha were planted with this transgenic potato in USA. Another GM potato with resistance to potato leaf roll virus (PRLV) was approved in 1998 and a variety resistant to PVY in 1999. Moreover the *Bt* trait was stacked with PRLV and/or PVY resistance. From 1995 to 1998 the area with GE (genetically engineered) potato increased to 20,000 ha representing 3.5% of total area of potato crop in USA. But, in 2000 the area planted with GE potato declined sharply, a decline attributed to lack of acceptance by some consumers, the fast-food chain refusal of GE potato use and the incapacity of potato industry to test and segregate GE from non-GE potato. In these conditions, growers were concerned that their GE potato will no more be purchased by their buyers. The farmers, on the other hand, were purchasing a new insecticide for CPB and other pests rather than using GE varieties. In 2001 Monsanto decided to close its potato division [69, 228, 229]. Another story is about Amflora potato in Europe. After authorization procedure and favourable scientific opinions the European Commission approved the cultivation of BASF Amflora starch potato in 2010. This was the first GE plant approved for cultivation in EU in 12 years. The Amflora potato was not intended to be authorised for food, only for industrial use in starch production and its by-products as feed. Many member states were reacting against the GE potato authorization. In 2013, the EU General Court annulled the authorization of Amflora potato. In 2012 the BASF Company decided to move it's headquarter in USA (North Carolina) and halted the production of Amflora potato from EU market. Although, new breeding technologies and particularly CRISPR-Cas technology does not leave any foreign DNA into targeted mutagenized crops, EU has decided to consider edited crops under GM low in 2018, but there is hope that these modified crops will be accepted for cultivation and commercialization in the near future. The acceptance of modified crops by consumers varies from one country to another, depending on culture, history, environmental pressure *etc.*, but it seems that the benefits of transgenic and editing methods will at the end extend at scientific level, because cost benefits, CO_2 reduction and reduction on pesticides use will override the consumers unscientific doubts. But until then there are other effective biotechnological tools that are not considered as GMOs, for instance, somatic fusion and production of somatic hybrids as presented above can also address many resistance traits and be included in breeding. We have proposed a new strategy of biotechnological results integration in potato breeding called combinatorial biotechnology and already gave some good examples for the SHs of potato varieties (4x) with the diploid wild species *S. bulbocastanum* and *S. chacoense* [145, 180]. For instance in the case of somatic hybrids potato + *S. cahacoense*, presented above, transgenesis using *AtMSH2* gene, somatic hybridization, molecular analysis and stress selection were combined. For further integration in breeding somatic hybrids have to be back-crossed with cultivars and embryo rescue will be applied for BCs regeneration. Moreover, to remove the transgenes another strategy has to be applied as: gene segregation, RNAi or CRISPR-Cas. Finally, these genotypes with very interesting traits: resistance to CPB (antibiosis and antixenosis), tolerance to drought and salt would be integrated in breeding. The adoption of these biotechnological tools coupled with new knowledge on potato genomics and phenome's will most probably change the ways how the biotechnology is integrated in potato breeding for resilient potato, which is indispensable in today's challenging agriculture.

6. Conclusions

There are many tools in potato biotechnology which could be applied to improve potato resistance to biotic and abiotic stresses and to increase the quality of potato tubers for different application as food, feed, industrial use of even medicinal applications (**Figure 3**). These tools coupled with the new knowledge of genomics

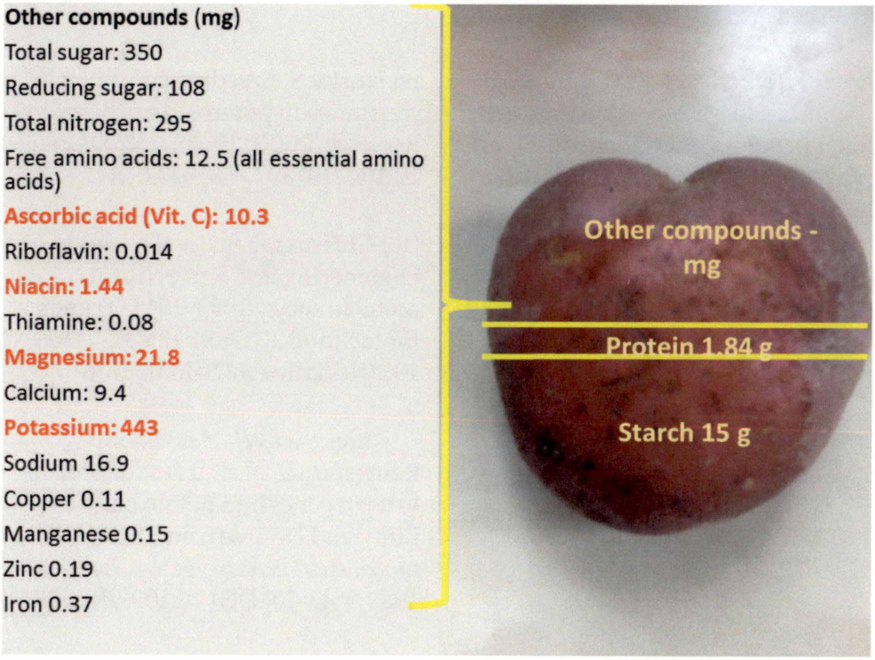

Figure 3.
Starch, protein and other valuable compound content of a potato raw tuber (detailed on the left (mg)) (modified data for cv. Russet Burbank https://www.researchgate.net/publication/265480176_27_).

and phenomics will be more and more accepted in improving potato crop for actual and future agriculture. Combinatorial biotechnology that in our opinion will use all advantages of potato genome manipulation, tissue culture techniques, and next generation biotechnologies along with genome, transcriptome and metabolome research will contribute to resilient potato crop.

Acknowledgements

I. M. was supported by funds from the Young Researcher Grants 2019-2020 (GTC-2020).

Author details

Elena Rakosy-Tican* and Imola Molnar
Plant Genetic Engineering Group, Department of Molecular Biology and Biotechnology, Faculty of Biology and Geology, Babeş-Bolyai University, Cluj-Napoca, Romania

*Address all correspondence to: arina5744@yahoo.com

References

[1] Barrell PJ, Meiyalaghan S, Jeanne ME et al. Applications of biotechnology and genomics in potato improvement. Plant Biotech J. 2013; 11:907-920. DOI: 10.1111/pbi.12099

[2] Bradshaw JE. Potato Breeding: Theory and Practice. Dordrecht: Springer International Publishing; 2021. 563 p. DOI 10.1007/978-3-030-64414-7

[3] Zaheer K, Akhtar MH. Potato production, usage, and nutrition–a review. Crit Rev Food Sci Nutr. 2016; 56:711-721. DOI: 10.1080/10408398.2012.724479

[4] Kehoe MA, Jones RAC. Improving Potato virus Y strain nomenclature: Lessons from comparing isolates obtained over a 73-year period. Plant Pathol. 2016; 65: 322-333. DOI: 10.1111/ppa.12404

[5] Vreugdenhil D, Bradshaw J, Gebhardt C, et al. editors. Potato Biology and Biotechnology. Advances and Perspectives. 1st ed. (2007) Amsterdam: Elsevier; 2007. 856 p.

[6] Halterman D, Guenthner J, Collinge S, et al. Biotech potatoes in the 21st century: 20 years since the first biotech potato. Am J Potato Res. 2016; 93:1-20. DOI: 10.1007/s12230-015-9485-1

[7] CIP Position Paper [Internet]. 2015. Available from: https://cip-cgiar-bucket1.s3.amazonaws.com/wp-content/uploads/2018/05/Biotechnology-Position-Paper-18Nov15.pdf [Accessed: 2021-04-15]

[8] Thieme R, Rakosy-Tican E. Somatic cell genetics and its application in potato breeding. In: Chakrabarti SK, Xie C, Tiwari JK, editors. The Potato Genome. AG Cham: Springer International Publishing; 2017. p. 217-269. DOI: 10.1007/978-3-319-66135-3_13

[9] Jansky S. Breeding for disease resistance in potato. Plant Breeding Reviews. 2000; 19: 69-155. DOI: 10.1002/9780470650172.ch4

[10] Christou P. Plant genetic engineering and agricultural biotechnology 1983-2013. Trends Biotechnol. 2013; 31:125-127. DOI: 10.1016/j.tibtech.2013.01.006

[11] Stiekema WJ, Heidekamp F, Louwerse JD, et al. Introduction of foreign genes into potato cultivars Bintjeand Désirée using an *Agrobacterium tumefaciens* binary vector. Plant Cell Rep. 1988; 7:47-50. DOI 10.1007/BF00272976

[12] Sheerman S, Bevan MW. A rapid transformation method for *Solanum tuberosum* using binary *Agrobacterium tumefaciens* vectors. Plant Cell Rep. 1988; 7:13-16. DOI: 10.1007/BF00272967

[13] Rakosy-Tican L, Aurori CM, Dijkstra C, et al. The usefulness of reporter gene *gfp* for monitoring *Agrobacterium*-mediated transformation of potato dihaploid and tetraploid genotypes. Plant Cell Rep. 2007; 26:661-671. DOI: 10.1007/s00299-006-0273-8

[14] Directive 2001/18/EC of the European Parliament and of the Council [Internet]. 2001. Available from: https://eur-lex.europa.eu/LexUriServ/LexUriServ.do?uri=CONSLEG:2001L0018:20080321:EN:PDF [Accessed: 2021-04-02]

[15] Evaluation of new breeding technology (EFSA) [Internet]. 2011. Available from: https://www.efsa.europa.eu/en/topics/topic/gmo [Accessed: 2021-04-03]

[16] Spooner DM, Hijmans RJ. Potato systematics and germplasm collecting, 1989-2000. Am J Potato Res. 2001; 78: 237-268. DOI: 10.1007/BF02875691

[17] Visser RGF, Bachem CWB, de Boer JM, et al. Sequencing the potato genome: Outline and first results to come from the elucidation of the sequence of the World's third most important food crop. Am J Potato Res. 2009; 86:417-429. DOI: 10.1007/s12230-009-9097-8

[18] De Block M. Genotype-independent leaf disc transformation of potato (*Solanum tuberosum*) using *Agrobacterium tumefaciens*. Theor Appl Genet. 1988; 76(5):767-74. DOI: 10.1007/BF00303524

[19] Gimenez-Ibanez S. Designing disease-resistant crops: From basic knowledge to biotechnology. Mètode Science Studies Journal. 2021; 11:47-53. DOI: 10.7203/metode.11.15496

[20] Dong OX, Roland PC. Genetic engineering for disease resistance in plants: Recent progress and future perspectives. Plant Physiology. 2019; 180:26-38. DOI: 10.1104/pp.18.01224

[21] Potato Genome Sequencing Consortium. Genome sequence and analysis of the tuber crop potato. Nature. 2011; 475:189-195. DOI: 10.1038/nature10158

[22] Aversano R, Contaldi F, Ercolano MR, et al. The *Solanum commersonii* genome sequence provides insights into adaptation to stress conditions and genome evolution of wild potato relatives. Plant Cell. 2015; 27:954-968. DOI: 10.1105/tpc.114.135954

[23] Leisner CP, Hamilton JP, Crisovan E, et al. Genome sequence of M6, a diploid inbred clone of the high-glycoalkaloid-producing tuber-bearing potato species *Solanum chacoense*, reveals residual heterozygosity. The Plant Journal. 2018; 94:562-570. DOI: 10.1111/tpj.13857

[24] Lorenc-Kukuła K, Jafra S, Oszmiański J, et al. Ectopic expression of anthocyanin 5-*O*-Glucosyltransferase in potato tuber causes increased resistance to bacteria. J Agric Food Chem. 2005; 53(2):272-281. DOI: 10.1021/jf048449p

[25] Almasia NI, Bazzini AA, Esteban Hopp H, et al. Overexpression of *snakin*-1 gene enhances resistance to *Rhizoctonia solani* and *Erwinia carotovora* in transgenic potato plants. Mol Plant Pathol. 2008; 9(3):329-338. DOI: 10.1111/j.1364-3703.2008.00469.x

[26] Bidondo LF, Almasia N, Bazzini A, et al. The overexpression of antifungal genes enhances resistance to *Rhizoctonia solani* in transgenic potato plants without affecting arbuscular mycorrhizal symbiosis. Crop Protection. 2019; 124: 104837. DOI: 10.1016/j.cropro.2019.05.031

[27] Jones JDG, Witek K, Verweij W, et al. Elevating crop disease resistance with cloned genes. Phil Trans R Soc B. 2014; 369:20130087. DOI: 10.1098/rstb.2013.0087

[28] Jo KR, Kim CJ, Kim SJ, et al. Development of late blight resistant potatoes by cisgene stacking. BMC Biotechnology. 2014; 14:50. DOI: 10.1186/1472-6750-14-50

[29] Listanto E, Riyanti EI, Santoso TJ, et al. Genetic stability analysis of *RB* gene in genetically modified potato lines tolerant to *Phytophthora infestans*. Indonesian J Agric Sci. 2015; 16(2):51-58. DOI: 10.21082/ijas.v16n2.2015.pp.51-58

[30] Zhu S, Li Y, Vossen JH, et al. Functional stacking of three resistance genes against *Phytophthora infestans* in potato. Transgenic Res. 2012; 21:89-99. DOI: 10.1007/s11248-011-9510-1

[31] Bouwmeester K, Han M, Blanco-Portales R, et al. The *Arabidopsis* lectin receptor kinase LecRK-I.9 enhances resistance to *Phytophthora infestans* in *Solanaceous* plants. Plant Biotechnol J. 2014; 12(1):10-6. DOI: 10.1111/pbi.12111

[32] Zhang Z, Yang F, Na R, et al. AtROP1 negatively regulates potato resistance to

Phytophthora infestans via NADPH oxidase-mediated accumulation of H_2O_2. BMC Plant Biol. 2014; 14:392. DOI:10.1186/s12870- 014-0392-2.44

[33] Jahan SN, Asman AK, Corcoran P, et al. Plant-mediated gene silencing restricts growth of the potato late blight pathogen *Phytophthora infestans*. Journal of Experimental Botany. 2015; 66: 2785-2794. DOI: 10.1093/jxb/erv094

[34] Ghislain M, Byarugaba AA, Magembe E, et al. Stacking three late blight resistance genes from wild species directly into African highland potato varieties confers complete field resistance to local blight races. Plant Biotechnol J. 2019; 17:1119-1129. DOI: 10.1111/pbi.13042

[35] Webi EN, Kariuki D, Kinyua J, et al. Extreme resistance to late blight disease by transferring 3 R genes from wild relatives into African farmer-preferred potato varieties. Afr J Biotechnol. 2019; 18(29):845-856. DOI: 10.5897/AJB2019.16856

[36] Osusky M, Osuska L, Kay W, et al. Genetic modification of potato against microbial diseases: *in vitro* and *in planta* activity of a dermaseptin B1 derivative, *MsrA2*. Theor Appl Genet. 2005; 111:711-722. DOI: 10.1007/s00122-005-2056-y

[37] Goyal RK, Hancock REW, Mattoo AK, et al. Expression of an engineered heterologous antimicrobial peptide in potato alters plant development and mitigates normal abiotic and biotic responses. PLoS ONE. 2013; 8(10):e77505. DOI: 10.1371/journal.pone.0077505

[38] Buziashvili A, Cherednichenko L, Kropyvko S, et al. Obtaining transgenic potato plants expressing the human Lactoferrin gene and analysis of their resistance to phytopathogens. Cytol Genet. 2020; 54:179-188. DOI: 10.3103/S0095452720030020

[39] Zhu W, Bai X, Li G, et al. SpCYS, a cystatin gene from wild potato (*Solanum pinnatisectum*), is involved in the resistance against *Spodoptera litura*. Theor Exp Plant Physiol. 2019; 31:317-328. DOI: 10.1007/s40626-019-00148-8

[40] Koropacka KB. Molecular contest between potato and the potato cyst nematode *Globodera pallida*: modulation of Gpa2-mediated resistance [thesis]. Wageningen: Wageningen University; 2010.

[41] Paal J, Henselewski H, Muth J, et al. Molecular cloning of the potato Gro1-4 gene conferring resistance to pathotype Ro1 of the root cyst nematode *Globodera rostochiensis*, based on a candidate gene approach. The Plant Journal. 2004; 38:285-297. DOI: 10.1111/j.1365-313X.2004.02047.x

[42] Green J, Wang D, Lilley CJ, et al. Transgenic potatoes for potato cyst nematode control can replace pesticide use without impact on soil quality. PLoS ONE. 2012; 7(2):e30973. DOI: 10.1371/journal.pone.0030973

[43] Missiou A, Kalantidis K, Boutla A, et al. Generation of transgenic potato plants highly resistant to potato virus Y (PVY) through RNA silencing. Mol Breed. 2004; 14:185-197. DOI: 10.1023/B:MOLB.0000038006.32812.52

[44] Bukovinszki A, Diveki Z, Csanyi M, et al. Engineering resistance to PVY in different potato cultivars in a marker-free transformation system using a 'shooter mutant' *A. tumefaciens.* Plant Cell Rep. 2007; 26:459-465. DOI: 10.1007/s00299-006-0257-8

[45] Dusi AN, Lopes de Oliveira C, de Melo PE, et al. Resistance of genetically modified potatoes to Potato virus Y under field conditions. Pesq Agropec Bras Brasília. 2009; 44(9):1127-1130. DOI: 10.1590/S0100-204X2009000900009

[46] Rakosy-Tican E, Aurori CM, Dijkstra C, et al. Generating marker free transgenic potato cultivars with a hairpin

construct of PVY coat protein. Romanian Biotech Lett. 2010; 15(1suppl):63-71.

[47] Tabassum B, Nasir IA, Khan A, et al. Short hairpin RNA engineering: In planta gene silencing of potato virus Y. Crop Protection. 2016; 86:1-8. DOI: 10.1016/j.cropro.2016.04.005

[48] Duan H, Richael C, Rommens C. Overexpression of the wild potato eIF4E-1 variant Eva1 elicits Potato virus Y resistance in plants silenced for native eIF4E-1. Transgenic Research. 2012; 21:929-938. DOI: 10.1007/s11248-011-9576-9

[49] Arcibal E, Morey Gold K, Flaherty S, et al. A mutant eIF4E confers resistance to potato virus Y strains and is inherited in a dominant manner in the potato varieties Atlantic and Russet Norkotah. Am J Potato Res. 2015; 93(1):64-71. DOI: 10.1007/s12230-015-9489-x

[50] Pooggin MM. RNAi-mediated resistance to viruses: a critical assessment of methodologies. Curr Opin Virol. 2017; 26:28-35. DOI: 10.1016/j.coviro.2017.07.010

[51] Zhan X, Zhang F, Zhong Z, et al. Generation of virus-resistant potato plants by RNA genome targeting. Plant Biotech J. 2019; 17:1814-1822. DOI: 10.1111/pbi.13102

[52] Osmani Z, Sabet MS, Nakahara KS, et al. Identification of a defense response gene involved in signaling pathways against PVA and PVY in potato. GM Crops & Food. 2021; 12(1):86-105. DOI: 10.1080/21645698.2020.1823776

[53] Naimov S, Dukiandjiev S, de Maagd RA. A hybrid *Bacillus thuringiensis* deltaendotoxin gives resistance against a coleopteran and a lepidopteran pest in transgenic potato. Plant Biotechnol J. 2003; 1:51-57. DOI: 10.1046/j.1467-7652.2003.00005.x

[54] Davidson MM, Takla MFG, Jacobs JME, et al. Transformation of potato (*Solanum tuberosum*) cultivars with a *cry1Ac9* gene confers resistance to potato tuber moth (*Phthorimaea operculella*). New Zealand J Crop Hort Sci. 2004; 32(1):39-50. DOI: 10.1080/01140671.2004.9514278

[55] Outchkourov NS, de Kogel WJ, Schuurman-de Bruin A, et al. Specific cysteine protease inhibitors act as deterrents of western flower thrips, *Frankliniella occidentalis* (Pergande), in transgenic potato. Plant Biotech J. 2004; 2:439-448. DOI: 10.1111/j.1467-7652.2004.00088.x

[56] Zhu F, Xu J, Palli R, et al. Ingested RNA interference for managing the populations of the Colorado potato beetle, *Leptinotarsa decemlineata*. Pest Manag Sci. 2011; 67:175-182. DOI:10.1002/ps.2048

[57] Zotti MJ, Smagghe G. RNAi technology for insect management and protection of beneficial insects from diseases: lessons, challenges and risk assessments. Neotrop Entomol. 2015; 44:197-213. DOI: 10.1007/s13744-015-0291-8

[58] Zhang J, Khan SA, Hasse C, et al. Full crop protection from an insect pest by expression of long double-stranded RNAs in plastids. Science. 2015; 347:991-994. DOI:10.1126/science.1261680

[59] Hussain T, Aksoy E, CalisKan MEC, et al. Transgenic potato lines expressing hairpin RNAi construct of molting-associated EcR gene exhibit enhanced resistance against Colorado potato beetle (*Leptinotarsa decemlineata*, Say). Transgenic Res. 2019; 28:151-164. DOI: 10.1007/s11248-018-0109-7

[60] Gui S, Taning CNT, Wei D, et al. First report on CRISPR/Cas9-targeted mutagenesis in the Colorado potato beetle, *Leptinotarsa decemlineata*. J Insect Physiol. 2020; 121:104013. DOI: 10.1016/j.jinsphys.2020.104013

[61] Sharma HC, Sharma KK, Crouch JH. Genetic transformation of crops for

insect resistance: Potential and limitations. Crit Rev Plant Sci. 2004; 23(1):47-72. DOI: 10.1080/07352680490273400

[62] Du J, Verzaux E, Chaparro-Garcia A, et al. Elicitin recognition confers enhanced resistance to *Phytophthora infestans* in potato. Nat Plants. 2015; 1(4):15034. DOI: 10.1038/nplants.2015.34

[63] Kawchuk LM, Hachey J, Lynch DR, et al. Tomato *Ve* disease resistance genes encode cell surface-like receptors. Proc Natl Acad Sci *USA*. 2001; 98:6511-6515. DOI: 10.1073/pnas.091114198

[64] Haverkort AJ, Boonekamp PM, Hutten R, et al. Durable late blight resistance in potato through dynamic varieties obtained by cisgenesis: scientific and societal advances in the DuRPh project. Potato Res. 2016; 59:35-66. DOI 10.1007/s11540-015-9312-6

[65] Waltz E. USDA approves next-generation GM potato. Nature bio-technology. 2015; 33(1):12-13. DOI: 10.1038/nbt0115-12

[66] Ridler K. Canada approves three types of genetically engineered potatoes [Internet]. 2017. Available from: https://www.ctvnews.ca/health/canada-approves-three-typesof-genetically-engineered-potatoes-1.3531998 [Accessed: 2021-04-22]

[67] Bendahmane A, Kohm BA, Dedi C, et al. The coat protein of potato virus X is a strain-specific elicitor of Rx1-mediated virus resistance in potato. Plant J. 1995; 8:933-941. DOI: 10.1046/j.1365-313x.1995.8060933.x

[68] Danan S, Veyrieras J-B, Lefebvre V. Construction of a potato consensus map and QTL meta-analysis offer new insights into the genetic architecture of late blight resistance and plant maturity traits. BMC Plant Biology. 2011; 11:16. DOI: 10.1186/1471-2229-11-16

[69] James C. (2011) Global Status of Commercialized Biotech/GM Crops, Vol 44. New York: ISAAA; 2011.

[70] Ferro DN, Logan JA, Voss RH, et al. Colorado potato beetle (Coleoptera: *Chrysomelidae*) temperature-dependent growth and feeding rates. Environ Entomol. 1985; 14:343-348. DOI: 10.1093/ee/14.3.343

[71] Balaško MK, Mikac KM, Bažok R, et al. Modern techniques in Colorado potato beetle (*Leptinotarsa decemlineata* Say) control and resistance management: History review and future perspectives. Insects. 2020; 11(9):581. DOI: 10.3390/insects11090581

[72] Balbyshev NF, Lorenzen JH. Hypersensitivity and egg drop: A novel mechanism of host plant resistance to Colorado potato beetle (Coleoptera: *Chrysomelidae*). J Econ Entomol. 1997; 90:652-657. DOI: 10.1093/jee/90.2.652

[73] Perlak FJ, Stone TB, Muskopf YM, et al. Genetically improved potatoes: Protection from damage by Colorado potato beetles. Plant Mol Biol. 1993; 22:313-321. DOI: 10.1007/BF00014938

[74] Grafius EJ, Douches DS. The present and future role of insect-resistant genetically modified potato cultivars in IPM. In: Romeis J, Shelton AM, Kennedy GG, editors. Integration of Insect-Resistant Genetically Modified Crops within IPM Programs. 1st ed. Dordrecht: Springer; 2008. p. 195-221. DOI: 10.1007/978-1-4020-8373-0_7

[75] Coombs JJ, Douches DS, Li W, et al. Combining engineered (Bt-cry3A) and natural resistance mechanisms in potato for control of Colorado potato beetle. J Am Soc Hortic Sci. 2002; 127:62-68. DOI: 10.21273/JASHS.127.1.62

[76] Baum J, Bogaert T, Clinton W, et al. Control of coleopteran insect pests through RNA interference. Nat Biotechnol. 2007; 25:1322-1326. DOI: 10.1038/nbt1359

[77] Joga MR, Zotti MJ, Smagghe G, et al. RNAi efficiency, systemic properties, and novel delivery methods for pest insect control: what we know so far. Front Physiol. 2016; 7: 553. DOI: 10. 3389/ fphys. 2016. 00553

[78] Mishra S, Dee J, Moar W, et al. Selection for high levels of resistance to double stranded RNA (dsRNA) in Colorado potato beetle (*Leptinotarsa decemlineata* Say) using non-transgenic foliar delivery. Sci Rep. 2021; 11:6523. DOI: 10.1038/s41598-021-85876-1

[79] Schoville SD, Chen YH, Andersson MN, et al. A model species for agricultural pest genomics: The genome of the Colorado potato beetle, *Leptinotarsa decemlineata* (Coleoptera: *Chrysomelidae*). Sci Rep. 2018; 8:1-18. DOI: 10.1038/s41598-018-20154-1

[80] Gaddelapati SC, Kalsi M, Roy A, et al. Cap'n'collar C regulates genes responsible for imidacloprid resistance in the Colorado potato beetle, *Leptinotarsa decemlineata*. Insect Biochem Mol Biol. 2018; 99:54-62. DOI: 10.1016/j.ibmb.2018.05.006

[81] Ochoa-Campuzano C, Martínez-Ramírez AC, Contreras E, et al. Prohibitin, an essential protein for Colorado potato beetle larval viability, is relevant to *Bacillus thuringiensis* Cry3Aa toxicity. Pestic Biochem Physiol. 2013; 107:299-308. DOI: 10.1016/j.pestbp.2013. 09.001

[82] Singh KB, Foley RC, Oñate-Sánchez L. Transcription factors in plant defense and stress responses. Curr Opin Plant Biol. 2002; 5:430-436. DOI: 10.1016/S1369-5266(02)00289-3

[83] Youm JW, Jeon JH, Choi D, et al. Ectopic expression of pepper *CaPF1* in potato enhances multiple stresses tolerance and delays initiation of *in vitro* tuberization. Planta. 2008; 228:701-708. DOI: /10.1007/s00425-008-0782-5

[84] Dou H, Xv K, Meng Q, et al. Potato plants ectopically expressing *Arabidopsis thaliana* CBF3 exhibit enhanced tolerance to high-temperature stress. Plant Cell Environ. 2015; 38:61-72. DOI: 10.1111/pce.12366

[85] Trapero-Mozos A, Morris WL, Ducreux LJM, et al. Engineering heat tolerance in potato by temperature-dependent expression of a specific allele of HEAT-SHOCK COGNATE 70. Plant Biotechnol J. 2018; 16:197-207. DOI: 10.1111/pbi.12760

[86] Behnam B, Kikuchi A, Celebi-Toprak F, et al. *Arabidopsis rd29A::DREB1A* enhances freezing tolerance in transgenic potato. Plant Cell Rep. 2007; 26:1275-1282. DOI: 10.1007/s00299-007-0360-5

[87] Stiller I, Dulai S, Kondrak M, et al. Effects of drought on water content and photosynthetic parameters in potato plants expressing the *trehalose-6-phosphate synthase* gene of *Saccharomyces cerevisiae*. Planta. 2008; 227:299-308. DOI: 10.1007/s00425-007-0617-9

[88] Zhang N, Si HJ, Wen G, et al. Enhanced drought and salinity tolerance in transgenic potato plants with a BADH gene from spinach. Plant Biotechnol Rep. 2011; 5:71-77. DOI: 10.1007/s11816-010-0160-1

[89] Kondrak M, Marincs F, Antal F, et al. Effects of yeast trehalose-6-phosphate synthase 1 on gene expression and carbohydrate contents of potato leaves under drought stress conditions. BMC Plant Biol. 2012; 12:74. DOI: 10.1186/1471-2229-12-74

[90] Pal AK, Acharya K, Vats SK, et al. Over-expression of *PaSOD* in transgenic potato enhances photosynthetic performance under drought. Biol Plant. 2013; 57(2):359-364. DOI: 10.1007/s10535-012-0277-x

[91] Iwaki T, Guo L, Ryals JA, et al. Metabolic profiling of transgenic potato

tubers expressing *Arabidopsis* dehydration response element-binding protein 1A (DREB1A). J Agric Food Chem. 2013; 61:893-900. DOI:10.1021/jf304071n

[92] Pieczynski M, Marczewski W, Hennig J, et al. Down-regulation of CBP80 gene expression as a strategy to engineer a drought-tolerant potato. Plant Biotechnology Journal. 2013; 11:459-469. DOI: 10.1111/pbi.12032

[93] Li S, Zhang N, Zhu X, et al. Enhanced drought tolerance with artificial microRNA-mediated *StProDH1* gene silencing in potato. Crop Sci. 2020; 60(3):1462-1471. DOI: 10.1002/csc2.20064

[94] Hmida-Sayari A, Gargouri-Bouzid R, Bidani A, et al. Overexpression of Δ1-pyrroline-5-carboxylate synthetase increases proline production and confers salt tolerance in transgenic potato plants. Plant Sci. 2005; 169(4):746-752. DOI: 10.1016/j.plantsci.2005.05.025

[95] Bayat F, Shiran B, Belyaev DV, et al. Potato plants bearing a vacuolar Na+/H+ antiporter HvNHX2 from barley are characterized by improved salt tolerance. Russ J Plant Physiol. 2010; 57(5):696-706. DOI: 10.1134/S1021443710050134

[96] Wang L, Liu Y, Li D, et al. Improving salt tolerance in potato through overexpression of AtHKT1 gene. BMC Plant Biol. 2019; 19(1):357. DOI: 10.1186/s12870-019-1963-z

[97] Liu M, Li Y, Li G, et al. Overexpression of StCYS1 gene enhances tolerance to salt stress in the transgenic potato (*Solanum tuberosum* L.) plant. J Integrative Agri. 2020; 19(9):2239-2246. DOI: 10.1016/S2095-3119(20)63262-2

[98] Lee HE, Shin D, Park SR, et al. Ethylene responsive element binding protein 1 (*StEREBP*1) from *Solanum tuberosum* increases tolerance to abiotic stress in transgenic potato plants. Biochem Biophys Res Commun. 2007; 353(4):863-868. DOI: 10.1016/j.bbrc.2006.12.095

[99] Tang Li, Kwon S-K, Kim S-H, et al. Enhanced tolerance of transgenic potato plants expressing both superoxide dismutase and ascorbate peroxidase in chloroplasts against oxidative stress and high temperature. Plant Cell Rep. 2006; 25:1380-1386. DOI: 10.1007/s00299-006-0199-1

[100] Movahedi S, Tabatabaei BES, Alizade H, et al. Constitutive expression of *Arabidopsis DREB1B* in transgenic potato enhances drought and freezing tolerance. Biol Plant. 2012; 56:37-42. DOI: 10.1007/s10535-012-0013-6

[101] Bouaziz D, Pirrello J, Amor HB, et al. Ectopic expression of dehydration responsive element binding proteins (StDREB2) confers higher tolerance to salt stress in potato. Plant Physiol Bioch. 2012: 60:98-108. DOI: 10.1016/j.plaphy.2012.07.029

[102] Bouaziz D, Pirrello J, Charfeddine M, et al. Overexpression of *StDREB1* transcription factor increases tolerance to salt in transgenic potato plants. Mol Biotechnol. 2013; 54:803-817. DOI: 10.1007/s12033-012-9628-2

[103] Sievers N, Muders K, Henneberg M, et al. Establishing glucosylglycerol synthesis in potato (*Solanum tuberosum* l. cv. Albatros) by expression of the *ggpPS* gene from *Azotobacter vinelandii*. J Plant Sci & Mol Breed. 2013; 2:1. DOI: 10.7243/2050-2389-2-1

[104] Ahmad R, Hussain J, Jamil M, et al. Glycinebetaine synthesizing transgenic potato plants exhibit enhanced tolerance to salt and cold stresses. Pak J Bot. 2014; 46(6):1987-1993

[105] Shahzad R, Harlina PW, Cong-hua X, et al. Overexpression of potato transcription factor (StWRKY1) conferred resistance to *Phytophthora*

infestans and improved tolerance to water stress. Plant Omics. 2016; 9(2):149-158. DOI: 10.21475/poj.160902.p7649x

[106] Muñiz García MN, Cortelezzi JI, Fumagalli M, et al. Expression of the Arabidopsis *ABF4* gene in potato increases tuber yield, improves tuber quality and enhances salt and drought tolerance. Plant Mol Biol. 2018; 98:137-152. DOI: 10.1007/s11103-018-0769-y

[107] Lehretz GG, Sonnewald S, Lugassi N, et al. Future-proofing potato for drought and heat tolerance by overexpression of Hexokinase and SP6A. Front Plant Sci. 2021; 11:614534. DOI: 10.3389/fpls.2020.614534

[108] Ahmad R, Kim MD, Back KH, et al. Stress-induced expression of choline oxidase in potato plant chloroplasts confers enhanced tolerance to oxidative, salt, and drought stresses. Plant Cell Rep. 2008; 27:687-698. DOI: 10.1007/s00299-007-0479-4

[109] Ahmad R, Kim YH, Kim MD, et al. Simultaneous expression of choline oxidase, superoxide dismutase and ascorbate peroxidase in potato plant chloroplasts provides synergistically enhanced protection agains various abiotic stresses. Physiol Plant. 2010; 138:520-533. DOI: 10.1111/j.1399-3054.2010.01348.x

[110] Gangadhar BH, Sajeesh K, Venkatesh J, et al. Enhanced tolerance of transgenic potato plants over-expressing non-specific lipid transfer Protein-1 (StnsLTP1) against multiple abiotic stresses. Front Plant Sci. 2016; 7:1228. DOI: 10.3389/fpls.2016.01228

[111] Cho K-S, Han E-H, Kwak S-S, et al. Expressing the sweet potato orange gene in transgenic potato improves drought tolerance and marketable tuber production. C R Biologies. 2016; 339:207-213. DOI: 10.1016/j.crvi.2016.04.010

[112] Gong L, Zhang H, Gan X, et al. Transcriptome profiling of the potato (*Solanum tuberosum* L.) plant under drought stress and water-stimulus conditions. PLoS ONE. 2015; 10(5): e0128041. DOI: 10.1371/journal.pone.0128041

[113] van Zelm E, Zhang Y, Testerink C. Salt tolerance mechanisms of plants. Ann Rev Plant Biol. 2020; 71:403-433. DOI: 10.1146/annurev-arplant-050718-100005

[114] Fidalgo F, Santos A, Santos I, et al. Effects of long-term salt stress on antioxidant defense systems, leaf water relations and chloroplast ultrastructure of potato plants. Ann Appl Biol. 2004; 145:185-192. DOI: 10.1111/j. 1744-7348.2004.tb00374.x

[115] Odemis B, Caliskan ME. Photosynthetic response of potato plants to soil salinity. Turk J Agric Nat Sci. 2014; 2:1429-1439.

[116] Dwelle RB, Kleinkopf GE, Pavek JJ. Stomatal conductance and gross photosynthesis of potato (*Solanum tuberosum* L.) as influenced byirradiance, temperature and growth stage. Potato Res. 1981; 24:49-59. DOI: 10.1007/bf02362016

[117] Dahal K, Li X-Q, Tai H, et al. Improving potato stress tolerance and tuber yield under a climate change scenario – A current overview. Front Plant Sci. 2019; 10:563. DOI: 10.3389/fpls.2019.00563

[118] Tang R, Zhu W, Song X, et al. Genome-wide identification and function analyses of heat shock transcription factors in potato. Front Plant Sci. 2016; 19(7):490. DOI: 10.3389/fpls.2016.00490

[119] Schenk PM, Kazan K, Wilson I, et al. Coordinated plant defense responses in *Arabidopsis* revealed by microarray analysis. Proc Natl Acad Sci USA. 2000; 97:11655-11660. DOI: 10.1073/pnas.97.21.11655

[120] Bouaziz D, Charfeddine M, Jbir R, et al. Identification and functional characterization of ten *AP2/ERF* genes in potato. Plant Cell Tiss Organ Cult. 2015; 123:155-172. DOI: 10.1007/s11240-015-0823-2

[121] Gangadhar BH, Yu JW, Sajeesh K, et al. A systematic exploration of high-temperature stress-responsive genes in potato using large-scale yeast functional screening. Mol Genet Genomics. 2014; 289:185-201. DOI: 10.1007/s00438-013-0795-z

[122] Jansky S. Breeding, genetics and cultivar development. In: Singh J, Kaur L, editors. Advances in potato chemistry and technology. Burlington: Academic Press; 2009. p. 27-62. DOI: 10.1016/B978-0-12-374349-7.00002-7

[123] Ross H. Potato breeding-problems and perspectives. Berlin: Springer; 1986.

[124] Muthoni J, Kabira J, Shimelis H, et al. Tetrasomic inheritance in cultivated potato and implications in conventional breeding. Aust J Crop Sci. 2015; 9(3): 185-190

[125] Haverkort AJ, Struik PC, Visser RGF, et al. Applied biotechnology to combat late blight in potato caused by *Phytophthora infestans*. Potato Res. 2009; 52:249-264. DOI: 10.1007/s11540-009-9136-3

[126] Thieme R, Darsow U, Gavrilenko T, et al. Production of somatic hybrids between *S. tuberosum* L. and late blight resistant Mexican wild potato species. Euphytica. 1997; 97:189-200. DOI: 10.1023/A:1003026125623

[127] Rokka VM. Protoplast technology in genome manipulation of potato through somatic cell fusion. In: Li X-Q, Donnelly D, Jensen TJ, editors. Somatic genome manipulation—advances, methods and applications. New-York: Springer; 2015. p. 217-235. DOI: 10.1007/978-1-4939-2389-2_10

[128] Davey MR, Anthony P, Power BJ, et al. Plant protoplasts: status and biotechnological perspectives. Biotechnol Adv. 2005; 23(2):131-171. DOI: 10.1016/j.biotechadv.2004.09.008

[129] Thieme R, Rakosy-Tican E, Gavrilenko T, et al. Novel somatic hybrids (*Solanum tuberosum* L. + *Solanum tarnii*) and their fertile BC1 progenies express extreme resistance to potato virus Y and late blight. Theor Appl Genet. 2008; 116:691-700. DOI: 10.1007/s00122-007-0702-2

[130] Thieme R, Rakosy-Tican E, Nachtigall M, et al. Characterization of the multiple resistance traits of somatic hybrids between *Solanum cardiophyllum* Lindl. and two commercial potato cultivars. Plant Cell Rep. 2010; 29:1187-1201. DOI: 10.1007/s00299-010-0905-x

[131] Rakosy-Tican E, Thieme R, Nachtigall M, et al. The recipient potato cultivar influences the genetic makeup of the somatic hybrids between five potato cultivars and one cloned accession of sexually incompatible species *Solanum bulbocastanum* Dun. Plant Cell Tiss Organ Cult. 2015; 122:395-407. DOI: 10.1007/s11240-015-0777-4

[132] Shepard JF, Totten RE. Mesophyll cell protoplasts of potato. Isolation, proliferation and plant regeneration. Plant Physiol. 1977; 60:313-316. DOI: 10.1104/pp.60.2.313

[133] Puite KJ, Roest S, Pijnacker LP. Somatic hybrid potato plants after electrofusion of diploid *Solanum tuberosum* and *Solanum phureja*. Plant Cell Rep. 1986; 5:262-265. DOI: 10.1007/BF00269817

[134] Wenzel G. Biotechnology in potato improvement. In: Gopal J, Khurana SM, editors. Handbook of potato production, improvement and postharvest management. New York: The Haworth Press Inc; 2006. p. 109-146. DOI: 10.1017/S0014479707005133

[135] Binding H, Nehls R, Schieder O, et al. Regeneration of mesophyll protoplasts isolated from dihaploid clones of *Solanum tuberosum* L. Plant Physiol. 1978; 43:52-54. DOI: 10.1111/j.1399-3054.1978.tb01565.x

[136] Binding H, Nehls R. Regeneration of isolated protoplasts to plant *Solanum dulcamara* L. Z Pflanzenphysiology. 1977; 85:279-280. DOI: 10.1016/S0044-328X(77)80255-9

[137] Behnke M. Regeneration in Gewebekulturen einiger dihaplider *Solanum tuberosum*-Klone. Z Pflanzenzücht. 1975; 75:262-265

[138] Rakosy-Tican E, Aurori A. Green fluorescent protein (GFP) supports the selection based on callus vigorous growth in the somatic hybrids *Solanum tuberosum* L. + *S. chacoense* Bitt. Acta Physiol Plant. 2015; 37:201. DOI: 10.1007/s11738-015-1946-0

[139] Senda M, Takeda J, Abe S, et al. Induction of cell fusion of plant protoplasts by electrical stimulation. Plant Cell Physiol. 1979; 20:1441-1443. DOI: 10.1093/oxfordjournals.pcp.a075944

[140] Zimmermann U, Scheurich P. High frequency fusion of plant protoplasts by electric fields. Planta. 1981; 151:26-32. DOI: 10.1007/BF00384233

[141] Rakosy-Tican L, Turcu I, Lucaciu MC. Quantitative evaluation of yield and lysis in electrofusion of oat and wheat mesophyll protoplasts. Rom Agri Res. 1998; 9-10:27-33

[142] Smyda-Dajmund P, Sliwka J, Wasilewicz-Flis I, et al. Genetic composition of interspecific potato somatic hybrids and autofused 4x plants evaluated by DArT and cytoplasmic DNA markers. Plant Cell Rep. 2017; 35:1345-1358. DOI: 10.1007/s00299-016-1966-2

[143] Lössl A, Adler N, Horn R, et al. Chondriome-type characterization of potato: mt α, β, γ, δ, ε and novel plastidmitochondrial configurations in somatic hybrids. Theor Appl Genet. 1999; 98:1-10. DOI: doi.org/10.1007/s001220051202

[144] Lössl A, Götz M, Braun A, et al. Molecular markers for cytoplasm in potato: male sterility and contribution of different plastid-mitochondrial configurations to starch production. Euphytica. 2000; 116:221-230. DOI: 10.1023/A:1004039320227

[145] Rakosy-Tican E, Thieme R, König J, et al. Introgression of two broad-spectrum late blight resistance genes, Rpi-Blb1 and Rpi-Blb3, from *Solanum bulbocastanum* Dun plus race-specific R genes into potato pre-breeding lines. Front Plant Sci. 2020; 11:699-716. DOI: 10.3389/fpls.2020.00699

[146] Rokka VM, Laurila J, Tauriainen A, et al. Glycoalkaloid aglycone accumulations associated with infection by *Clavibacter michiganensis* ssp. Sepedonicus in potato species *Solanum acaule* and *Solanum tuberosum* and their interspecific somatic hybrids. Plant Cell Rep. 2005; 23(10-11):683-691. DOI: 10.1007/s00299-004-0868-x

[147] Tek AL, Stevenson WR, Helgeson JP, et al. Transfer of tuber soft rot and early blight resistances from *Solanum brevidens* into cultivated potato. Theor Appl Genet. 2004; 109:249-254. DOI: 10.1007/s00122-004-1638-4

[148] Chen L, Guo X, Xie C, et al. Nuclear and cytoplasmic genome components of *Solanum tuberosum* + *S. chacoense* somatic hybrids and three SSR alleles related to bacterial wilt resistance. Theor Appl Genet. 2013; 126(7):1861-1872. DOI: 10.1007/s00122-013-2098-5

[149] Chen L, Guo X, Wang H, et al. Tetrasomic inheritance pattern of the pentaploid *Solanum chacoense* (+) *S. tuberosum* somatic hybrid (resistant to bacterial wilt) revealed by SSR detected

alleles. Plant Cell Tiss Organ Cult. 2016; 127:315-323. DOI: 10.1007/s11240-016-1051-0

[150] Wang H, Cheng Z, Wang B, et al. Combining genome composition and differential gene expression analyses reveals that SmPGH1 contributes to bacterial wilt resistance in somatic hybrids. Plant Cell Rep. 2020; 39:1235-1248. DOI: 10.1007/s00299-020-02563-7

[151] Fock I, Collonnier C, Lavergne D, et al. Evaluation of somatic hybrids of potato with *Solanum stenotomum* after a long-term in vitro conservation. Plant Physiol Biochem. 2007; 45(3-4):209-215. DOI: 10.1016/j.plaphy.2007.02.004

[152] Ahn YK, Park TH. Resistance to common scab developed by somatic hybrids between *Solanum brevidens* and *Solanum tuberosum*. Acta Agric Scand Sect B. 2013; 63(7):595-603. DOI: 10.1080/09064710.2013.829867

[153] Thomson AJ, Taylor RT, Pasche JS, et al. Resistance to *Phytophthora erythroseptica* and *Pythium ultimum* in a potato clone derived from *S. berthaultii* and *S. etuberosum*. Am J Potato Res. 2007; 84:149-160. DOI: 10.1007/BF02987138

[154] Colton LM, Groza HI, Wielgus SM, et al. Marker-assisted selection for the broad-spectrum potato late blight resistance conferred by gene RB derived from a wild potato species. Crop Sci. 2006; 46:589-594. DOI: 10.2135/cropsci2005.0112

[155] Iovene M, Savarese S, Cardi T, et al. Nuclear and cytoplasmic genome composition of *Solanum bulbocastanum* (+) *S. tuberosum* somatic hybrids. Genome. 2007; 50:443-450. DOI: 10.1139/g07-024

[156] Chen Q, Li HY, Shi YZ, et al. Development of an effective protoplast fusion system for production of new potatoes with disease and insect resistance using Mexican wild potato species as gene pools. Can J Plant Sci. 2007; 88(4):611-619. DOI: 10.4141/CJPS07045

[157] Chandel P, Tiwari JK, Ali N, et al. Interspecific potato somatic hybrids between *Solanum tuberosum* and *S. cardiophyllum*, potential sources of late blight resistance breeding. Plant Cell Tiss Organ Cult. 2015; 123:579-589. DOI: 10.1007/s11240-015-0862-8

[158] Espejo R, Cipriani G, Rosel G, et al. Somatic hybrids obtained by protoplast fusion between *Solanum tuberosum L. subsp*. tuberosum and the wild species *Solanum circaeifolium* Bitter. Rev Peru Biol. 2008; 15(1):73-78.

[159] Szczerbakowa A, Tarwacka J, Oskiera M, et al. Somatic hybridization between the diploids of *S.* x *michoacanum* and *S. tuberosum*. Acta Physiol Plant. 2010; 32:867-873. DOI: 10.1007/s11738-010-0472-3

[160] Smyda P, Jakuczun H, Dębski K, et al. Development of somatic hybrids *Solanum* X *michoacanum* Bitter. (Rydb.) (+) *S. tuberosum* L. and autofused 4 X *S.* X *michoacanum* plants as potential sources of late blight resistance for potato breeding. Plant Cell Rep. 2013; 32:1231-1241. DOI: 10.1007/s00299-013-1422-5

[161] Szczerbakowa A, Maciejewska U, Zimnoch-Guzowska E, et al. Somatic hybrids *Solanum nigrum* (+) *S. tuberosum*: morphological assessment and verification of hybridity. Plant Cell Rep. 2003; 21:577-584. DOI: 10.1007/s00299-002-0555-8

[162] Polzerova P, Patzak J, Greplova M. Early characterization of somatic hybrids from symmetric protoplast electrofusion of *Solanum pinnatisectum* Dun. and *Solanum tuberosum* L. Plant Cell Tiss Organ Cult. 2011; 04:163-170. DOI: 10.1007/s11240-010-9813-6

[163] Sarkar D, Tiwari JK, Sharma S, et al. Production and characterization of

somatic hybrids between *Solanum tuberosum* L. and *S. pinnatisectum* Dun. Plant Cell Tiss Organ Cult. 2011; 107:427-440. DOI: 10.1007/s11240-011-9993-8

[164] Tiwari JK, Poonam, Kumar V, et al. Evaluation of potato somatic hybrids of dihaploid *S. tuberosum* (+) *S. pinnatisectum* for late blight resistance. Potato J. 2013; 40 (2):176-179

[165] Tiwari JK, Luthra SK, Devi S, et al. Development of advanced back-cross progenies of potato somatic hybrids and linked ISSR markers for late blight resistance with diverse genetic base-First ever produced in Indian potato breeding. Potato Journal. 2018; 45:17-27.

[166] Tiwari JK, Rawat S, Luthra SK, et al. Genome sequence analysis provides insights on genomic variation and late blight resistance genes in potato somatic hybrid (parents and progeny). Mol Biol Rep. 2021; 48: 623-635. DOI: 10.1007/s11033-020-06106-x

[167] Greplova M. Isolation, cultivation and fusion of protoplasts of *Solanum* genera [thesis]. Olomouc: Palacký University Olomouc; 2010

[168] Tarwacka J, Polkowska-Kowalczyk L, Kolano B, et al. Interspecific somatic hybrids *Solanum villosum* (+) *S. tuberosum*, resistant to *Phytophthora infestans*. J Plant Physiol. 2013; 170(17):1541-1548. DOI: 10.1016/j.jplph.2013.06.013

[169] Nouri-Ellouz O, Gargouri-Bouzid R, Sihachakr D, et al. Production of potato intraspecific somatic hybrids with improved tolerance to PVY and *Pythium aphanidermatum*. J Plant Physiol. 2006; 163(12):1321-1332. DOI: 10.1016/j.jplph.2006.06.009

[170] Kim-Lee H, Moon JS, Hong YJ, et al. Bacterial wilt resistance in the fusion hybrids between haploid of potato and *Solanum commersonii*. Am J Potato Res. 2005; 82:129-137. DOI: 10.1007/BF02853650

[171] Novy RG, Gillen AM, Whitworth JL. Characterization of the expression and inheritance of potato leafroll virus (PLRV) and potato virus Y (PVY) resistance in three generations of germplasm derived from *Solanum etuberosum*. Theor App Genet. 2007; 114:1161- 1172. DOI: 10.1007/s00122-007-0508-2

[172] Novy RG, Alvarez JM, Corsini DL, et al. Resistance to PVY, PLRV, PVX, green peach aphid, Colorado potato beetle, and wireworm in the progeny of a tri-species somatic hybrid. Am J Potato Res. 2004; 81(1):77-78

[173] Gavrilenko T, Thieme R, Heimbach U, et al. Fertile somatic hybrids of *Solanum etuberosum* (+) dihaploid *Solanum tuberosum* and their backcrossing progenies: relationships of genome dosage with tuber development and resistance to potato virus Y. Euphytica. 2003; 131:323-332. DOI: 10.1023/A:1024041104170

[174] Tiwari JK, Poonam Sarkar D, Pandey SK, et al. Molecular and morphological characterization of somatic hybrids between *Solanum tuberosum* L. and *S. etuberosum* Lindl. Plant Cell Tiss Organ Cult. 2010; 103: 175-187. DOI: 10.1007/s11240-010-9765-x

[175] Molnár I, Besenyei E, Thieme R, et al. Mismatch repair deficiency increases the transfer of antibiosis and antixenosis properties against Colorado potato beetle in somatic hybrids of *Solanum tuberosum + S. chacoense*. Pest Manag Sci. 2017; 73:1428-1437. DOI:10.1002/ps.4473

[176] Brown CT, Mojtahedi H, James S, et al. Development and evaluation of potato breeding lines with introgressed resistance to Columbia root-knot nematode (*Meloidogyne chitwoodi*). Am J Potato Res. 2006; 83(1):1-8. DOI: 10.1007/BF02869604

[177] Zhang LH, Mojtahedi H, Kuang H, et al. Marker-assisted selection of

Columbia root-knot nematode resistance introgressed from *Solanum bulbocastanum*. Crop Sci. 2007; 47:2021-2026. DOI: 10.2135/cropsci2007.01.0003

[178] Novy RG, Alvarez JM, Sterret SB, et al. Progeny of a tri-species potato somatic hybrid express resistance to wireworm in eastern and western potato production regions of the US. Am J Potato Res. 2006; 83(1):126

[179] Kammoun M, Bouallous O, Ksouri MF, et al. Agro-physiological and growth response to reduced water supply of somatic hybrid potato plants (*Solanum tuberosum* L.) cultivated under greenhouse conditions. Agricultural Water Management. 2018; 203:9-19. DOI: 10.1016/j.agwat.2018.02.032

[180] Rakosy-Tican E, Lörincz-Besenyei E, Molnár I, et al. New phenotypes of potato co-induced by mismatch repair deficiency and somatic hybridization. Front Plant Sci. 2019; 10:3. DOI: 10.3389/fpls.2019.00003

[181] Tu W, Dong J, Zou Y, et al. Interspecific potato somatic hybrids between *Solanum malmeanum* and *S. tuberosum* provide valuable resources for freezing-tolerance breeding. Research Square. 2021. DOI: 10.21203/rs.3.rs-307076/v1

[182] Bidani A, Nouri-Ellouz O, Lakhoua L, et al. Interspecific potato somatic hybrids between *Solanum berthaultii* and *Solanum tuberosum* L. showed recombinant plastome and improved tolerance to salinity. Plant Cell Tiss Organ Cult. 2007; 91:179-189. DOI: 10.1007/s11240-007-9284-6

[183] Jbir-Koubaa R, Charfeddine S, Bouaziz D, et al. Enhanced antioxidant enzyme activities and respective gene expressions in potato somatic hybrids under NaCl stress. Biologia plantarum. 2019; 63:633-642. DOI: 63. 10.32615/bp.2019.075

[184] Kumar A, Cocking EC. Protoplast fusion, a novel approach to organelle genetics in higher plants. Am J Bot. 1987; 74:1289-1303. DOI: 10.2307/2444164

[185] Rakosy-Tican L, Aurori A, Aurori CM, et al. Transformation of wild *Solanum* species resistant to late blight by using reporter gene *gfp* and *msh2* genes. Plant Breed Seed Sci. 2004; 50:119-128

[186] Thach NQ, Frei U, Wenzel W. Somatic fusion for combining virus resistance in *Solanum tuberosum*. L. Theor Appl Genet. 1993; 85:863-867. DOI: 10.1007/bf00225030

[187] Austin S, Lojkowska E, Ehlenfeldt MK, et al. Fertile interspecific somatic hybrids of *Solanum*: a novel source of resistance to *Erwinia* soft rot. Phytopathology. 1988; 78:1216-1220. DOI: 10.1094/Phyto-78-1216

[188] Mattheij WM, Eijlander R, De Koning JRA, et al. Interspecific hybridization between the cultivated potato *Solanum tuberosum* subspecies tuberosum L. and the wild species *S. ciraeifolium* subsp ciraeifolium Bitter exhibiting resistance to *Phytophthora infestans* (Mont.) de Bary and *Globodera pallida* (Stone) Behrens. I. Somatic hybrids. Theor Appl Genet1992; 83:459-466. DOI: DOI: 10.1007/BF00226534

[189] Helgeson JP, Pohlman JD, Austin S, et al. Somatic hybrids between *Solanum bulbocastanum* and potato: a new source of resistance to late blight. Theor Appl Genet. 1998; 96:738-742. DOI: 10.1007/s001220050796

[190] Song J, Bradeen JM, Naess SK, et al. Gene RB cloned from *Solanum bulbocastanum* confers broad spectrum resistance to potato late blight. Proc Natl Acad Sci USA. 2003; 100:9128-9133. DOI: 10.1073/pnas.1533501100

[191] Lozoya-Saldana H, Belmar-Diaz C, Bradeen JM, et al. Characterization of

Phytophthora infestans isolates infecting transgenic and somatic hybrid potatoes resistant to the pathogen in the Toluca Valley, Mexico. Am J Potato Res. 2005; 82:79

[192] Naess S, Bradeen J, Wielgus S, et al. Analysis of the introgression of *Solanum bulbocastanum* DNA into potato breeding lines. Mol Genet Genom. 2001; 265:694-704. DOI: 10.1007/s004380100465

[193] Van der Vossen EAG, Sikkema A, Lintel Hekkert B, et al. An ancient R gene from the wild potato species *Solanum bulbocastanum* confers broad-spectrum resistance to *Phytophthora infestans* in cultivated potato and tomato. Plant J. 2003; 36:867-882. DOI: 10.1046/j.1365-313x.2003.01934.x

[194] Oosumi T, Rockhold D, Maccree M, et al. Gene Rpi-bt1 from *Solanum bulbocastanum* confers resistance to late blight in transgenic potatoes. Am J Potato Res. 2009; 86:456-465. DOI: 10.1007/s12230-009-9100-4

[195] Lokossou AA, Park T-H, van Arkel G, et al. Exploiting knowledge of R/Avr genes to rapidly clone a new LZ-NBS-LRR family of late blight resistance genes from potato linkage group IV. Mol Plan-Microbe Interact. 2009; 22(6):630-641. DOI: 10.1094/MPMI-22-6-0630

[196] Orbegozo J, Roman ML, Rivera C, et al. Rpi-blb2 gene from *Solanum bulbocastanum* confers extreme resistance to late blight disease in potato. Plant Cell Tiss Organ Cult. 2016; 125(2):169-281. DOI: 10.1007/s11240-016-0947-z

[197] Fock I, Collonnier C, Luisetti J, et al. Use of *Solanum stenotomum* for introduction of resistance to bacterial wilt in somatic hybrids of potato. Plant Physiol Biochem. 2001; 39:899-908. DOI: 10.1016/S0981-9428(01)01307-9

[198] Rokka VM. Potato Haploids and Breeding. In: Touraev A, Forster BP, Jain SM, editors. Advances in haploid production in higher plants. Heidelberg: Springer; 2009. p. 199-208. DOI: 10.1007/978-1-4020-8854-4_17

[199] Cooper-Bland S, DeMaine MJ, Fleming ML, et al. Synthesis of intraspecific somatic hybrids of *Solanum tuberosum*: assessments of morphological, biochemical and nematode (*Globodera pallida*) resistance characteristics. J Exp Bot. 1994; 45:1319-1325. DOI: 10.1093/jxb/45.9.1319

[200] Helgeson JP, Haberlach GT. Somatic hybrids of *Solanum tuberosum* and related species. In: Altman A, Ziv M, Izhar S, editors. Plant biotechnology and in vitro biology in the 21st century, current plant science and biotechnology in agriculture, vol 36. Heidelberg: Springer; 1999. p. 151-154. DOI: 10.1007/978-94-011-4661-6_35

[201] Gavrilenko T. Potato cytogenetics. In: Vreugdenhil D, editor. Potato biology and biotechnology: Advances and perspectives. Amsterdam: Elsevier BV; 2007. p. 203-216. DOI: 10.1016/B978-044451018-1/50052-X

[202] Orczyk W, Przetakiewicz J, Nadolska-Orczyk A. Somatic hybrids of *Solanum tuberosum*— application to genetics and breeding. Plant Cell Tiss Organ Cult. 2003; 74:1-13. DOI: 10.1023/A:1023396405655

[203] Lakshmanan PS, Eeckhaut T, Deryckere D, et al. Asymmetric somatic plant hybridization: status and applications. Am J Plant Sci. 2013; 4:1-10. DOI: 10.4236/ajps.2013.48A001

[204] Grosser JW, Gmitter FG. Protoplast fusion for production of tetraploids and triploids: applications for scion and rootstock breeding in citrus. Plant Cell, Tissue Organ Cult. 2011; 104:343-357. DOI: 10.1007/s11240-010-9823-4

[205] Dudits D, Maroy E, Praznovszky T, et al. Transfer of resistance traits from carrot into tobacco by asymmetric

somatic hybridization: regeneration of fertile plants. Proc Natl Acad Sci USA. 1987; 84:8434-8438. DOI: 10.1073/pnas.84.23.8434

[206] Oberwalder B, Schilde-Rentschler L, Ruo B, et al. Asymmetric protoplast fusion between wild species and breeding lines of potato—effect of recipients and genome stability. Theor Appl Genet. 1998; 97:1347-1354. DOI: 10.1007/s001220051028

[207] Hall RD, Rouwendal GJA, Krens FA. Asymmetric somatic cell hybridization in plants. I. The early effects of (sub) lethal doses of UV and gamma irradiation on the cell physiology and DNA integrity of cultured sugarbeet (Beta vulgarris L.) protoplasts. Mol Gen Genet. 1992; 234:306-314.

[208] Gleba YY, Hinnisdaels S, Sidorov VA, et al. Intergeneric asymmetric hybrids between *Nicotiana plumbaginifolia* and *Atropa belladonna* obtained by gamma-fusion. Theor Appl Genet. 1998; 76(5):760-766. DOI: 10.1007/BF00303523

[209] Ramulu KS, Dijkhuis P, Famelaer I, Cardi T, Verhoeven HA. Cremart: a new chemical for efficient induction of micronuclei in cells and protoplasts for partial genome transfer. Plant Cell Rep. 1994; 13:687-691. DOI: 10.1007/BF00231625

[210] Valkonen JPT, Xu Y-S, Rokka V-M, et al. Transfer of resistance to potato leaf roll virus, potato virus Y and potato virus X from *Solanum brevidens* to *S. tuberosum* through symmetric and designed asymmetric somatic hybridization. Ann Appl Biol. 1994; 124:351-362. DOI: 10.1111/j.1744-7348.1994.tb04139.x

[211] Wolters AMA, Koornneef M, Gilissen LJW. The chloroplast and mitochondrial DNA type are correlated with the nuclear composition of somatic hybrid calli of *Solanum tuberosum* and *Nicotiana plumbaginifolia*. Curr Genet. 1993; 24:260-267. DOI: 10.1007/BF00351801

[212] Ali SNH, Juigen DJ, Ramanna MS, et al. Genomic in situ hybridization analysis of a trigenomic hybrid involving *Solanum* and *Lycopersicon* species. Genome. 2000; 44:299-304. DOI: 10.1139/g00-114

[213] Lusser M, Parisi C, Plan D, et al. Deployment of new biotechnologies in plant breeding. Nat Biotechnol. 2012; 30:231-239 DOI: 10.1038/nbt.2142

[214] Jinek M, Chylinski K, Fonfara I, et al. A programmable dual-RNA-guided DNA endonuclease in adaptive bacterial immunity. Science. 2012; 337:816-821. DOI: 10.1126/science.1225829

[215] Kawall K. New possibilities on the horizon: Genome editing makes the whole genome accessible for changes. Frontiers in Plant Sci. 2019; 10: 525. DOI: 10.3389/fpls.2019.00525

[216] Wang S, Zhang S, Wang W, et al. Efficient targeted mutagenesis in potato by the CRISPR/Cas9 system. Plant Cell Rep. 2015; 34:1473-1476. DOI: 10.1007/s00299-015-1816-7

[217] Butler NM, Atkins PA, Voytas DF, et al. Generation and inheritance of targeted mutations in potato (*Solanum tuberosm* L.) using the CRISPR/Cas system. PLoS ONE. 2015; 10(12):e0144591. DOI: 10.1371/journal.pone.0144591

[218] Butler NM, Baltes NJ, Voytas DF, et al. Geminivirus mediated genome editing in potato (*Solanum tuberosum* L.) using sequence-specific nucleases. Front Plant Sci. 2016; 7:1045. DOI: 10.3389/fpls.2016.01045

[219] Tang L, Zeng Y, Du H, et al. CRISPR/Cas9-mediated gene editing in human zygotes using Cas9 protein. Mol Genet Genomics. 2017; 292(3):525-533. DOI: 10.1007/s00438-017-1299-z

[220] Andersson M, Turesson H, Olsson N, et al. Genome editing in potato

via CRISPR-Cas9 ribonucleoprotein delivery. Physiol Plant. 2018; 164:378-384. DOI: 10.1111/ppl.12731

[221] Nadakuduti SS, Buell CR, Daniel F, et al. Genome editing for crop improvement – Applications in clonally propagated polyploids with a focus on potato (*Solanum tuberosum* L.). Front Plant Sci. 2021; 9: 1607. DOI: 10.3389/fpls.2018.01607

[222] Voytas DF. Plant genome engineering with sequence-specific nucleases. Annu Rev Plant Biol. 2013; 64:327-350. DOI: 10.1146/annurev-arplant-042811-105552

[223] Draffehn A, Meller S, Li L, et al. Natural diversity of potato (*Solanum tuberosum*) invertases. BMC Plant Biol. 2010; 10: 271. DOI: 10.1186/1471-2229-10-271

[224] Clasen BM, Stoddard TJ, Luo S, et al. Improving cold storage and processing traits in potato through targeted gene knockout. Plant Biotechnol J. 2015; 14(1):169-176. DOI: 10.1111/pbi.12370

[225] González MN, Massa GA, Andersson M, et al. Reduced enzymatic browning in potato tubers by specific editing of a polyphenol oxidase gene via ribonucleoprotein complexes delivery of the CRISPR/Cas9 system. Front Plant Sci. 2019; 10:1649. DOI: 10.3389/fpls.2019.01649

[226] Kieu NP, Lenman M, WangES, et al. Mutations introduced in susceptibility genes through CRISPR/Cas9 genome editing confer increased late blight resistance in potatoes. Scientific Reports. 2021; 11:4487. DOI: 10.1038/s41598-021-83972-w

[227] Genome editing database [Internet]. Available from: http://plantcrispr.org/cgi-bin/crispr/index.cgi [Accessed: 2021-04-30]

[228] Committee on Genetically Engineered Crops. Genetically Engineered Crops: Experiences and Prospects [Internet]. 2016. Available from: http://nap.edu/23395 [Accessed: 2021-04-28]

[229] Fresh Fruit Portal [Internet]. 2021. Available from: https://www.freshfruitportal.com/news/2021/04/26/usda-deregulates-simplots-genetically-engineered-potato/ [Accessed: 2021-04-26]

Section 5

Potato Seeds

Chapter 15

Use of Quality Potato Seeds in Family Farming Systems in the Highlands Zones of Peru

Rember Pinedo-Taco, Percy Rolando Egusquiza-Bayona and Dylan Anderson-Berens

Abstract

In the Andean region of Peru, the predominant production system for potatoes is family farming, oriented towards self-consumption, seed provision, and the sale of surplus production. Labor force activities for land preparation, sowing, maintenance, harvest and postharvest are under the responsibility of the family and eventually they hire farm laborers, when parcels are of a considerable size. Approximately 95% of the cultivated surface of potato crops is located in the high Andean zone, from 3000 to 4200 meters above sea level (masl), employing native varieties of tuber seeds and modern seeds introduced to production systems in the past 50 years. Potato systems in Peru, like the majority of underdeveloped countries, are characterized by the co-existence of formal and informal systems. Formal systems prioritize production and commercialization of seeds of just a few varieties positioned in modern markets which are regulated and accredited by a certification body according to the current legislation, while in the informal system the guarantee of seed quality falls under the responsibility of the very producers and users of those seeds.

Keywords: certified seed, family farming, seed regulation, *Solanum tuberosum*, traditional seed

1. Introduction

Peru is considered to be a center of origin and diversity of edible food species, and among them the potato (*Solanum tuberosum* L), a crop initially domesticated in the northern area of Lake Titicaca where among wild species the first cultivated forms were selected 7,000–10,000 years AD [1–4]. According to the International Potato Center (CIP), in Peru there can be found nine cultivated species of potatoes that originate from wild species of the group with *S. candolleanum* as a potential ancestor [5]. Other studies indicate that, in the Peruvian Andes 8 cultivated species can be found and around 200 wild species, which give rise to a large diversity of cultivated species included in polyploid series (2n = 24, 36, 48, and 60) with approximately 4200 varieties or morphotypes of native potatoes, recognized worldwide for their high nutritional value and potential for genetic improvement [4, 6, 7].

Starting in the second half of the 16th century, the Spanish began a process of world expansion of potato cultivation. Initially the first tubers were brought from Peru to the Canary Islands and thereafter to Spain, the United Kingdom, the Netherlands, India, Ethiopia, and Saudi Arabia [6]. Currently the potato on a global level is the crop with the fourth largest cultivated surface area after maize, wheat and rice, and therefore, forms a critical part of the global food system [8, 9]. In the high Andean zone in Peru, for the majority of farmers, the potato continues to be a basic food crop and is produced through traditional techniques such as the use of numerous different varieties and cultivated species with different spatial and temporal distribution.

It is the main transitory crop with approximately 367,000 ha of area planted, directly involving 710,000 families [6–8, 10, 11]. They are grown from sea level to 4200 meters above sea level depending on the adaptation of the varieties, the production system (traditional or conventional) and the destination of production for self-consumption or sale to the market [6, 12]. Modern varieties such as INIA 303 Canchan and UNICA adapt very well to the agro climatic conditions of the coast and were 'Andeanized' or adapted by family farming and they are cultivated from sea level, in the inter-Andean valleys and part of the plateau up to 3800 meters above sea level. The native varieties of various cultivated species are sown under the rainfall regime of 3,500 to 3800 meters above sea level, and bitter varieties of the species *S.* X *juzepczukii* y *S.* x *curtilobum* from 3800 to 4200 meters above sea level.

According to data from the last National Agricultural Census (CENAGRO-2012), family farming represents 97% of the more than 2.2 million agricultural units (AU), concentrated mainly in the Sierra region [13]. According to the Encuesta Nacional de Hogares (ENAHO - National Household Survey), Family Farming (FF) generates about 80% of the food products consumed in the national market. In order to recognize and strengthen small famers in rural areas, the government promulgated the Estrategia Nacional de Agricultura Familiar (ENAF- National Strategy for Family Farming); however, it does not include plans for the sustainability of potato cultivation based on genetic and ecological potential, to favor access and use of quality seeds, ENAF-2015 [14].

Other temporary interventions were implemented by some non-governmental organizations (NGOs), international technical cooperation organizations, the church and local governments; the most notable being Semillas Andinas (Andean Seeds) implemented by MINAGRI and FAO between 2011 and 2016. The project, with a high component of capacity building in seed technology, legislation and business plans, managed to overcome paradigms that small farmers would not be able to produce high quality certified seed. Family farmer system producer organizations that were trained in the production and use of quality seed through the field school methodology (ECA), managed to increase potato crop yields by 64%, exceeding the national average [15]. However, failure to follow-up by the agricultural sector and the lack of strategies for the continuity of successful models impeded the consolidation and autonomy of seed producing organizations [10, 11, 15].

Various studies carried out on the sustainability of potato cultivation in the high Andean zone agree that the use of low-quality seeds is the main factor that explains the low yields [15, 16]. For the year 2018, the national average yield was 15.76 t/ha; while, at the level of small farmers in the High Andean zone, yields are below 8; considering that almost all potato production is located in the Sierra (90%) it deserves immediate attention [8]. This situation is less overwhelming for potato producers in Latin American countries such as Argentina, Brazil, Colombia, Chile and Venezuela, which reach average yields of 20.86 t/ha. These differences may be due to environmental conditions, technology, management, but are mainly due to the use of quality seed [6, 8].

Much of the yield gap that currently limits productivity in low-income countries is attributed to poor seed quality. The availability, access and use of quality seeds of adaptable crop varieties are of vital importance to improve agricultural productivity, ensure food security and improve farmers' livelihoods. However, despite the advances in research and development of varieties, the rate of use of quality seed is low in traditional systems [9].

Formal systems promote the production and use of certified seed, generally of modern or hybrid varieties, preferred in conventional production systems that use external inputs such as pesticides, fertilizers and intensive soil tillage. Likewise, this system is governed by seed legislation supervised by the seed authority in charge of public entities. On the other hand, informal potato seed systems use native seeds or the so-called local, artisanal or 'personal seeds', with an ancestral dynamic such as the exchange of seeds, seed production in high areas to reduce the risk of virus infection and, therefore, for various reasons, they often produce relatively high quality seeds. The informal seed system can be complex; however, there are many links between the two prevalent systems in Peru [9, 11].

The coexistence of formal and informal seed systems is evident in the case of potato cultivation, which began with the enactment of the Ley General de Semillas (LGS- General Seed Law) in order to promote research, production, marketing and the use of quality seeds. The LGS regulations prioritize the production and commercialization of certified seeds with a clear objective of formalizing all seed production; however, in the last 30 years the rate of use of certified potato seed has not exceeded 1%. Consequently, this process has favored technology transfer processes, including the introduction of modern potato varieties that were adopted with relative success in family farming systems.

Family farming can contribute significantly to the development of the formal seed sector, not only in increasing the rate of use of certified seed, but also in the production of seed of native and modern varieties [10, 15]. However, the sustainability of the system will depend on a favorable environment, based on adequate seed legislation, efficient services, technical assistance, training, and technology transfer both at the level of producers and seed users [11, 15].

2. Family farming systems

Family farming is of high importance for food security, generation of agricultural employment, poverty alleviation, conservation of biodiversity and cultural traditions of communities in Latin America and throughout the world [10]. They are agricultural holdings with a predominance of use of the family labor force, where the administration of the economic-productive unit is assigned to the head of the household [17]. Another distinctive characteristic of other forms of agriculture is the limited access to land, water and capital resources; multiple-income survival strategy and heterogeneity [14].

Unlike other production systems, family farming presents a high degree of flexibility, dedicating efforts to work according to the situation and especially according to market prices. The management of production systems using the logic of crop diversification allows this flexibility, and is a factor that contributes to the economic stability of the sector. Likewise, it includes on-farm and off-farm activities (temporary work on other farms, mining and other activities), generating economic income in rural or urban areas. These activities are carried out in dynamic interrelation with the social, economic, cultural and environmental circumstances. Hence, it is inseparable from the family production unit, since it has the same resources at its disposal, and decisions about employment influence both in the family and the productive unit.

In family farming, the size of the farm and/or agricultural production is a determining factor for its classification [14, 17]. This classification includes subsistence family farming, families without land and family farmers fulfilling existing demand and generating surpluses. Family Farming (**Table 1**), depending on its land resources, cultivated area, technology and access to the market, can be classified as Subsistence Family Farming (SFF), intermediate or transitional Family Farming (IFF) or Consolidated Family Farming (CFF).

2.1 Potato cultivation in high Andean production systems

Peru is considered one of the main centers of genetic diversity and variability of cultivated and wild potato species and it is known that an intensive process of

Variable	Subsistence family farming	Intermediate family farming	Consolidated family farming
Production system	Traditional, use of supplies and local technology	Mixed production system: traditional and conventional	Conventional production
Farming system	Diversified crops, more than two crops in one plot and more than two species per crop	Monoculture, one variety	Single crop monoculture
Workforce	Family or community	Family and local farmers	Local laborers contractors or from other localities
Type of exploitation	(150–200 wages/ha)	80–120 wages/ha	60–100 wages/ha
Water dependency	Under temporary regime, it depends on rains	Temporary regime and irrigation	Temporary regime and irrigation
Seed system	Informal, the use of traditional seed	Informal/formal	Formal
Environmental conditions	Vulnerability to water extremes (drought, frost, hail or excess of rain)	Variable but less adverse conditions	Variable but less adverse conditions
Seed use	Native; wide variability available, “mixed sowing” (Chagro, Huachuy)	Native and modern selected by market demand	Native and modern selected by market demand, urban distributors, or processing companies
Type of Production and exploitation of land	Smallholding, planting area less than 1 ha	Small and medium farmers of 1 to 4 ha	Medium to large farmers from 10 to 150 ha
Efficiency	2–10 t / ha (extremes from 0 to 15 t / ha)	15–25 t / ha (extremes from 8 to 30 t / ha)	20–35 t / ha (extremes from 10 to 50 t / ha)
Market	Far from urban centers or markets	Medium proximity to urban centers or markets	Close proximity to urban markets
Cost of production	US $ 600–1200 / ha	US $ 1500–2500 / ha	US $ 3000–4500 / ha
Endpoints production	Production for self-consumption, exchange and less than 20% sale to the market	80% to the market	To the market (100%)

Table 1.
Characteristics of family farming potato production systems and their relationship with seed production and use.

domestication of cultivated species took place from species that gave rise to various potato morphotypes called native varieties that are characterized by their various shapes, colors, and size. Archeological remains suggest that Pre-Columbian and Inca cultures were possibly those that contributed to selecting and developing the cultivated species that are currently known. Among the species cultivated in Peru are; *Solanum tuberosum sp. andigena*, *Solanum goniocalyx*, *Solanum stenotomum*, *Solanum x chaucha*, *Solanum x ajanhuiri*, *Solanum x juzepczuki*. As a result, farmers have a great diversity of potatoes that contribute to their resilience and food security strategies. For example, the diversified multi-variety planting with more than two cultivated species of potato in a single plot in the high Andean zone called Chaqro, is a strategy that allows them to manage and mitigate the effects of climatic risks such as frost or the incidence of pests and play an important role in the in-situ conservation of native potato diversity [18].

According to the national agricultural census carried out in 2012 (CENAGRO-2012), in Peru there are approximately 2,160,000 farmers, of which 90% have less than 10 ha and correspond to the category of small and medium farmers [13]. In the case of the high Andean zone of Peru, the average landholding per family is 0.80 ha where small farmers, especially traditional ones, practice diversified agriculture with potato crops as head crops, corn (*Zea mays* L.), wheat (*Triticum aestivum* L.), barley (*Hordeum vulgare* L.), broad bean (*Vicia faba* L.), tarwi (*Lupinus mutabilis* Sweet) and some vegetables [17, 19, 20]. Potato cultivation in Peru generates permanent work for approximately 710,000 families with 33 million daily wages in all tasks and production activities, harvest and post-harvest.

Potato production generates approximately, a corresponding demand for 4.5 million tons of seed tubers annually, thus constituting an important activity in the generation of income and the basis of their food security [6]. The average per capita consumption is 90 kg/person/year and continues with a growing trend despite the introduction of other foods, both those from government social programs and the flow of industrialized or processed products that are gradually causing changes in the rural communities diet [18].

3. Seed systems in potato cultivation

Seeds are the principal input of agriculture, regardless of the production system, the technology used, and the end product point. Consequently, the supply of seeds is a function of the predominant seed systems in a region where formal systems (certified seeds) and informal systems (non-certified seeds) coexist. Various actors from public and private institutions, producer organizations, plant breeders, service providers, technical assistants, agricultural innovation centers and universities participate in both systems; likewise, commercial potato and seed producers and traders. Seed inspectors also participate in the certification, supervision, marketing and distribution of seeds [3, 20, 15].

The coexistence of informal and formal systems is unavoidable and possibly the same actors participate in both systems; however, each has its own characteristics. Formal systems produce and use generally modern certified seed from conventional production systems (**Figure 1**); it typically works for a limited sector of farmers [3, 21]. In contrast, informal potato seed systems use native seeds or so-called local, artisanal or personal seeds, with a natural dynamic such as the exchange of seeds that dates back several centuries and, for various reasons, often produces quality seeds. Although the informal seed system is complex, there are many links between the two systems [9, 11, 22].

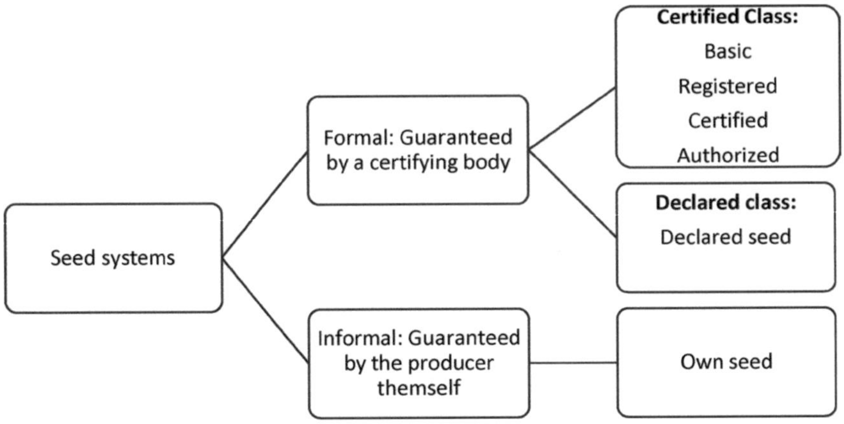

Figure 1.
Potato seed systems.

Family farming contributes significantly to the production of both certified and non-certified seeds; Currently, modern varieties are registered in the national registry of commercial cultivars, as well as native varieties with greater demand in the national market, which may be subject to certification [15].

3.1 Access to seeds categorized as higher quality

The production of elite seeds such as pre-basic and basic potato is in the charge of public and private institutions that have infrastructure and laboratories for the accelerated multiplication of potato cultivars with the highest demand in the market. Generally, the maintenance and management of a seed program has high costs because it is a highly specialized activity. In many developing countries, these functions are often performed by public sector breeding programs [10, 11, 15]. Consequently, the sustainability of certified seed production in family farming systems is weak, because it is almost impossible to maintain improvement programs with their own resources that do not allow them to access new varieties and/or produce seeds of higher categories (genetic, basic or registered).

For this reason, the producers of seeds of certified, authorized and declared categories depend on the operational and logistical capacity of the institutions or companies that produce the pre-basic and basic seeds. Delays in the availability of adequate quantity and quality of seeds can cause large bottlenecks in the production of quality seeds. Therefore, it is necessary to promote the establishment of agreements or contracts between seed producer organizations and public improvement programs that establish commitments to deliver seeds with quantity, quality and punctuality [10]. Likewise, it is necessary to strengthen the capacities of producers and users of quality seeds with training methodologies where knowledge is gained through participatory and experiential practices.

3.2 Seed quality in family farming systems

Four key characteristics have a big impact on the quality of the harvest of a potato field: physiological age, genetic purity of the seed, size and phytosanitary appearance of the seed [6, 23]. When referring to seed tubers, in the case of high Andean farmers, quality refers to the level of response of the seed to adverse agro-climatic situations. When a local variety does not respond to the expectations of the farmer who uses the seed, distrust of the seed producer begins and it is classified as poor quality or degenerate and is generally discarded. Potato degeneration is the main cause of reduction in yield or quality caused by accumulation of pathogens and pests in planting material due to successive cycles of vegetative propagation and it has been a long-standing production challenge for farmers and potato growers around the world [24]. In the Andes, seed degeneration caused by virus infection is much slower at altitudes greater than 2800 masl and even more reduced at altitudes greater than 3500 masl [6, 25].

Native cultivars managed by high Andean farmers have been maintained for several years with relatively low rates of disease spread. This is supported by studies carried out in traditional Andean potato seed systems during the last 30 years; they often found relatively low frequencies of virus-infected tubers limiting yield [6]. Farmers maintain local knowledge to define whether a seed is of quality or not; among them, knowledge of the phytosanitary status by verifying the presence of pest insects such as the Andean weevil (*Premnotryphes spp.*) that damage seed tubers; fungal diseases such as wart (*Synchytrium endobioticum*), powdery scab (*Spongospora subterranea*); and bacteria such as potato smut (*Tecaphora solani*) or blackleg (*Pectobacterium ssp.*) and finally the discoloration or dwarfism caused by viruses [16]. When farmers observe tuber deformations, they associate that the seed is tired, degenerated or aged, due to the use of a batch of seeds for several planting seasons, generally for more than five years [24].

Factors such as genetic material, agronomic management and cultural practices contribute to maintaining the quality of potato seed in the informal systems of the Andes. Agricultural practices, location of seed production fields at high altitudes, types of tillage, crop rotation, diversified planting, sectoral management and the number and height of hills are factors which can reduce the phases of diseases transmitted in the soil leading to degeneration of seed tubers [6, 26]. Other intrinsic factors of each variety such as resistance to Potato virus Y (PVY) and Potato leafroll virus (PLRV) in *S. tuberosum* ssp. *andigena* may contribute to reducing the spread of these diseases between plants or their replication within plants [6, 25].

3.3 Perception of the concept of quality seeds in the informal sector

In informal sector seed production systems, perceptions of seed quality differ from those of formal system farmers; while farmers who use their own seed rate their seeds as good, bad or fair, farmers with formal systems trust some certifying body that guarantees the quality of the seeds [3, 15]. Farmers in the high Andean areas associate quality through color, shape, size and the presence or absence of deformations in the tubers [16, 27].

Currently, with the introduction of modern varieties, this can be observed in rural communities, where good farmers are those who meet the minimum requirements for quality seed production. In the perception of quality, small farmers in the Andean highlands prioritize personal values such as safety, health and well-being and based on this logic they prefer seed tubers that reflect the characteristics of the variety in combination with seed quality signals that reflect altitude, soil and low input management and do not associate the latter with tuber seeds from the formal seed sector [28]. Seed certification cards do not substitute for perceptions of quality that farmers have learned to use and seeds they trust for generations.

A very important aspect is the leading role played by women in family farming systems in the management of agricultural units. They are not only responsible for family care and feeding the children but also for the maintenance of basic resources for family food security. Women play a key role in maintaining the genetic diversity of potatoes, particularly in the seed selection stages of the complex varieties, storage and planning of future plantings [15, 29, 30]. In the higher altitudes of Andes of Peru and Bolivia, women act as conservationists preserving the greatest possible diversity of bitter potatoes (*Solanum Jueepczukii* and *S. curtilobum*) that can be grown at temperatures as low as −3°C and can be dried to obtain traditional products such as *chuño* and *moraya* consisting of frozen, dried and dehydrated potatoes.

The selection of seeds of the native varieties is carried out based on the morphological interpretation and the in-situ yield of the crops, the culinary quality, the crop yield, the quality of the processing and the resistance to diseases, drought or flood. Consequently, controlling genetic diversity through careful management of variety and combinations allows communities to manage risks, particularly where climatic stress is more frequent and intense.

3.4 Certified seed potato production in the formal system

The production of certified seeds starts from *in vitro* seedlings with origins fully proven under specific production guidelines or standards and these stocks move through a series of steps following clear regulations and result in high-quality certified seeds [6, 31]. The aforementioned procedure includes the production of virus-free seedlings for both native and modern cultivars. However, the exclusion from the registry of commercial cultivars of other native varieties with high potential in quality and yield are limiting in the development of the seed system in the high Andean zone; this may influence the fact that a lack of virus cleaning in relegated varieties can result in low yields and low profitability [11].

In Peru, rice, cotton and hard yellow corn crops have the highest rate of use of certified seed; while, for crops such as potatoes and starchy maize, the seed use rate does not exceed 1% [8]. Therefore, it is explained that formal systems respond to the expectations of large commercial farmers and companies that prioritize a very limited number of crops with modern or hybrid varieties [21]. In the case of the cultivation of potatoes and other Andean grains, the flow of seeds works with its own dynamics and operates at different scales in local contexts which guarantee the supply of seeds [20].

In the past 10 years, there have been some favorable changes in seed legislation, improving the production, access and use of quality seeds which has gone some way towards strengthening local seed companies in relation to supporting the sustainability of family farming [21]. The renewed legislation in the regulation of specific potato seeds incorporates the declared class that is not subjected to the certification process, so the guarantee of its quality is the responsibility of the producer; additionally, it includes traditional seed categories from native potato biodiversity [31]. However, there are several important challenges to improve the institutional framework of the seed authority and to apply what is indicated in the Seed Law, which states the production and use of quality seeds is of national interest [32].

The limited supply or flow of premium seeds is unfavorable to the sustainability of individual seed producers as well as organized ways to meet local demand for quality seeds. Another aspect that influences the sustainability of the formal sector is the slow development of technology for harvesting and post-harvest aimed at family farmers, which is why some work has to be carried out manually or with rudimentary tools, reducing the efficiency of the process [11].

The Ministry of Agriculture (as the governing body of the agricultural sector), international cooperation and other private initiatives have made great efforts and investment to promote the adoption of quality seed; however, the use of certified seed has had little penetration into informal seed systems, as is currently the case in most developing countries. The rate of use of certified seed of the formal system in Peru is 0.5%; while in China and India it can reach up to 20% [8, 24]. Small farmers in developing and particularly high Andean countries continue to use seed tubers acquired through the informal seed system, that is, produced on the farm or purchased from neighbors or local markets. The formal certification system in Peru is a much smaller percentage of total seeds used (less than 5%), but it is useful to introduce genetic material free of viruses and other key pathogens that affect the quality of the seeds. This explains why in recent years the national average yield has had a significant increase from 8.5 to 15.5 t ha^{-1}, thanks to the growing adoption, access and use of quality seed tubers.

It is possible to improve the access and use of quality seeds by advocating for the capacity building of the official sector (Ministry of Agriculture), extension agents, producers and seed users. Between 2011 and 2015, FAO, together with the Ministry of Agriculture and the National Institute of Agrarian Innovation (INIA), implemented a joint initiative to promote the use of certified seeds of potato, starch corn and quinoa with farmers of family farming systems. The training method through the FFSs facilitated the process and 32 producer organizations managed to integrate into the formal sector for the production of certified seeds. The yields of the three crops indicated above increased by 50%. Agricultural organizations and individual producers that chose to use certified seed managed to increase yields by 64% in potatoes, 56% in quinoa and 31% in starchy corn respectively, a fact that has contributed to improving food security in the high Andean zone [15].

3.5 Agronomic management of potato seed production in family farming systems

Informal potato seed production systems are typically traditional, incorporating components and technologies from the local environment, as well as projects promoted by the public and private sectors. The informal system has its own characteristics and strategies to produce and supply the demand for seeds of local communities. It involves the appropriate seed size, the correct cultivars for the particular niche, utilizing certain production areas (typically in higher mountainous areas to reduce the degeneration of stocks from virus infections), a complex distribution system of seeds and good coordination between regions in Peru according to planting seasons and seed needs.

In family farming systems, the supply of seeds does not represent a great difficulty, applying traditional techniques and not necessarily using certified seed, producers can achieve a multiplication rate of 1:32, supposing that an average seed tuber of a native cultivar generates six stems. However, the seeds can have high rates of infection by bacteria, fungi, viruses and viroids. The huge reserves of native potatoes cannot always be of quality during successive cycles of vegetative propagation [9, 24, 33].

The flow of potato seeds in Peru is characterized by a horizontal and vertical distribution, a fact that contributes to the active exchange of genetic material between communities and farmers [16]. Seed exchange spaces occur at local fairs, religious festivals and community anniversaries, where not only seeds of native varieties are exchanged, but also modern varieties that were incorporated into their production systems; with however, the risk of displacing their local varieties for the best yield and demand in the markets.

3.6 Agronomic management of potato seed production

The main form of multiplication of the potato crop is through clonal propagation; however, this potato seed production system can be laborious, expensive and time consuming to accelerate seed multiplication [6, 9, 33]. Due to the low multiplication rate (generally 1: 6), it takes many years to produce significant quantities of seed to meet the demand. The clonal seed is voluminous, delicate and constitutes a vehicle for the transmission of diseases, increasing the cost of producing, handling and transporting the seed; consequently, seeds represent a high percentage of production costs: 20–25% is typical in developing countries [25].

In family farming systems, the fields for the production of potato seeds should be located between 3,000 to 4,200 masl; in addition to being part of the traditional food safety and quality management strategies. In the case of certified seed production, it is a mandatory requirement and its non-compliance may be grounds for rejection in accordance with the specific regulation of potato seed. Seed produced in cold climates exhibits a broader growth curve and greater productive potential than seed produced in hot climates [16]. In the higher altitude areas, the bitter potato varieties that belong to *S. curtilobum* y *S. juzepczukii* predominate. In the intermediate zones very diverse varieties are cultivated belonging to *S. tuberosum* subsp. *andigena, S. goniocalyx, S. stenostomum* and *S. chaucha*. The areas of lower altitude are characterized by a greater presence of modern or hybrid varieties (*S. tuberosum* subsp. *tuberosum* x *S. tuberosum* subsp. *andigena*).

Fallow systems, locally known as *laymes*, predominate in higher altitude areas. These involve communal lands subdivided into well-defined sectors that are managed under the decision of the local authorities. Rotation commonly begins with potato cultivation during the first year, followed by rest or barley cultivation in the second year [5]. The rest period fluctuates between 3 to 6 years and is characterized by the rotating use of the sectors. This applies mainly to subsistence family farming.

Soil preparation is an important practice given the requirements of the potato plant to facilitate good root development and facilitate complementary agricultural tasks such as hilling and weed management. The use of agricultural machinery and tools in Andean farming currently depends on the location and physiography of the land. In areas with steep slopes, for primary tillage, the ancestral tool called "*chakitaclla*" is commonly used, a manual tool adapted for the difficult geographical conditions of the Andes and consisting of a 1.5-inch wooden arm and metal teeth.

The selection and classification of seeds are practices that consist of selecting healthy potatoes in the first instance, then classifying by size; it is generally customary to classify seed tubers into large, medium, and small size. Due to the volume of crops, their delicate and perishable condition and the possibility of transmitting diseases during harvesting, post-harvest and seed processing, it represents a high percentage of production costs [25].

In traditional systems, sowing is done manually, where 15 to 20 daily wages per hectare are used. The amount of seed varies depending on the size of the seed used. Medium-sized seeds (40 to 60 g) are equivalent to 1800 to 2000 kg of seed per hectare. Some farmers prefer small tubers whose weight does not exceed 60 g, therefore, adjustments are made in the spacing between plants and between rows. With a distance of 20 cm between plants and 80 cm between rows, a population of 37,000 plants per ha is estimated. Whole seeds are used as it is considered that the risk of disease transmission and virus infection is reduced.

The production of seed tubers from potato botanical seed progeny is a technological alternative for obtaining abundant first-generation seed tubers of high phytosanitary and physiological quality in a short period of time. This technology can help reduce production costs and increase the multiplication rate of high-quality seeds, in addition to being kept for five to seven years with a high germination percentage equal to or greater than 70%. However, for family farming systems that mostly use native varieties, the use of botanical seed is not common due to its unstable behavior in terms of risk of segregation [6]. In addition, the true seed management process is a challenge for the environmental conditions of the Peruvian Andes due to sudden changes in temperature, precipitation, drought and other adverse factors. A seedling from true seed when affected by frost for example is totally lost, while a stem from true seed in its early stages can be recovered by a regrowth of the seed tuber.

'Hilling' is a complementary task that consists of accumulating soil in the form of a ridge or ridges at the base of the stem of each plant when the plants reach between 30 to 40 cm in height (or 55 days after sowing), in order to give the plant support, protect tubers from pests and damage by biophysical agents such as low temperatures [6, 26]. The number of hills depends on the characteristics of the soil, the variety, the sowing season and the production system; therefore, in the high Andean areas for some varieties hilling may be one or two ridges [26].

Regarding pest management, farmers integrate agronomic labor, physical, mechanical and chemical control methods. In traditional production systems they are less dependent on external inputs such as industrial pesticides. Agronomic practices are part of traditional strategies in potato production to reduce damage from biotic and agricultural factors such as crop rotation, variety selection, use of good quality seed tubers, sowing density, timing sowing depth, sowing depth, number and timing of hilling and pest management and harvest [6]. A physical control strategy includes the use of barriers with species such as Tarwi (*Lupinus mutabilis* L.) and the use of polyethylene as physical barriers to prevent the entry of *Prennotrypes ssp*. For disease control they mostly use chemical products as in the case of the oomycete *Phytopthora insfestan* (Mont.) de Bary.

To manage weeds in seed fields, farmers combine agricultural practices and mechanical and chemical methods. In family farming systems tillage is an important activity for weed control in potatoes, regardless of region or production system, and includes a wide variety of tactics ranging from simple manual weeding to the use of complicated implements of cultivation [6, 16]. Chemical control is mostly carried out in conventional production systems, but the options for active ingredients of herbicides recommended to control potato weeds are limited; while in traditional production systems it is carried out during the production practices of preparing the land and hilling. The aforementioned activities directly or indirectly involve the concept of integrated weed management where control methods with agronomic work, as well as physical and mechanical work are combined with local knowledge of weed biology [34].

Research in Peru also indicates potentially complex host-pathogen interactions in degenerative diseases. The incidence of the virus could decrease in subsequent generations (that is, not passing from an infected mother plant to all the tubers of the progeny) and that this phenomenon is strongly favored by the production of seeds in high altitude areas. The effectiveness of management practices, such as symptom-free planting selection and clearing, is highly dependent on disease detection.

When the potato crop reaches harvest maturity, farmers carry out the harvests that can be manual or mechanical in some cases, depending on the production system; however, due to the difficult geographical conditions, it is usually manual. The seed tuber storage systems are carried out according to the characteristics of the cultivar, the harvest time, and the dynamics of the seed market. In subsistence family farming systems [14], the seeds are stored indoors or on pallets, covered with straw and muña (*Minthostachys mollis*) or eucalyptus (*Eucalyptus globulus*) leaves. In intermediate family farming systems that are in the process of articulation to the markets, the harvesting and processing of seed consists of the selection, classification, bagging and labeling as long as it is certified seed. For informal systems this process may differ. Seed producers make large piles of tuber seeds and are protected from frost and sunny days with a thick layer of straw. Under these conditions storage can be successful for up to three months.

4. Conclusions

In family farming potato seed systems, informal or non-accredited seed systems predominate over formal systems with small farmers maintaining their native or traditional varieties with seeds selected from their own harvest to guarantee future planting seasons. On the other hand, in formal systems, priority is given to the production of certified seeds of modern varieties of higher categories such as basic, registered and certified -- which are regulated by a certifying body. In this context, in family farming, the use of certified seeds is not a priority or a conditioning factor for potato production; farmers maintain strategies such as the exchange of seeds that since ancient times guarantee the flow of seeds. However, the quality of the seed is not always the best, due to the high levels of genetic degeneration of its seed matter due to the low rotation or refreshment with quality seed. Consequently, the use of low-quality seeds has an impact on yield and are not self-sustaining.

Certified seeds cannot always be considered quality seeds; regulatory and institutional factors such as non-compliance with the quality parameters established in the Seed Law and the specific regulation for each class and category of seeds, adulteration of certification cards and the lack of internal control in institutions and companies that are responsible for generating seeds of the genetic and certified class weaken the seed system, generating distrust in users of quality seeds.

Author details

Rember Pinedo-Taco[1*], Percy Rolando Egusquiza-Bayona[1] and Dylan Anderson-Berens[2]

1 La Molina National Agrarian University, Lima, Peru

2 Alliance for Bioversity International and CIAT, Cali, Colombia

*Address all correspondence to: rpinedo@lamolina.edu.pe

References

[1] INIA. 2019. Sistematización de la experiencia de los subproyectos de papa financiados por el programa nacional de innovación agraria. MINAGRI. 45 p.

[2] Rosen, C.J. Gupta S. K and Souza, E.F.C. 2020. Potato. IN Souza Wien, H. C., Stützel, H. (Edits) The physiology of vegetable crops

[3] Forbes G.A., Charkowski A., Andrade-Piedra J., Parker M.L., Schulte-Geldermann E. (2020) Potato Seed Systems. In: Campos H., Ortiz O. (eds) The Potato Crop. Springer, Cham. DOI: 10.1007/978-3-030-28683-5_12

[4] Spooner DM, Ghislain M, Simon R, Jansky SH, Gavrilenko T (2014) Systematics, diversity, genetics, and evolution of wild and cultivated potatoes. Bot Rev 80:283-383. https://doi.org/10.1007/s12229-014-9146-y

[5] Centro Internacional de la Papa (CIP), y la Federación Departamental de Comunidades Campesinas (FEDECH), 2006.Catálogo de variedades de papa nativa de Huancavelica-Perú. 205 p.

[6] Egúsquiza, R. La papa en el Perú (2nd ed.). Universidad Nacional Agraria La Molina. 2014; 84 p.

[7] Soto, J., Medina, T., Aquino, Y., & Estrada, R. (2013). Diversidad genética de papas nativas (Solanum spp.) conservadas en cultivares nativos del Perú. Revista Peruana de Biología, 20(3), 215-222

[8] MINAGRI (2020). Análisis de mercado de papa. Sierra y selva exportadora. 85 p.

[9] Singh RK, Buckseth T, Tiwari JK, Sharma A K, Singh V, Kumar D, Chakrabarti, S K. Seed potato (Solanum tuberosum) production systems in India: A chronological outlook. Indian Journal of Agricultural Sciences. 2019; 89(4): 578-587.

[10] Santivañez, T, Tejada G, Aguilera, J, Mastrocola N, Pinedo R. Retos y oportunidades de la producción de semillas certificadas para la agricultura familiar en la zona andina. En Salcedo, S., Guzmán, L (Ed): Agricultura Familiar en América Latina y el Caribe: Recomendaciones de Política. FAO. 2014; 473p.

[11] Tejada G, Santivañez T, Aguilera J, Mastrocola N, Pinedo R. Experience on Certified Seed Production with Associated Family Farmers in the Highlands of Bolivia, Ecuador and Peru. En: FAO y ICRISAT. Ojiewo CO, Kugbei S, Bishaw Z.; Rubyogo, J.C. (Eds). Community Seed Production on Workshop Proceedings, 9-11 December 2013. Roma, Italia. 2015; 176 p.

[12] Rosen CJ, Gupta SK, Souza EFC. Potato. IN Souza Wien, H. C., Stützel, H. (Edits) The physiology of vegetable crops. 2020.

[13] INEI. Resultados definitivos. IV Censo Nacional Agropecuario 2012. Instituto Nacional de Estadística e Informática. Lima, Perú. 2013; 62 p.

[14] MINAGRI (Ministerio de Agricultura). ENAF 2015 Ministerio de Agricultura y Riego. Estrategia nacional de agricultura familiar 2015-2021. 2015.

[15] Pinedo R. Innovaciones tecnológicas con metodología de ECA en producción y adopción de uso de semilla certificada en sistemas de Agricultura Familiar. Tierra Nuestra. 2019; 13(1): 77-86.DOI: 10.21704/rtn.v13i1.1294

[16] Ezeta F.N. 2001. Producción de Semilla de Papa en Latinoamérica. Revista Latinoamericana de la Papa. 2001; 12:1-14

[17] Maletta, H. La pequeña agricultura familiar en el Perú. Una tipología micro

regionalizada. En IV Censo Nacional Agropecuario 2012: Investigaciones para la toma de decisiones en políticas públicas. Libro V, FAO. Lima, Perú. 2017; 179 p.

[18] Caycho-Ronco J, Arias-Mesia A, Oswald A, Esprella-Elias R, Rivera A, Yumisac F, Andrade-Piedra J. Tecnologías sostenibles y su uso en la producción de papa en la región altoandina. Revista Latinoamericana de la Papa. 2009; 15(1): 20-37

[19] Pinedo R, Gómez L, Julca A. Sostenibilidad ambiental de la producción de quinua (Chenopodium quinoa Willd.) en los valles interandinos del Perú. Cienc. Tecnol. Agropecuaria. 2020; 21 (3): e1309

[20] Almekinders CJM, Walsh S, Jacobsen KS et al. Why interventions in the seed systems of roots, tubers and bananas crops do not reach their full potential. Food Sec. 2019; 11: 23-42. DOI: 10.1007/s12571-018-0874-4

[21] Hruska, A. Sustainable Seed Systems for Family Farming: Promoting More Inclusive Public Institutions – Lessons Learned from Mesoamerica. En: FAO y ICRISAT. Ojiewo CO, Kugbei S, Bishaw Z.; Rubyogo, J.C. (Eds). Community Seed Production on Workshop Proceedings, 9-11 December 2013. Roma, Italia. 2015: 176 p.

[22] FAO/ICRISAT. Community Seed Produc on, by Ojiewo CO, Kugbei S, Bishaw Z & Rubyogo JC, eds. Workshop Proceedings, 9-11 December 2013. FAO, Rome & ICRISAT, Addis Ababa. 2015: 176 pp

[23] Merk EM. A Growers Guide to Quality Potato Seed. Creative Components. 2019: 411. https://lib.dr.iastate.edu/creativecomponents/411

[24] Thomas-Sharma S, Abdurahman A, Ali S, Andrade-Piedra JL, Bao S, Charkowski AO, Crook D, Kadian M, Kromann P, Struik PC. Torrance L, Garrett KA, Forbes GA. Seed degeneration in potato: the need for an integrated seed health strategy to mitigate the problem in developing countries. Plant Pathol. 2016; 65: 3-16. DOI: doi.org/10.1111/ppa.12439

[25] Thiele G. 1999. Sistemas Informales de Semilla de Papa en los Andes: ¿Por qué Son Importantes y qué Hacer con Ellos? 1999; 49 p.

[26] Rodríguez G, Pinedo R, Sullca F. Efecto del aporque en el rendimiento y la rentabilidad en cultivares nativos de papa. Ciencia Tecnología Agropecuaria. 2020; 21 (3): e1798

[27] Aboites G, Martínez F, Torres G. El negocio de la producción de semillas mejoradas y su rol en el proceso de privatización de la agricultura mexicana. Espiral, VI. 1999; 16:151-185

[28] Urrea-Hernandez C, Almekinders CJM, van Dam YK. Understanding perceptions of potato seed quality among small-scale farmers in Peruvian highlands, NJAS - Wageningen Journal of Life Sciences. 2016; 76: 21-28. DOI: 10.1016/j.njas.2015.11.001

[29] GRAIN. Potato: a fragile gift from the Andes GRAIN|15 September 2000|SeedlingSeptemb. 2000; https://grain.org/en/article/220-potato-a-fragile-gift-from-the-andes

[30] Sarapura S. Gender and agricultural innovation in peasant production of native potatoes in the Central Andes of Peru. PhD Thesis, University of Guelph, Canada. 2013.

[31] Davidson RD, Xie K. Seed Potato Production. IN: Navarre, R., Pavek, M. J. editors The potato: botany, production and uses. 2014; 115-132. DOI: 10.1079/9781780642802.0115

[32] INIA, 2013. Ley General de Semillas (Ley N° 27262, modificada con Decreto

Legislativo N° 1080). Reglamento General de la Ley General de Semillas (DS N° 006-2012-AG).

[33] Buckseth T, Sharma AK, Pandey KK, Singh B P, Muthuraj R. Methods of pre-basic seed potato production with special reference to aeroponics-A review. Scientia Horticulturae, 2016; 204: 79-87. DOI: 10.1016/j.scienta.2016.03.041

[34] Burke, C, Everman, WJ. Weed Control Strategies for Potato (Solanum tuberosum). IN: Navarre R, Pavek MJ editors. The potato: botany, production and uses; 2014. p. 225-236. DOI: 10.1079/9781780642802.0225

Chapter 16

Apical Rooted Cuttings Revolutionize Seed Potato Production by Smallholder Farmers in the Tropics

Peter VanderZaag, Tung Xuan Pham, Victoria Escobar Demonteverde, Cynthia Kiswa, Monica Parker, Shadrack Nyawade, Pieter Wauters and Alex Barekye

Abstract

Potato apical rooted cuttings (ARC) originating from juvenile simple rounded leaf mother plants are a significant new way of transplanting and field growing of seed potatoes under smallholder field conditions in the tropical highlands. The aim of this paper is to highlight the development of the technology by researchers and farmers in Vietnam, Philippines, Kenya and Uganda. The development of cultivars with late blight resistance for which no source of tuber seed was available stimulated the creation of using ARC. The demystification of tissue culture by the 1980s greatly aided this development. The key hurdle was to multiply tissue culture plants in beds of growing media and maintain the physiological young stage of the mother plants from which apical cuttings could be repeatedly taken for several months to produce ARC for sale to farmers who demanded the new cultivars (cvs) with all the desirable attributes. The technology was first developed in warmer climates at lower elevations of less than 1,500 meters above mean sea level (mamsl) but gradually it was successfully developed at cooler climates in East Africa. The technology is well established in the highlands of Vietnam and Philippines. The largest family operation is producing over 4 million ARC annually. These high-quality ARC along with improved cvs have markedly improved yields of smallholder farmers, improving food security and increasing their income levels. In Kenya and Uganda there is a rapid adoption of ARC by seed producers, smallholder farmers and youths. The ARC revolution is bringing a great deal of excitement and promise of prosperity to remote poor highland communities.

Keywords: Apical Rooted Cuttings, tissue culture, juvenile mother plants, seed potatoes, smallholder farmers, food security

1. Introduction

Potatoes, the third most important food crop globally, is a major crop for both food security as well as income in many tropical countries. Millions of smallholder resource poor farmers now grow potatoes. Productivity measured as harvested yields remains low in comparison to the yields obtained by European and North American farmers. The USA has an average yield of 49.8 t/ha as compared to less than 11 t/ha for countries in East Africa and 15 t/ha for the Philippines and Vietnam [1]. One of the primary reasons for this disparity is the lack of quality seed potatoes of suitable cvs. The continual reuse of seed stocks for many generations leads to severe virus infection. Bacterial Wilt infected seed caused by *Ralstonia solanacearum* further hinders potato production in the mid elevation tropical settings [2].

Traditional western seed potato production systems have been copied in many tropical countries with minimal success. Prevalence of soil borne diseases and virus diseases build up over several generations of seed propagation coupled with prolonged seed storage at warm ambient temperatures are major constraints. The amounts of tuber seed produced are a small fraction of the actual seed requirements, the cost is prohibitive and often seeds of desired cultivars are not available at the right time. Seed importation fills a small gap in the seed requirement of wealthier farmers in some countries.

Over the past 40 years, there has been a major push towards developing suitable seed systems in various countries with donor support and expertise from the International Potato Center (CIP) as well as experts from numerous national and international agencies. China has strongly supported a diverse set of measures using tissue culture, large scale greenhouses for mini tuber production both in substrate as well as in aeroponic systems. Government subsidies were a key part of this success both for capital investments as well as subsidized pricing for the buyers of the mini tubers. India has developed a large-scale seed production system which primarily benefits larger landowners who have the financial means to provide all the needed inputs to grow higher yielding crops from certified or higher standard seed potatoes grown in the northern isolated areas of the country.

Unfortunately, majority of smallholder farmers in most tropical countries did not benefit from these seed systems and yet grew potatoes as a food and income security crop. Potato production, although yields were exceptionally low, still fit into their cropping systems and was also considered a valuable cash crop. Two factors started the change in the opportunity for smallholder farmers to get better quality planting materials. During the 1970s there was an international focus stimulated by CIP on evaluating cvs with late blight (*Phytophthora infestans*) resistance in many parts of the tropical world. Cvs with superior resistance to late blight proved to be a large benefit to farmers who could not afford or access fungicides. Farmers expressed high demand in these new cultivars but there was no immediate source of seed potatoes. The second major development was the modernization and simplification of the use of tissue culture. Rapidly, the technical aspects of tissue culture were being learned and adopted by young scientists, technicians, and private businesses in many developing countries. CIP started distributing these new late blight resistant cvs for evaluation in different countries of South East Asia and Eastern Africa.

2. Principles for successful utilization of apical rooted cuttings as transplants for tuber production

The physiology of the transplant is the most important factor in maximizing the growth and productivity in the greenhouse or field for tuber production. During the 1970s and 80s the use of stem cuttings became a rapid way of producing mini tubers in substrate in greenhouses. These stem cuttings generally originated from a sprouted tuber and always had compound leaves indicating physiological maturity thus yielding only a 2–5 tubers per stem cutting. With the popularization of tissue culture, the plants *in vitro* had simple rounded juvenile leaves and such tissue cultured plants, when transplanted to a substrate, produced many tubers, often exceeding 15 per plant. This leads to the thinking that if the juvenile plantlets with simple rounded leaves could be maintained in a substrate and from which apical cuttings could be taken periodically over an extended period, a large number of ARC could be obtained. Once transplanted to the field, many tubers could be obtained as first field generation seed tubers (G_1). It is critical to maintain the mother plants in a juvenile simple rounded leaf state and then the apical cuttings would develop into excellent ARC for transplanting to the field [3, 4]. **Figure 1a** and **b** illustrates the desired physiology of the mother plants and the ARC. **Figure 1c** shows the initial phase of compound leaf formation. Depending on the genetic background of the cv, these would be discarded as these would not develop into a high tuber producing plant.

Environmental factors are also important to consider in a successful maintenance of rapidly growing juvenile mother plants in substrate filled beds. The tropical environments are generally in the 11 to 13 hour photoperiod range, which is conducive for tuberization. Extending day length to 16 hours improved apical

Figure 1.
Mother plants and rooted cuttings: a. Tissue culture origin mother plants maintaining the juvenile simple round leaf stage; b. Apical rooted cutting with simple rounded leaves; c. Apical rooted cutting with undesirable trait of already developing compound pointed leaves.

cutting production [5]. Photoperiod is easily prolonged by turning some lights on during the evenings over the mother plant nurseries. A light break of 1–2 hours is sufficient to simulate the effect of extended day length. Warmer air temperatures encourage rapid growth of the plantlets and their regrowth after harvesting apical cuttings. Minimum air temperatures above 16°C are ideal such as in Dalat [3] at 800 to 1,500 mamsl and in the Philippines at 800 mamsl [5].

At higher elevations, a proper greenhouse may be needed. At lower elevations, productivity is substantially reduced due to overall heat stress with minimum and maximum temperatures above 23°C and 31°C, respectively [6]. Reducing the light intensity through shading also helps to delay physiological aging and naturally delays tuberization.

Most cultivars can be maintained in a young vegetative stage up to 6–9 months with appropriate nursery management with extended day length, warm temperature, and appropriate spacing and fertilization [7]. Gibberellic Acid (GA_3) application was helpful for some cultivars but was later not being practiced. The use of hormones is not necessary when rooting ARC.

Interplay of factors would affect the survival, growth and yield of ARC in the field. Availability of water at transplanting and during the growth cycle is crucial for better survival and yield of transplants. In earlier studies, the most important factors that would influence growth and yield were determined. Cuttings perform better when planted in raised beds compared to planting in flat beds or furrows. Hilling up improved tuber yield (size, number, and weight) and reduced greening of tubers. Pruning of the apical shoot after transplanting did not stimulate more branching nor improve yields. Plant populations greater than 40,000 transplants/hectare did improve total yields but reduced average tuber size. The highest yields were at 100,000 plants/hectare. The cost benefit ratio and the goal of the field production will determine the ideal spacing in each season for most varieties. In terms of fertilization, nitrogen supply was the most important, as deficiency or excessive nitrogen supply reduced yield [4].

Genetic background of the cultivars being utilized is a major consideration. Generally, the cultivars from *Solanum andigena and Solanum phureja* backgrounds are most prolific and remain longer physiologically young. This is partially because tuber induction in these species is not so strong at 11–13 hours of photoperiod. Most *Solanum tuberosum* cultivars, are adapted to the longer days of Europe or other temperate climates and are strongly induced to tuberize under the short-day tropical conditions. Cv Igorota, a cultivar bred with Peruvian germplasm produced 3 times as many apical cuttings as compared to the European cv Granola in a trial in the Philippines [2].

Knowing the cultivars as well as the environment and managing them appropriately are essential for successfully launching a program of mass production of ARC and their transplanting in the field for further multiplication.

3. Vietnamese farmers demystify sophisticated tissue culture technology for the successful utilization of apical rooted cuttings as transplants for tuber production

3.1 Background

Potato production was severely negatively impacted by the post war period in Southern Vietnam with a lack of clean healthy seed potatoes for viable potato production. That, coupled with a lack of good cultivars, placed the farmers and researchers in a difficult situation. The Center for Experimental Biology requested

CIP for new cultivars with resistance to Late Blight. A total of 16 cvs were received as virus free tissue culture plants. After some quick field evaluations there were 3 cultivars that the farmers demanded. With no source of seed tubers available, farmers took it upon themselves to take the tissue cultured plants and multiply them *in vitro* and then to establish the plantlets in a substrate as mother plants from which cuttings could be taken and after rooting, be transplanted to the field [3, 8]. With the support of researchers, farmers quickly learned how to maintain *in vitro* plants with the use of coconut water and some supplemental nutrients in a corner of their bedrooms with plastic walls and a Bunsen burner for sterilization. Then through trial and error, a suitable substrate media was developed using subsoil, sand and fine coconut husk material, composted manure and other materials to grow the mother plants.

Initially, 10 farming families established small outdoor mother plant nurseries on less than 100 m^2 of land next to their homes. Nine of those farmers maintained tissue culture plants in their homes [3]. These multipliers were soon selling over 2.5 million ARC in total annually to smallholder farms in Dalat area [8]. With the rapid adoption of the new cultivars, the demand for ARC was reduced and soon only 3 farmers remained with tissue culture plants and mass production of ARC.

3.2 Recent developments in the utilization of ARC by farmers in Dalat and Lamdong Province

Production of G_1 tuber seed directly in the field from ARC is a routine practice in Lam Dong province (1,000–1,500 mamsl). In 2020, there were 4 major ARC propagators who supplied around 5 million ARC to potato growers in the area. The largest multiplier sells 3 to 4 million ARC a year depending on the growers' demand. Some half a million marble size mini tubers were also harvested from the mother plant beds and sold as quality planting materials to the growers. With good agronomic management and plant density of 50–55 thousand per hectare, the average tuber yield of 20–25 tons per hectare (t/ha) has been obtained from the ARC with over half of that being ware potatoes. There were cases that yields of over 40 tons t/ha were obtained and up to 70% were sold as ware potatoes.

Production of the ARC normally starts in May with tissue cultured disease-free plantlets in Dalat. Those are raised in stock plant beds on good substrate, under net house conditions with about 50% shade. The first two or three cuttings harvested from a stock plantlet are used for establishment of mother plant beds (**Figure 2**).

Further apical cuttings are then harvested for production of ARC for the tuber seed production in Dalat and lower elevation areas of Don Dzuong and Duc Trong (1,000 mamsl). The whole process goes through the months of May to October with rigorous application of hygiene measures and crop protection. The transplanting of ARC into the field takes place mostly in November and December with sprinkler irrigation as rainfall is minimal. The ARC are arranged in wooden trays holding 1,000 each (**Figure 3**), and when ready for planting, are delivered to the growers' farms by the multiplier on a truck or taken directly by the grower himself on a motorbike.

Not as in the past, when an apical cutting was rooted in a small hand-made banana leaf pot of the size approximately 2 x 3 cm for planting into the field, ARC are now rooted in a substrate cube of the size 3 x 3 cm (**Figure 4**). The cubes are mechanically produced in mass by a uniquely designed motorized machine. The substrate is basically the same, i.e. a mixture of fine and clean clay loam subsoil, coconut fiber dust and some composted manure. Much of the substrate production operation has also been mechanized to minimize the costs on screening, mixing the component materials and making the rooting cubes.

Figure 2.
Taking apical cuttings from mother plants.

Figure 3.
Rooting of cuttings in wooden trays.

Figure 4.
Rooting of apical cuttings in substrate cubes 3 x 3 cm in size.

Tissue culture is currently carried out in a research institution and three private farmers' laboratories. Expensive test tubes and/or Erlenmeyer flasks have been replaced by plastic bags and boxes or reused serum bottles in raising the *in vitro* plantlets. This helps reduce the cost on culture vessels and labor for washing.

For the last 25 years, only two potato cvs are used for production for fresh market held by farmers, locally namely 07 (internationally known as Utatlan) and PO3 (released as Igorota in the Philippines). Seed stocks are customarily renewed by planting ARC after several generations in field production. Cv PO3 is resistant to several major viruses, thus seed tubers could be retained from ware fields up to 5–6 generations provided no fungal and/or bacterial infections occur, while cv 07 could hold only for 2–3 generations due to its high susceptibility to viruses.

Though many operational techniques have been improvised, the basic steps in the technological line remain the same through the years since the early 1980s: disease free *in vitro* plantlets – stock mother plants trays/beds – mother plants beds – apical cuttings rooted in pots – G_1 seed fields. The improvements take account for the better laboratory equipment, greenhouse facilities, irrigation system, mechanization in the preparation of rooting substrate and pots/cubes and delivery system.

These all, coupled with the great availability of quality pesticides and fertilizers, help assure stability and success of the novel seed potato system in Dalat.

From an economical point of view, the ARC propagation is a profitable business. With the current price of USD 1.75 per 100 ARC, the total revenue of all the multipliers would be USD 87,500 for the 5 million ARC a year. Their net profit is estimated to be about 30% of the amount. A little income would be added from the mini tubers harvested from the mother plant beds. Though the unit price has increased by 2.5 since the early 1990s the proportion of net profit seems unchanged mainly due to the situation that transplanting of the ARC in the field (sale of ARC), take place primarily only during two months (November and December). Whereas the mother plants could be maintained for apical cuttings harvesting for up to 4–5 months [9].

Partly, inflation of inputs, especially the labor cost, adds some strain to improvement of net profit. However, the scenario shows that most of the benefit from the ARC technology has been going to the small holder farmers who make much higher profitability from their potato production owing to the higher quality of planting materials. The fact that the average potato yield in Dalat has been lifted from less than 10 t/ha during the late 1970s to around 15 during the 1990s and 25 at the present is largely attributable to the use of disease-free ARC for seed propagation.

4. Apical rooted cuttings permitted rapid adoption and maintenance of new cultivars for smallholder farms in Philippines

4.1 Background

The Regional Germplasm and Training Center of CIP in the Philippines was established in 1982 to multiply new sets of cvs for distribution and evaluation by institutions in Southeast Asia and the Pacific. ARC were successfully used for quick multiplication of tissue culture plants and the subsequent evaluation of diverse cvs in five different environments in the country [4].

In 1985–1986, ARC field trials were initiated with highland farmers. In some field trials yields were up to 38 t/ha. Based on these preliminary results some farmers were willing to buy ARC at nominal price [10]. In 1990, 3 out of 8 new clones with superior late blight resistance were selected by farmers. LBR1–5 was the most prolific in terms of cutting production while I-1039 and LBR-9 were highest yielders. Four of the clones from cuttings out yielded Granola from tubers (control check) [11]. LBR1–5, later coded as PO3 and officially released as cv Igorota in 2004, was selected for its early maturity, good tuber skin and shape, and high yields with large sized tubers. Cv Igorota was initially multiplied and sold for seed by Mr. Peter Raymundo, a farmer from Buguias, Benquet [2]. The use of ARC made possible the rapid multiplication of pathogen tested materials and introduction of new cvs for on farm evaluations.

4.2 Adoption of ARC and new cvs for commercial potato production

The commercialization of ARC for seed potato production was a concerted effort of government, private sector and farmers' organizations. The Department of Agriculture provided funds for research and infrastructure development; Land Bank provided loans; while Northern Philippine Root Crops Research and Training Center (NPRCRTC) of Benquet State University conducted research, on-farm trials and dissemination of ARC and new cvs. Farmer organizations were multipliers. They bought ARC or *invitro* plants as source of mother plants. The multipliers produced G_0-G_3 seed tubers for other farmers. NPRCRTC and a farmer cooperator were accredited by the Bureau of Plant Industry to produce certified ARC and mini tubers for farmers in Benquet and beyond.

Mr. Nelio Compelio is an outstanding multiplier. He started using ARC in 1992 and *invitro* plants as source of mother plants in 2010. He replaces his seed stocks in 2 years or whenever decrease in yields is observed. He started selling Igorota seed potatoes to farmers in other municipalities in 1995. Now, farmers have the option to buy from him either ARC or seed tubers. He was able to build a big house in Atok and another in Baguio out of his production venture. They call these "Houses that Igorota Built".

Several farmer organizations followed the aforementioned example and supported their members through multiplication of ARC of their preferred cvs. On average, tubers harvested from an ARC field-grown crop can be multiplied 4 times as seed before selling them to other farmers for table or processing potato production. All cooperators attest to the fact that seed production using ARC is efficient and profitable [12].

4.3 Impacts of ARC on dispersal of new/recommended cvs

The use of ARC and the informal seed systems by farmers had greatly contributed to the selection and dispersal of new and improved potato cvs. The dispersal of clean planting materials from NPRCRTC accelerated this process. As example, cvs Montañosa, Dalisay and Igorota were propagated in nurseries and in farmer's fields prior to its release as official Seed Board Varieties. NPRCRTC had sold 353,000 ARC of cv Igorota from 1987 to 1994. To date, 7.8 million pathogen tested ARC and minitubers were sold. The number of farmers served over the years also increased (**Table 1**).

Distribution of different cvs continues to the present. The most popular is Igorota with 3,440,000 ARC and 371,000 G_0 mini tubers sold to farmers. The demand for other varieties varied depending on farmers' preferences and market demand (**Table 2**).

Year(s)	No. of Individual Clients	Planting Material ('000's) Distributed to Farmers			
		ARC	G_0 tubers	Total	No. per Farmer
1987–1994		353		353	
1995–1999	305	596	136	732	2.4
2000–2004	226	828	94	922	4.1
2005–2009	752	1,276	121	1,397	1.9
2010–2014	757	1,224	150	1,374	1.8
2015–2019	1,085	1,887	479	2,366	2.2
2020	156	139	152	291	1.9
April 2021	143	138	227	365	2.6
Total	3,424	6,441	1359	7,800	
Mean					2.3

Table 1.
Planting materials distributed by NPRCRTC from 1987 to April 2021.

Cultivar	ARC and Mini tubers Sold ('000's)				
	2006–2010	2013–2017	2018–2020	January to April-21	Total
Granola					
a. ARC	94	184	140	5	423
b. G_0 Tubers		62	185	194	441
Igorota					
a. ARC	1,280	1,544	487	129	3,440
b. G_0 Tubers		189	167	15	371
Other cvs					
a. ARC	30	26	76	4	136
b. G_0 Tubers		14	45	17	76
Total	1,404	2,019	1,100	364	4,887
a. ARC	1,404	1,754	703	138	3,999
b. G_0 Tubers		265	397	226	888

Table 2.
ARC and G_0 mini tubers sold by NPRCRTC from January 2006 to April 2021.

The NPRCRTC estimated production cost for ARC is 0.0210 USD/piece while the selling price is 0.031 USD/piece. The production cost and selling price per piece of *invitro* plant is 0.23 USD and 0.314 USD, respectively. Whereas, the G_0 is produced at 0.0858 USD and sold at 0.104 USD per piece.

5. Kenyan private and public sectors develop and rapidly promote ARC technology for farmers

5.1 Background

In Kenya, potato ranks the second most important crop after maize and is cultivated in the high-altitude areas between 1,500 and 3,000 mamsl [13]. Production is mainly rain-fed and is done in the long and short rains that occur in the period of March–July and October–January, respectively. The public institutions mandated to produce breeder seed or pre-basic seed potatoes can only meet about 1 to 2% of the certified seed demand in the country [14, 15]. These institutions mainly adopt the conventional clonal multiplication in which a set of disease-free tubers is repeatedly multiplied to bulk the seed. However, this method has low multiplication ratios of 3–6 and cannot meet the national seed demand. Certified seed is therefore highly priced in Kenya (at a cost of USD 30 to 40 per 50 kg bag) and is estimated to account for 20–70% of the total production costs. This compels farmers to use seed tubers obtained from informal sources. Such seeds are often infected with bacterial wilt and potato cyst nematode, thus resulting in low yields, poor quality produce, and spread of pests and diseases. The obtained potato yields are generally low with an average of 10 t/ha [16] against the potential of 30–40 t/ha [1, 13].

It is against this backdrop that CIP in partnership with the private sector Stokman Rozen Kenya (SRK), Kenya Agriculture and Livestock Organization (KALRO), Kenya Plant Health Inspectorate Service (KEPHIS- a body regulating seed certification in the country), and private businesses, sought to explore ARC as a complimentary rapid seed multiplication technique. The new improved CIP cvs: Unica, Wanjiku, Nyota, Chulu, Lenana and Kongo have been trialed against the conventional cv Shangi. Yields obtained from ARC have been robust and average 8–18 tubers per plant depending on the cv and management (**Table 3**). Subsequently, ARC has been endorsed by KEPHIS into the seed certification protocol to bulk up seed potato for multiplication and distribution to potato growers.

5.2 Taking ARC technology to farmers' doorstep

At least 10 satellite nurseries investing in apical cutting production have been set up in Meru, Nakuru and Nyandarua counties. Four of these nurseries were initiated by Farm Inputs Promotions Africa and CIP, and each has a capacity to produce 100–200,000 ARC annually. World Food Program in partnership with CIP has supported additional 2 satellite nurseries. Two other commercial private nurseries (Grace Rock Ltd. and SRK) have incurred great investments in ARC production with capacity to each produce up to 1 million ARC annually.

5.3 Juvenility of the mother plant- Kenya's case

As noted across the nurseries, mother plant productivity is highly dependent on the prevailing environmental conditions and routine management. Maximum productivity, vigor and mother plant juvenility has been associated with temperatures of about 18-25°C, relative humidity of 60–85%; regular fertigation with adequate nitrogen applied at 2–3 day intervals and use of clean media that exhibits low salt concentration (EC <0.5 dS/m), pH 6.0–6.5, and can retain good moisture and nutrients). Cocopeat is the most commonly used media in Kenya. However, if not properly washed, cases of stunted growth, yellowing and even death of the ARC have been reported. As observed by most multipliers, frequent cutting of the mother plant extends juvenility up to 9 months. Cuts should be obtained as soon as apical shoots have grown 5–7 cm high, or if the shoot has 4–7 complete leaves with 2–3 internodes. This is a key management protocol to attain prolific and juvenile mother plants. With regular proper training, nursery multipliers can vividly differentiate a physiologically young mother plant which they measure by simple round leaves, dark green vigorous shoot, delayed tuber formation/shoot senescence, and stems which are soft and easy to root in the absence of rooting hormones (**Figure 1a** and **b**). The commercial apical cutting derived from a juvenile mother plant has the bottommost leaves round and simple as opposed to the compound leaf characterizing the cutting derived from a mature mother/stem cutting (**Figure 1c**). A mature mother plant literally gives stem cuttings with low multiplication ratio of 3–7 tubers/cutting; thus, the multipliers are made aware that only juvenile plants are propagated to result in high rates of ARC production. To rescue a mature mother, the multipliers cut back the plant as soon as compound leaves are evident. Once the mother plant is too old to be rejuvenated, it is transferred for mini tuber production.

County	#of farmers sampled	#of tubers >20 mm					
		Shangi	Unica	Wanjiku	Nyota	Konjo	Chulu
Kiambu	96	11.5	8.8	18.2	13.9	10.9	10.8
Nakuru	24	12.8	9.3	18.2	14.5	10.3	11.6
Uasin Gishu	16	14.4	9.7	17.6	13.2	9.5	10.5

Table 3.
Average tubers per plant obtained from six potato cvs sampled from a total of 136 farmers. Sampling of at least 140 plants was randomly done from farmers who received ARC in Kiambu, Nakuru and Uasin Gishu in the long rains 2020.

5.4 Submothering for rapid ARC multiplication

With the technical guidance from CIP, the nurseries have adopted the practice of submothering to enhance rapid multiplication. The practice derives 1 to 2 submother plants from the first 1 to 2 cuts obtained from an *in vitro* plantlet. The submothers are only derived from a very juvenile tissue culture derived mother plants (<1 month) that exhibits dark-green vigorous shoot with all round and simple leaves. Submothering allows rapid multiplication rates of high potential *invitro* derived mother plants and can help attain multiplication rates up to 70 ARC per tissue culture plant in a 4 month period just like the case of Grace Rock Farm Ltd. (**Table 4**).

5.5 Making ARC a demand driven technology

Building market demand becomes key to accelerate the uptake of ARC. Thus, over the last few years, CIP in partnership with KALRO and county governments have built capacity of more than 200 extension agents to train farmers on ARC production. Women and youth groups have been supported to develop into small businesses. Additionally, CIP has built capacity of public agriculture training centers (ATCs) and supported the formation and registration of potato cooperative societies and private seed multipliers. Some of the cooperatives have been licensed as seed merchants and have acquired over 100 acres of land for seed multiplication. A few seed potato businesses now operate as out growers using ARC as starter seed material.

Increased sales of ARC have been observed across businesses licensed to produce and sell ARC. For example, farmers privately purchased 417,311 ARC in 2020 from 8 nurseries valued USD 42,000, while CIP purchased 168,000 for training and promotional purposes. The purchase made by the farmers were directly related to the trainings and field demos conducted in the preceding period, and generally indicated high preferences and acceptance of the new improved CIP potato cvs (Unica, Nyota, Wanjiku and Chulu). Benard Mwaura, a crop officer in Kiambu County, reported that "farmers are able to appreciate the rapid seed multiplication rates with the ARC technology and are so impressed that some of these cvs have good resistance to late blight and can tolerate water stress". Particularly, Wanjiku cv has gained popularity in Meru, Nakuru and Kiambu counties due to its good table qualities, and high multiplication rates ranging up to 50 ARC/tissue culture plantlet (**Table 5**). Cv Chulu was noted by farmers to be resistant to late blight while cv Unica has been preferred for its fast maturity and ability to tolerate heat and water stress.

Cultivar	#in-vitro plantlets used	#sub mothers produced	Total mother plants (*invitro* + sub mothers)	Total cuttings produced	Multiplication Ratio
Shangi	250	388	638	45,342	71
Unica	100	147	247	17,697	72
Wanjiku	150	132	282	16,617	59
Chulu	100	98	198	9,900	50
Nyota	50	33	83	3,513	42
Konjo	100	90	190	5,242	28
				98,311	

Table 4.
Total quantity of ARCs produced from combined mothers and sub mothers, and average multiplication ratio by six potato cvs obtained by Grace Rock Farm Ltd. in the period of March to December 2020.

Decentralized ARC producers	Location	Elevation (masl)	Cultivar			
			Shangi		Wanjiku	
			# invitro **plantlets used**	**Ratio**	***# invitro*** **plantlets used**	**Ratio**
Cecinta Nduru	Meru	2,360	200	50	720	13
Mary Nkatha	Meru	2,360	250	37	500	29
Robert Kimathi	Meru	2,023	325	19	300	24
Paul Munene	Meru	1,710	413	7	697	28
Faith Kajuju	Meru	2,222	270	13	—	
Erick Bittok	Nandi	2,006	2,196	3	1,071	4
Potato Empire	Nakuru	2,790	379	46	200	15

Table 5.
Invitro *plants used and average multiplication ratio for two cvs by privately owned satellite nurseries in the period of September 2020 – April 2021.*

6. Ugandan private and public sectors develop and rapidly promote ARC technology for farmers

6.1 Background

Potato is a key food and cash crop in Uganda, grown primarily by smallholder farmers in the eastern and southwestern highlands of the country. There are two major potato growing seasons (March–July; September–January), however, some off-season production also occurs in swamps, valley bottoms and irrigated areas.

National potato production has grown steadily over time, responding to an increasing demand and consumption [17]. This increase in production has been obtained by expanding the land cultivated rather than by increasing productivity [1]. Currently, Ugandan potato farmers harvest an average of 3–12 t/ha [18, 19]. The primary reason for this low yield is the use of low-quality seeds, that farmers recycle from previous harvests or purchase from other farmers or in the local markets [20]. Such seeds are often infected with seed-borne pathogens.

To improve farmers' access to high-quality seed of desired cvs, and to unlock the yield potential of potato in Uganda, CIP and other development partners have supported decentralized seed multiplication for more than a decade. Decentralized seed multipliers (DSMs) are based in potato growing communities and use early generation seed (EGS) – mostly basic seed – as starter material for onward field multiplication and bulking. DSMs have the advantage of making seed available in proximity to ware potato farmers at affordable prices. Quality assurance is the main risk to seed production under a DSM approach, but can be addressed with effective system management.

DSMs in Uganda, however, are frequently confronted with a lack of quality basic seed because public and private seed producers do not generate enough seeds to assure these demands. The current seed potato production system in Uganda is relying on the production of mini tubers from *invitro* plantlets in the screenhouse, followed by two seasons of field multiplication to produce basic seed. Public agricultural research institutes are currently leading the country's seed potato production. Private sector investment in commercial seed potato production is small and consists mainly of a few farmers in potato producing areas managing small screenhouses producing EGS and bulking seed in the field [21].

6.2 Recent developments in the utilization of ARC by farmers in Eastern and Southwestern Uganda

In 2019, CIP partnered with public agricultural research institutes, farmer managing screenhouses and a private commercial company to promote production and field multiplication of ARC. This initiative was to address the challenge of accessing tuber seed for farmers. The partners were trained and supported to produce ARC. The locations of their operations varied from 1,200 to over 2,200 mamsl (**Table 6**).

Nursery	Type of nursery	Location	Elevation (mamsl)	# *invitro* plantlets used	Ratio	Number of cvs	Sale price (USD)	ARC Sold
Agromax Ltd.	Private	Kampala	1200	8,824	12.5	8	0.2–0.25	81,510
Farmer Greenhouse managers	Private	South western and Eastern Uganda	1,665-2,433	11,335	12.9	8	0.11–0.25	51,902
KaZARDI	Public	Kabale district (South western Uganda)	2,223	5,119	18.9	2	0.28	2,550
BugiZARDI	Public	Bulambuli district (Eastern Uganda)	1,760	4,250	5.9	4	—	0

Table 6.
Different nurseries in Uganda producing ARC and their results during 2019 and 2020.

Figure 5.
Seed multiplier in a seed plot with potato plants from ARC.

DSMs were trained in field multiplication techniques to produce tuber seed of different cvs from ARC as illustrated by a smallholder field crop growing from ARC in **Figure 5**.

The NGO Self Help Africa was a key partner in implementing and monitoring field activities. The ARC multiplication rates varied among nurseries and cvs planted. The highest multiplication rate of producing ARC/tissue culture plant was at the KaZARDI in Kachwekano. The most popular cvs demanded by the smallholder farmers were: Victoria, Kachpot, Rwangume and Kinigi.

Despite the immense advantages of ARC for seed production, five major challenges are being addressed: to sell ARC at an affordable price depends largely on improving the multiplication rate of mother plants from tissue culture; improved cost-effective, easy-to-manipulate, locally acceptable, and environmentally-friendly packing materials that can maintain the quality of the ARCs during transport; improve transplants survival and tuber yields; a better coordination among potato value chain stakeholders to develop and stabilize the market demand for ARC and avoid oversupply; and quality assurance mechanisms need to be developed and implemented to use ARC in certified seed potato production that is aligned with the seeds and plants regulations of Uganda.

7. Discussion

The development and utilization of ARC in the 4 countries have some significant commonalities and differences. The initial primary driver of adopting the use of ARC is the highly desirable characteristics of the cvs sought by the smallholder farmers in each country. Improving the access of seeds of cultivars resistant to late blight was the most significant initial reason for the trial and adoption of ARC. The cvs mentioned in each country were either derived from CIP shared germplasm or coming from the Mexican late blight resistance selection program prior to 1975. Cvs Igorota (PO3) and Utatlan are the prime examples for the Philippines and Vietnam. The cvs promoted in Kenya and Uganda are all from CIP germplasm with moderate late blight resistance. These cvs proved to be vastly superior for late blight resistance compared to European cvs for which seed could be imported. This factor was the primary driver for the rapid acceptance and utilization of tissue culture and the mass production of ARC.

The successful development of juvenile rounded simple leaf ARC after many months of maintaining mother plants in the juvenile stage allowed the transplants to develop in the field to large vigorous plant with only one primary stem producing on average 11 and 18 tubers/plant. This has been the case in all 4 countries. This productivity is equal to that of standard seed tuber planted crops with 3–4 stems/plant. The climatic conditions for maintaining juvenile mother plants were initially at elevations of less than 1,500 mamsl in Vietnam and the Philippines. These warmer conditions favored rapid growth and frequent harvesting of apical cuttings. Interestingly in both Kenya and Uganda the productivity of juvenile mother plants can be maintained at cooler higher elevations above 2,000 mamsl. Better greenhouses and seasonal considerations as well as adding temperature control measures improved productivity.

The demand for ARC has stabilized in the highlands of both southern Vietnam and the northern Philippines. The high level of virus resistance of cv Igorota (PO3) permits the G_1 seed to be regrown by farmers 5–7 times before replacing with new ARC. In the case of Dalat Vietnam, when bacterial wilt appears in a field crop, it will not be retained as seed for replanting. For the approximate 1,500 hectares of potatoes grown in the greater Dalat area a total of about 5 million ARC are sold annually.

In the Philippines highlands with an area of about 7,500 hectares in potatoes annually, approximately 350,000 ARC plus 100,000 mini tubers are demanded annually from NPRCRTC. Many small holder farmers multiply the ARC further.

In Kenya, there was a rapid growth phase in the acceptance of ARC as the technology allows to have virus free planting material of cvs with late blight resistance and other good attributes. Licensed seed merchants can now use ARC as starter materials to produce certified seeds. Uganda is at an earlier stage in the introduction and adoption of ARC.

There is a large difference in the selling price of 100 ARC among 4 countries. In Vietnam, the price is USD 1.75; Philippines 3.00; Kenya 10.00 and Uganda is 11.00–26.00. The producers of ARC in Vietnam calculate a return to labor and investment of 30% while in the Philippines it is 24%. The largest producer in Vietnam prepares 100,000 tissue culture plants in May. These are placed in the substrate beds and multiplied to 720,000 mother plants by September. Then for 2–3 months apical cuttings are harvested and rooted every 5 days. Over 4 million ARC will be sold to smallholder farmers by December when the transplanting season ends. The large-scale efficiency in the production process allows this producer and the others in Dalat to sell 100 ARC at USD 1.75, with a significant profit. The selling prices in the Philippines has stabilized at USD 3.00/100 ARC. ARC sold per single tissue culture plant is greater than 40. In Kenya, the multiplication rate is similar with some cvs such as Shangi and Unica reaching 70 ARC/tissue culture plant. In Uganda, these numbers are still generally less than 20. Production costs and selling prices will be lowered through efficiencies in Kenya and Uganda as the technology matures and competitors join in the business.

Impact of the ARC technology coupled with new desired cvs had made a marked difference in the level of food security and prosperity for the smallholder potato farmers in all countries. In Southern Vietnam, it is estimated that the average yield of potatoes has improved from 10 to now over 25 t/ha due to clean seed of late blight resistant cultivars. In the Philippines, a similar improvement has been observed over the past 20 years with yields of 40 t/ha frequently recorded. The impact is seen in the purchases of refrigerators, motorcycles, and other household amenities. One multiplier of ARC in Dalat even purchased a baby grand piano! In the Philippines, there are stories about houses built on the production and sale of ARC and G_0 mini tubers by some farmers. In general, all smallholder farmers who have opted to grow ARC and G_0 mini tubers have improved food production and generated more income.

The ARC technology will continue to face strong competition from European tuber seed for the larger seasonal lowland production systems after rice and in the semi-arid regions of North Africa. Upland mountain area of India and other parts of Africa will be conducive for ARC adoption, especially where the planting season can be prolonged for 2 months or more. Aeroponics is an established system in high temperate locations of China, India as well as in other localities in South America and some in Africa to produce mini tubers. The level of sophistication with the need for reliable electricity, however, limits its adoption and favors the use of ARC in most tropical environments where potatoes are grown.

Acknowledgements

Dr. Uyen Van Uyen, who was the dynamic leader of the ARC revolution in Dalat Vietnam to help alleviate the severe poverty of the farming communities after the civil war. All the early adopters of simplified tissue culture practices, ARC multipliers and smallholder farmers, who took the risk to adopt this technology in each of the 4 countries. CIP for supporting the technology development in each of the 4 countries as well as providing the improved potato germplasm or cvs. To all the International and national funding agencies who provided the start-up capital and other resources needed. The national institutions and local government extension services to ensure technical support. The decentralized and private seed companies for sharing the sales and production data.

Author details

Peter VanderZaag[1]*, Tung Xuan Pham[2], Victoria Escobar Demonteverde[3], Cynthia Kiswa[4], Monica Parker[5], Shadrack Nyawade[5], Pieter Wauters[6] and Alex Barekye[7]

1 Sunrise Potato, Alliston, Ontario, Canada

2 Institute of Agricultural Sciences for Southern Vietnam, Ho Chi Minh City, Vietnam

3 Potato Systems Research and Training Center, Canlaon City, Philippines

4 Northern Philippines Root Crops and Research Training Center, Benquet State University, Philippines

5 International Potato Center, Nairobi, Kenya

6 International Potato Center, Uganda

7 Kachwekano Zonal Agricultural Research and Development Institute, Uganda

*Address all correspondence to: peter@sunrisepotato.com

References

[1] FAO STAT 2018

[2] Chien DH, Ganga ZN, Simongo DK, Kiswa CG, Gonzales IC, Perez J, Tung PX, Hau NT, Artola K, Valkonen JPT, Li C, VanderZaag P. The adoption of cv Igorota in the Philippines and Vietnam. Am J of Potato Res. 2016; 93: 296-301

[3] Uyen NV, VanderZaag P. Vietnamese farmers use tissue culture for commercial potato production. Am Potato J. 1983; 60: 873-879

[4] Escobar V, VanderZaag P. Field performance of potato (*Solanum spp*) cuttings in the warm tropics: influence of planting system, hilling, density and pruning. Am Potato J. 1988; 65: 1-10

[5] VanderZaag P, Escobar V. Rapid multiplication of potatoes in the warm tropics: rooting and establishments of cuttings. Potato Res. 1990; 33: 13-21

[6] Minh TV, Uyen NV, VanderZaag P. Rapid multiplication of potatoes: influence of environment and management on growth of juvenile apical cuttings. Am Potato J. 1990; 67: 789-797

[7] Escobar V, Montierro C, VanderZaag P. Rapid multiplication of potatoes (*Solanum spp.*): influence of P, foliar nutrients, daylength, and genotype on cutting production by mother plants. CIP Region VII Working Paper. 1986; 86-84

[8] Uyen NV, VanderZaag P. Potato production using tissue culture in Vietnam: The status after four years. Am Potato J. 1985; 62: 237-241

[9] Uyen NV, Ho TV, Tung PX, VanderZaag P, Walker TS. Economic impact of the rapid multiplication of high-yielding, late blight resistant varieties in Dalat, Vietnam. In: Case studies of the economic impact of CIP-related technology. T Walker & C Crissman (Ed.), International Potato Center. 1996.

[10] Caringal EM, VanderZaag P. On-farm evaluation of rapid multiplication of potato (*Solanum spp.*) in Benquet. CIP Region VII Working Paper. 1986; 86-15

[11] Simongo DK, Demonteverde VE, Tandang LL, VanderZaag P, Chujoy E. An assessment of Potato clones using apical cuttings by farmers in Benguet. Philipp J Crop Sci. 1991;16(2):49-55

[12] Demonteverde VE. Rapid multiplication techniques for potato production in the Cordillera Administration Region: A collaborative program for sustainable development [Ed.D dissertation]. Open University Systems. Pangasinan State University; 2013.

[13] Muthoni J, Shimelis H, Mbiri DG, Schulte-Geldermann E. Assessment of national performance trials of potatoes in mid-altitude regions of Kenya. Journal of Agriculture and Crops. 2021; 7: 7-13

[14] KEPHIS. Seed potato production and certification guidelines – Kenya [Internet]. 2016. Available from http://www.seedsectorplatformkenya.com [Accessed 2021 May 5]

[15] International Finance Corporation. Diagnostic increasing seed potato availability in Kenya: Priority investments and policy actions [Internet]. 2019. Available from https://www.ifc.org/wps/wcm/connect/a1483e51-b9414da5bbfc577edfe00854/IFC+KSPID+Report_FINAL.pdf?MOD=AJPERES&CVID=npBxw4c [Accessed 2021 May 5]

[16] FAOSTAT [Internet]. 2019. Available from http://www.fao.org/faostat/en/#data/QC [Accessed 2021 May 5]

[17] Sebatta C, Mugisha J, Katungi E, Kasharu AK, Kyomugisha H. Adding Value at the Farm: The Case of Smallholder Potato Farmers in the Highlands of Uganda, Asian Journal of Agricultural Extension, Economics & Sociology. 2015; 4(3): 210-223

[18] Uganda Bureau of Statistics (UBOS). 2020.

[19] Uganda Annual Agricultural Survey 2018. Kampala, Uganda; UBOS.

[20] Aheisibwe AR, Barekye A, Namugga P, Byarugaba AA. Challenges and opportunities for quality seed potato availability and production in Uganda. Uganda Journal of Agricultural Sciences. 2015; 16(2): 149-159

[21] Byarugaba AA, John K, Aheisibwe RA, Deo T, Barekye A. Bridging the gap in quality and quantity of seed potatoes through farmer managed screen houses in Uganda. African Journal of Plant Science. 2017; Vol. 11(2) pp30-37